Elementary Decision Theory

Elementary Decision Theory

HERMAN CHERNOFF
Professor of Statistics, Harvard University

LINCOLN E. MOSES
Professor of Statistics, Stanford University

Dover Publications, Inc.
NEW YORK

Published in Canada by General Publishing Company, Ltd., 30 Lesmill Road, Don Mills, Toronto, Ontario.
Published in the United Kingdom by Constable and Company, Ltd., 10 Orange Street, London WC2H 7EG.

This Dover edition, first published in 1986, is an unabridged, corrected republication of the work first published by John Wiley & Sons, Inc., New York, in 1959.

Manufactured in the United States of America
Dover Publications, Inc., 31 East 2nd Street, Mineola, N.Y. 11501

Library of Congress Cataloging-in-Publication Data

Chernoff, Herman.
Elementary decision theory.

Reprint. Originally published: New York: Wiley, 1959. (Wiley publications in statistics)
Includes index.
1. Mathematical statistics. 2. Statistical decision. I. Moses, Lincoln E. II. Title.
QA279.4.C47 1986 519.5 86-16842
ISBN 0-486-65218-1

Dedicated to the memory
of M. A. Girshick

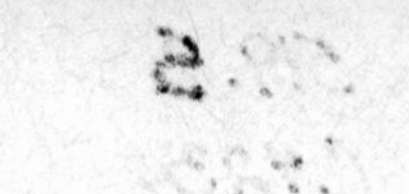

Preface

In recent years, Statistics has been formulated as the science of decision making under uncertainty. This formulation represents the culmination of many years of development and, for the first time, furnishes a simple and straightforward method of exhibiting the fundamental aspects of a statistical problem. Earlier representations had serious gaps which led to confusion and misunderstanding, especially on the part of elementary students without well-developed statistical intuition.

This book is the result of nine years of experience at Stanford in teaching a first course in Statistics from the decision making point of view. The course was first started by the late M. A. Girshick and this book may be regarded in part as an extension of his teaching.

A "first course" is contained in the first seven chapters. Our experience has been mainly with social science students in a five-unit, one-quarter course. Here we covered the seven chapters in detail, including the optional sections marked (†), and portions of the other chapters.

A background of high school mathematics suffices for this course, and considerable effort has been taken to by-pass involved computational reasonings which confuse the inexperienced. On the other hand, there has been no reluctance to use symbols, and the militantly non-mathematical student will not enjoy this book.

The mathematical novice should find in this book a well-motivated introduction to certain important and uncomplicated mathematical notions such as set, function, and convexity. Students who have a strong background in mathematics will find it profitable to spend time on the Appendixes which have the proofs of basic results in decision theory.

The reader will observe that new topics and ideas are introduced by examples. Also, certain exercises carry an essential burden in the development of the material. These exercises are starred and should be assigned. The teacher should take care to note that a few of the exercises in the first three chapters and in Chapter 7 are liable to be time-consuming.

In teaching the course, we have found it expedient to cover the material in Chapter 1 in one lecture and to spend another hour or two on the associated homework and classroom discussion. Chapter 2 was customarily disposed of rapidly, in about three hours. Considerable time was devoted to Chapter 5 which presents in detail most of the underlying decision theory.

The last three chapters may be studied in any order. Chapter 8 consists of a relatively informal discussion of model building. Chapters 9 and 10 treat classical statistical theory from the decision theory point of view and are rather technical. Ordinarily we have covered only portions of them. We regard it as inadvisable to attempt to do much in these chapters unless the students have some previous background in statistics or mathematics.

This book does not have a treatment of classical statistical methodology. It is our hope to follow this volume with a second one presenting existing methodology in the decision theory framework. Thus, this book will not be of much use to students who would like a single course devoted to the study of a few generally applicable statistical methods. Rather, we feel, that this book is well designed for those who are interested in the fundamental ideas underlying statistics and scientific method and for those who feel that they will have enough need for Statistics to warrant taking more than one course.

HERMAN CHERNOFF
LINCOLN E. MOSES

Stanford University
March 1959

Acknowledgments

Our greatest debt is to Sir Ronald A. Fisher, Jerzy Neyman, Egon S. Pearson, John von Neumann, and Abraham Wald. The present state of statistical theory is largely the result of their contributions which permeate this book. Particularly great is our debt to the late Abraham Wald who formulated statistics as decision making under uncertainty.

We acknowledge with warm thanks the keen and constructive comments of William H. Kruskal; we have acted upon many of his suggestions and have improved the book thereby. We thank Patrick Suppes for helpful comments on certain problems of mathematical notation. Harvey Wagner suggested that we incorporate the outline of a proof of the existence of the utility function as a handy reference for the mathematically sophisticated reader. He also proposed a proof which we found useful. Max Woods was especially effective in his careful preparation of the answers to the exercises and his detailed reading of the manuscript and proofs. Others who helped in reading the manuscript and proofs were Arthur Albert, Stuart Bessler, Judith Chernoff, Mary L. Epling, Joseph Kullback, Gerald Lenthall, Edward B. Perrin, Edythalena Tompkins, and Donald M. Ylvisaker. The completion of this manuscript would have long been delayed except for the unstinting cooperation of Carolyn Young, Charlotte Austin, Sharon Steck, and Max Woods. We thank our wives Judith Chernoff and Jean Moses for their support and encouragement.

The authors wish to express their appreciation to E. S. Pearson, C. Clopper, P. C. Mahalanobis, F. E. Croxton, and C. M. Thompson for the use of their tables. We are indebted to Professor Sir Ronald A. Fisher, Cambridge, to Dr. Frank Yates, Rothamsted, and to Messrs. Oliver and Boyd Ltd., Edinburgh, for permission to reprint Tables III, IV, and XXXIII from their book, Statistical Tables for Biological, Agricultural, and Medical Research.

H. C.
L. E. M.

Contents

CHAPTER 1

Introduction

1. INTRODUCTION

Beginning students are generally interested in what constitutes the subject matter of the theory of statistics. Years ago a statistician might have claimed that statistics deals with the processing of data. As a result of relatively recent formulations of statistical theory, today's statistician will be more likely to say that statistics is concerned with decision making in the face of uncertainty. Its applicability ranges from almost all inductive sciences to many situations that people face in everyday life when it is not perfectly obvious what they should do.

What constitutes uncertainty? There are two kinds of uncertainty. One is that due to *randomness*. When someone tosses an ordinary coin, the outcome is random and not at all certain. It is as likely to be heads as tails. This type of uncertainty is in principle relatively simple to treat. For example, if someone were offered two dollars if the coin falls heads, on the condition that he pay one dollar otherwise, he would be inclined to accept the offer since he "knows" that heads is as likely to fall as tails. His knowledge concerns the *laws of randomness* involved in this particular problem.

The other type of uncertainty arises when it is not known which laws of randomness apply. For example, suppose that the above offer were made in connection with a coin that was obviously bent. Then one could assume that heads and tails were not equally likely but that one face was probably favored. In statistical terminology we shall equate the laws of randomness which apply with the *state of nature*.

What can be done in the case where the state of nature is unknown? The statistician can perform relevant experiments and take observations. In the above problem, a statistician would (if he were permitted) toss the coin many times to estimate what is the state of nature. The decision on whether or not to accept the offer would be based on his estimate of the state of nature.

One may ask what constitutes enough observations. That is, how many times should one toss the coin before deciding? A precise answer would be difficult to give at this point. For the time being it suffices to say that the answer would depend on (1) the cost of tossing the coin, and (2) the cost of making the wrong decision. For example, if one were charged a nickel per toss, one would be inclined to take very few observations compared with the case when one were charged one cent per toss. On the other hand, if the wager were changed to $2000 against $1000, then it would pay to take many observations so that one could be quite sure that the estimate of the state of nature were good enough to make it almost certain that the right action is taken.

It is important to realize that no matter how many times the coin is tossed, one may never know for sure what the state of nature is. For example, it is possible, although very unlikely, that an ordinary coin will give 100 heads in a row. It is also possible that a coin which in the long run favors heads will give more tails than heads in 100 tosses. To evaluate the chances of being led astray by such phenomena, the statistician must apply the theory of probability.

Originally we stated that statistics is the theory of decision making in the face of uncertainty. One may argue that, in the above example, the statistician merely estimated the state of nature and made his decision accordingly, and hence, decision making is an overly pretentious name for merely estimating the state of nature. But even in this example, the statistician does more than estimate the state of nature and act accordingly. In the $2000 to $1000 bet he should decide, among other things, whether his estimate is good enough to warrant accepting or rejecting the wager or whether he should take more observations to get a better estimate. An estimate which would be satisfactory for the $2 to $1 bet may be unsatisfactory for deciding the $2000 to $1000 bet.

2. AN EXAMPLE

To illustrate statistical theory and the main factors that enter into decision making, we shall treat a simplified problem in some detail. It is characteristic of many statistical applications that, although real problems are too complex, they can be simplified without changing their essential characteristics. However, the applied statistician must try to keep in mind all assumptions which are not strictly realistic but are introduced for the sake of simplicity. He must do so to avoid assumptions that lead to unrealistic answers.

Example 1.1. The Contractor Example. Suppose that an electrical contractor for a house knows from previous experience in many communities that houses are occupied by only three types of families: those whose peak loads of current used are 15 amperes (amp) at one time in a circuit, those whose peak loads are 20 amp, and those whose peak loads are 30 amp. He can install 15-amp wire, or 20-amp wire, or 30-amp wire. He could save on the cost of his materials in wiring a house if he knew the actual needs of the occupants of that house. However, this is not known to him.

One very easy solution to the problem would be to install 30-amp wire in all houses, but in this case he would be spending more to wire a house than would actually be necessary if it were occupied by a family who used no more than 15 amp or by one that used no more than 20 amp. On the other hand, he could install 15-amp wire in every house. This solution also would not be very good because families who used 20 or 30 amp would frequently burn out the fuses, and not only would he have to replace the wire with more suitable wire but he might also suffer damage to his reputation as a wiring contractor.

TABLE 1.1

LOSSES INCURRED BY CONTRACTOR

States of Nature \ Actions	a_1 Install 15 amp	a_2 Install 20 amp	a_3 Install 30 amp
θ_1—family uses 15 amp	1	2	3
θ_2—family uses 20 amp	5	2	3
θ_3—family uses 30 amp	7	6	3

Table 1.1 presents a tabulation of the losses which he sustains from taking various actions for the various types of users.

The thetas (θ) are the possible categories that the occupants of a particular house fall into; or they are the possible states of nature. These are: θ_1—the family has peak loads of 15 amp; θ_2—the family has peak loads of 20 amp; and θ_3—the family has peak loads of 30 amp.[1]

The a's across the top are the actions or the different types of installations he could make. The numbers appearing in the table are his own estimates of the loss that he would incur if he took a particular action in the presence of a particular state.

For example, the 1 in the first row represents the cost of the 15-amp wire. The 2 in the first row represents the cost of the 20-amp wire, which is more expensive since it is thicker.[2]

In the second row we find a 5 opposite state θ_2, under action a_1. This reflects the loss to the contractor of installing 15-amp wire in a home with 20-amp peak loads; cost of reinstallation, and damage to his reputation, all enter into this number. It is the result of a subjective determination on his part; for one of his competitors this number might be, instead, a 6. Other entries in the table have similar interpretations.

Since he could cut down the losses incurred in wiring a house if he knew the value of θ for the house (i.e., what were the electricity requirements of the occupant), he tries to learn this by performing an experiment. His experiment consists of going to the future occupant and asking how many amperes he uses. The response is always one of four numbers: 10, 12, 15, or 20. From previous experience it is known that families of type θ_1 (15-amp users) answer z_1 (10 amp) half of the time and z_2 (12 amp) half of the time; families of type θ_2 (20-amp users) answer z_2 (12 amp) half of the time and z_3 (15 amp) half of the time; and families of type θ_3 (30-amp users) answer z_3 (15 amp) one-third of the time and z_4 (20 amp) two-thirds of the time. These values are shown in Table 1.2. In fact, the entries represent the probabilities of observing the z values for the given states of nature.

[1] We shall almost always use Greek letters to denote states of nature or special characteristics of states of nature.

[2] Apparently the contractor has already been paid off and he considers every dollar out of his pocket a loss.

TABLE 1.2

FREQUENCY OF RESPONSES FOR VARIOUS STATES OF NATURE IN THE CONTRACTOR EXAMPLE

States of Nature \ Observations	z_1 (10 amp)	z_2 (12 amp)	z_3 (15 amp)	z_4 (20 amp)
θ_1	1/2	1/2	0	0
θ_2	0	1/2	1/2	0
θ_3	0	0	1/3	2/3

The contractor now formulates a strategy (rule for decision making) which will tell him what action to take for each kind of observation. For instance, one possible rule would be to install 20-amp wire if he observes z_1; 15-amp wire if he observes z_2; 20-amp wire if he observes z_3; and 30-amp wire if he observes z_4. This we symbolize by $s=(a_2, a_1, a_2, a_3)$, where the first a_2 is the action taken if our survey yields z_1; a_1 is the action taken if z_2 is observed; the second a_2 corresponds to z_3; and a_3 corresponds to z_4.

Table 1.3 shows five of the 81 possible strategies that might be employed, using the above notation.

TABLE 1.3

STRATEGIES (RULES FOR DECISION MAKING)

Strategies \ Observations	z_1 (10 amp)	z_2 (12 amp)	z_3 (15 amp)	z_4 (20 amp)
s_1	a_1	a_1	a_2	a_3
s_2	a_1	a_2	a_3	a_3
s_3	a_3	a_3	a_3	a_3
s_4	a_1	a_1	a_1	a_1
s_5	a_3	a_3	a_2	a_1

Note that s_2 is somewhat more conservative than s_1. Both s_3 and s_4 completely ignore the data. The strategy s_5 seems to be one which only a contractor hopelessly in love could select.

How shall we decide which of the various strategies to apply?

First, we compute the average loss that the contractor would incur for each of the three states and each strategy. For the five strategies, these losses are listed in Table 1.4.

TABLE 1.4

AVERAGE LOSS IN CONTRACTOR EXAMPLE

States of Nature \ Strategies	s_1	s_2	s_3	s_4	s_5
θ_1—family uses 15 amp	1	1.5	3	1	3
θ_2—family uses 20 amp	3.5	2.5	3	5	2.5
θ_3—family uses 30 amp	4	3	3	7	6.67

They are computed in the following fashion:

First we compute the action probabilities for $s_1 = (a_1, a_1, a_2, a_3)$. If θ_1 is the state of nature, we observe z_1 half the time and z_2 half the time (see Table 1.2). If s_1 is applied, action a_1 is taken in either case, and actions a_2 and a_3 are not taken. If θ_2 is the state of nature, we observe z_2 half the time and z_3 half the time. Under strategy s_1, this leads to action a_1 with probability 1/2, action a_2 with probability 1/2, and action a_3 never. Similarly, under θ_3, we shall take action a_1 never, a_2 with probability 1/3, and a_3 with probability 2/3. These results are summarized in the *action probabilities* for s_1 (Table 1.5) which are placed next to the losses (copied from Table 1.1).

If θ_1 is the state of nature, action a_1 is taken all of the time, giving a loss of 1 all of the time. If θ_2 is the state of nature, action a_1 yielding a loss of 5 is taken half the time and action a_2 yielding a loss of 2 is taken half the time. This leads to an average loss of

$$5 \times 1/2 + 2 \times 1/2 = 3.5.$$

Similarly the average loss under θ_3 is

$$6 \times 1/3 + 3 \times 2/3 = 4.$$

Thus the column of *average losses* corresponding to s_1 has been computed. The corresponding tables for strategy s_2 are indicated in Table 1.5. The other strategies are evaluated similarly.

In relatively simple problems such as this one, it is possible to compute the average losses with less writing by juggling Tables 1.1, 1.2, and 1.3 simultaneously.

Is it clear now which of these strategies should be used? If we look at the chart of average losses (Table 1.4), we see that some of the strategies give greater losses than others. For example,

TABLE 1.5

LOSSES, ACTION PROBABILITIES, AVERAGE LOSS

States of Nature	Losses			Action Probabilities			Average Loss
				For $s_1 = (a_1, a_1, a_2, a_3)$			
	a_1	a_2	a_3	a_1	a_2	a_3	
θ_1	1	2	3	1	0	0	1
θ_2	5	2	3	1/2	1/2	0	3.5
θ_3	7	6	3	0	1/3	2/3	4
				For $s_2 = (a_1, a_2, a_3, a_3)$			
	a_1	a_2	a_3	a_1	a_2	a_3	
θ_1	1	2	3	1/2	1/2	0	1.5
θ_2	5	2	3	0	1/2	1/2	2.5
θ_3	7	6	3	0	0	1	3

if we compare s_5 with s_2, we see that in each of the three states the average loss associated with s_5 is equal to or greater than that corresponding to s_2. The contractor would therefore do better to use strategy s_2 than strategy s_5 since his average losses would be less for states θ_1 and θ_3 and no more for θ_2. In this case, we say "s_2 dominates s_5." Likewise, if we compare s_4 and s_1, we see that except for state θ_1, where they were equal, the average losses incurred by using s_4 are larger than those incurred by using s_1. Again we would say that s_4 is dominated by strategy s_1. It would be senseless to keep any strategy which is dominated by some other strategy. We can thus discard strategies s_4 and s_5. We can also discard s_3 for we find that it is dominated by s_2.

If we were to confine ourselves to selecting one of the five listed strategies, we would need now only choose between s_1 and s_2. How can we choose between them? The contractor could make this choice if he had a knowledge of the percentages of families in the community corresponding to states θ_1, θ_2, and θ_3. For instance, if all three states are equally likely, i.e., in the community one-third of the families are in state θ_1, one-third in state θ_2, and one-third in state θ_3, then he would use s_2, because for s_2 his average loss would on the average be

$$1.5 \times 1/3 + 2.5 \times 1/3 + 3 \times 1/3 = 2.33$$

whereas, for s_1 his average loss would on the average be

$$1 \times 1/3 + 3.5 \times 1/3 + 4 \times 1/3 = 2.83.$$

However, if one knew that in this community 90% of the families were in state θ_1 and 10% in θ_2, one would have the *average losses* of

$$1 \times 0.9 + 3.5 \times 0.1 = 1.25 \qquad \text{for } s_1$$
$$1.5 \times 0.9 + 2.5 \times 0.1 = 1.60 \qquad \text{for } s_2$$

and s_1 would be selected. Therefore, the strategy that should be picked depends on the relative frequencies of families in the three states. Thus, when the actual proportions of the families in the three classes are known, a good strategy is easily selected. *In the absence of such knowledge, choice is inherently difficult.* One principle which has been *suggested* for choosing a strategy is called the "minimax average loss rule." This says, "Pick that strategy for which the largest average loss is as small as possible, i.e., minimize the maximum average loss." Referring to Table 1.4, we see that, for s_1, the maximum average loss is 4 and for s_2 it is three. The minimax rule would select s_2. This is clearly a pessimistic approach since the choice is based entirely on consideration of the worst that can happen.

In considering our average loss table, we discarded some strategies as being "dominated" by other procedures. Those we rejected are called *inadmissible* strategies. Strategies which are not dominated are called *admissible.*

In our example it might turn out that s_1 or s_2 would be dominated by one of the 76 strategies which have not been examined; on the other hand, other strategies not dominated by s_1 or s_2 might be found. An interesting problem in the theory of decision making is that of finding all the *admissible* strategies.

Certain questions suggest themselves. For example, one may ask why we put so much dependence on the "average losses."

This question will be discussed in detail in Chapter 4 on utility. Another question that could be raised would be concerned with the reality of our assumptions. One would actually expect that peak loads of families could vary continuously from less than 15 amp to more than 30 amp. Does our simplification (which was presumably based on previous experience) lead to the adoption of strategies which are liable to have very poor consequences (large losses)? Do you believe that the assumption that the only possible observations are z_1, z_2, z_3, and z_4 is a serious one? Finally, suppose that several observations were available, i.e., the contracter could interview all the members of the family separately. What would be the effect of such data? First, it is clearly apparent that, with the resulting increase in the number of possible combinations of data, the number of strategies available would increase considerably. Second, in statistical problems, the intelligent use of more data generally tends to decrease the average losses.

In this example we ignored the possibility that the strategy could suggest (1) compiling more data before acting, or (2) the use of altogether different data such as examining the number of electric devices in the family's kitchen.

3. PRINCIPLES USED IN DECISION MAKING

Certain points have been illustrated in the example. One is that the main gap in present-day statistical philosphy involves the question of what constitutes a good criterion for selecting strategies. The awareness of this gap permits us to see that in many cases it really is not serious. First, in many industrial applications, the frequencies with which the state of nature is θ_1, θ_2, etc., is approximately known, and one can average the average losses as suggested in the example. In many other applications, the minimum of the maximum average loss is so low that the use of the minimax rule cannot be much improved upon.

Another point is that the statistician must consider the *consequences* of his actions and strategies in order to select a good rule for decision making (strategy). This will be illustrated further in Exercise 1.1 where it is seen that two reasonable people with different loss tables may react differently to the same data even though they apply the same criterion.

Finally, the example illustrates the relation between statistical theory and the scientific method. Essentially every scientist who designs experiments and presents conclusions is engaging in decision making where the costs involved are the time and money for his experiments, on one hand, and damage to society and his reputation if his conclusions are seriously in error, on the other hand. It is not uncommon to hear of nonsense about conclusions being "scientifically proved." In real life very little can be certain. If we tossed a coin a million times, we would not know the exact probability of its falling heads, and we could (although it is unlikely) have a large error in our estimate of this probability. It is true that, generally, scientists attach a large loss to the act of presenting a conclusion which is in error and, hence, they tend to use strategies which decide to present conclusions only when the evidence in favor is very great.

4. SUMMARY

The essential components in decision-making problems are the following:

1. *The available actions* $a_1, a_2, \cdots$. A problem exists when there is a choice of alternative actions. The consequence of taking one of these actions must depend on the state of nature. Usually the difficulty in deciding which action to take is due to the fact that it is not known which of

2. *the possible states of nature* $\theta_1, \theta_2, \cdots$ is the true one.

3. *The loss table* (*consequence of actions*) measures the cost of taking actions $a_1, a_2, \cdots$ respectively when the states of nature are $\theta_1, \theta_2, \cdots$ respectively.

Given the loss table, it would be easy to select the best action if the state of nature were known. In other words, the *state of nature* represents the underlying "facts of life," knowledge of which we would like in order to choose a proper action. A state of nature can be made to reveal itself partially through an

4. *experiment*, which leads to one of

4(a). *the possible observations* $z_1, z_2, \cdots$. The probabilities of these various observations depend upon what the state of nature actually is.

4(b). *The table of frequency responses* shows this dependence.

An informative experiment is one where the frequencies of responses depend heavily on the state of nature. Each of

5. *the available strategies* $s_1, s_2 \cdots$ is a recipe which tells us how to react (which action to take) to any possible data.

For example, Paul Revere's assistant was to hang lamps according to the strategy "One if by land, two if by sea" (and implicitly, none if the British did not come).

6. Finally, *the average loss table* gives us the consequence of the strategies. It is in terms of this table that we must determine what constitutes a good strategy. With a well-designed or informative experiment, there will be strategies for which the average loss will tend to be small.

An intermediate step in the computation of the average loss table consists of evaluating a table of *action probabilities* for each strategy. This table of action probabilities tells how often the strategy will lead to a given action when a certain state of nature applies.

**Exercise 1.1.* Suppose that our contractor is a "fly-by-night operator," and his loss table is not given by Table 1.1 but by the following table (Table 1.6):

TABLE 1.6

LOSSES FOR "FLY-BY-NIGHT OPERATOR"

States of Nature \ Action	a_1	a_2	a_3
θ_1	1	2	3
θ_2	1	2	3
θ_3	1	2	3

What would constitute a good strategy for him?

Exercise 1.2. Suppose that the contractor (the original respectable business man and not the fly-by-night operator) discovers a new type of wire available on the market. This is a 25-amp wire. Even though this wire would not be appropriate for any state of

* Starred exercises should be assigned to students.

nature if the state of nature were known, a reasonable strategy may call for its use occasionally. Extend Table 1.1 so that, for a_4 (installing 25-amp wire), the losses are 2.4, 2.4, and 4 for θ_1, θ_2, and θ_3 respectively. Introduce at least two new "reasonable" strategies which occasionally call for a_4. Evaluate the associated average losses. Are these strategies admissible? If one of them is admissible, describe the circumstances, if any, in which you would use it.

**Exercise 1.3.* Construct and carry through the details of a problem which exhibits all the characteristics of decision making summarized in Section 4. Present and evaluate some reasonable and unreasonable strategies. Indicate which, if any, of the simplifications introduced into this problem are liable to be serious in that they may strongly affect the nature of a good strategy.

Remarks. Among other interesting problems are those which deal with diagnosis of illness, the question of wearing rainclothes, and the decision between not shaving and arriving late on a date.

In the contractor problem the experiment yields only one out of four possible observations. If the contractor had supplemented his question with a count of the number of electric appliances in the house, his data would have consisted of two observations out of very many possible pairs of observations. Then the number of available strategies would have been multiplied considerably. For the sake of simplicity, it is suggested that you construct a problem where the experiment will yield only one *of a few* possible observations; for example, one temperature reading on a thermometer with possible readings "normal," "warm," and "hot." Please note that if there is only one possible observation, you have no experiment. Thus the executive who asks his "yes" man for criticism obtains no relevant information.

Exercise 1.4. The minimax average regret principle is a modification of the minimax average loss principle. In Table 1.1, we note that, if θ_1 were the state of nature, the least loss of all the actions considered would be 1. For θ_2 and θ_3, these minimum losses would be 2 and 3. These losses may be considered unavoidable and due to the state of nature. Each loss then represents this unavoidable loss plus a regret (loss due to ignorance of θ). Subtracting these unavoidable losses, we obtain the regret table, Table 1.7, and the average regret table, Table 1.8.

Table 1.8 could be obtained in either of two ways. First, we could construct it from Table 1.7 just as Table 1.4 was constructed from Table 1.1. Alternatively, we can subtract the minimum possible losses 1, 2, and 3 from the θ_1, θ_2, and θ_3 rows respectively, of Table 1.4.

TABLE 1.7

REGRET IN CONTRACTOR EXAMPLE

States of Nature \ Action	a_1	a_2	a_3
θ_1	0	1	2
θ_2	3	0	1
θ_3	4	3	0

TABLE 1.8

AVERAGE REGRET IN CONTRACTOR EXAMPLE

States of Nature \ Strategy	s_1	s_2	s_3	s_4	s_5
θ_1	0	0.5	2	0	2
θ_2	1.5	0.5	1.0	3.0	0.5
θ_3	1	0	0	4	3.67
Max. average regret	1.5	0.5	2	4	3.67

The strategy which minimizes the maximum regret is s_2. Apply these ideas in your example of Exercise 1.3 by evaluating the minimax average loss and minimax average regret strategies. Must these principles always yield the same strategy for all decision-making examples?

Exercise 1.5. When Mr. Clark passed through East Phiggins in its pioneer days, he was expected to bring back information to guide the coming settlers on whether or not to take along air conditioners. The available actions for the settlers were a_1, to take air conditioners, and a_2, to leave them behind. He considered three possible states of nature describing the summer weather in East Phiggins; θ_1—80% of the days are very hot and 20% hot; θ_2—50% of the days are very hot, 30% hot, and 20% mild; and θ_3—20% of the days are very hot, 30% hot, and 50% mild. Since he passed through Phiggins in one day, his data consisted of one of the following possible observations describing that day: z_1 very hot, z_2 hot, and z_3 mild. One of the settlers, after considerable introspection involving the difficulty of carrying air conditioners which would replace other important items and the discomfort without them, represented his losses by Table 1.9.

TABLE 1.9

LOSSES IN AIR-CONDITIONER PROBLEM

States of Nature \ Actions	a_1	a_2
θ_1	3	10
θ_2	3	4
θ_3	3	1

The first row of the frequency of response table is given by

	z_1	z_2	z_3
θ_1	0.8	0.2	0

(a) Complete the frequency of response table.

(b) List five strategies and evaluate their average losses.

(c) Among these five strategies, point out which is the minimax average loss strategy, and which is the best strategy if θ_1, θ_2, and θ_3 were equally likely.

(d) Mr. Clark passed through on a hot day. What actions do the strategies of (c) call for?

Exercise 1.6. For a decision-making problem with losses, frequency of responses, and strategies given in Tables 1.10 (a), (b), and (c), evaluate the average losses for strategies s_1, s_2, s_3, and s_4. How many possible strategies are there?

TABLE 1.10(a)

LOSSES

Actions / States of Nature	a_1	a_2	a_3	a_4
θ_1	0	2	4	8
θ_2	6	4	2	5
θ_3	3	2	1	0

TABLE 1.10(b)

FREQUENCY OF RESPONSE

Responses / States of Nature	z_1	z_2	z_3	z_4
θ_1	0	0.4	0.6	0
θ_2	0.4	0.2	0.2	0.2
θ_3	0.1	0.2	0.3	0.4

TABLE 1.10(c)

ACTIONS REQUIRED BY CERTAIN STRATEGIES

Responses / Strategies	z_1	z_2	z_3	z_4
s_1	a_1	a_2	a_3	a_4
s_2	a_2	a_3	a_4	a_3
s_3	a_3	a_3	a_3	a_3
s_4	a_4	a_4	a_1	a_1

Exercise 1.7. Jane Smith can cook spaghetti, hamburger, or steak for dinner. She has learned from past experience that if her husband is in a good mood she can serve him spaghetti and save money, but if he is in a bad mood, only a juicy steak will calm him down and make him bearable. In short, there are three actions:

a_1 prepare spaghetti;
a_2 prepare hamburger;
a_3 prepare steak.

Three states of nature:

θ_1 Mr. Smith is in a good mood;
θ_2 Mr. Smith is in a normal mood;
θ_3 Mr. Smith is in a bad mood.

The loss table is:

	a_1	a_2	a_3
θ_1	0	2	4
θ_2	5	3	5
θ_3	10	9	6

The experiment she performs is to tell him when he returns home that she lost the afternoon paper. She foresees four possible responses. These are:

z_1 "Newspapers will get lost";
z_2 "I keep telling you 'a place for everything and everything in its place'";
z_3 "Why did I ever get married?"
z_4 an absent-minded, far-away look.

The frequency of response table is

	z_1	z_2	z_3	z_4
θ_1	0.5	0.4	0.1	0
θ_2	0.2	0.5	0.2	0.1
θ_3	0	0.2	0.5	0.3

(a) List four strategies and evaluate their average losses.

(b) Point out the minimax average loss strategy.

(c) Which is the best of these strategies if Mr. Smith is in a good mood 30% of the time and in a normal mood 50% of the time?

Exercise 1.8. Replace Table 1.1 by

	a_1	a_2	a_3
θ_1	1	2	3
θ_2	10	2	3
θ_3	10	10	3

Evaluate the average losses for the five strategies of Table 1.3.

SUGGESTED READINGS

Problems in statistics and in game theory are very closely related. For example, statistical problems are sometimes referred to as games against nature. A very brief discussion will be found in Appendix F_1.

An elementary exposition of game theory is given in:

[1] Williams, J. D., *The Compleat Strategyst, Being a Primer on the Theory of Games of Strategy*, McGraw-Hill Book Co., New York, 1954, Dover Publications, Inc., New York, 1986.

A popularized version of decision making applied to statistics will be found in:

[2] Bross, I.D.J., *Design for Decision*, The Macmillan Co., New York, 1953.

CHAPTER 2

Data Processing

1. INTRODUCTION

In the preceding chapter we remarked that the number of available strategies increases rapidly with the amount of data available. When there are many observations, at least some rough methods of summarizing their content should be considered. Of course, what constitutes a good method of summarizing data is determined by the uses to which the data will be put. For example, if a manufacturer were interested in the probability that an item produced by a certain machine will be defective, this quantity could be estimated by taking the proportion of defectives obtained during the output of one day. If, however, he were interested in knowing whether his machine was "in control," i.e., whether the probability of producing a defective was constant during the day, then he might compare the proportion of defectives obtained during the morning with that for the afternoon.

In this chapter we shall very briefly present some standard procedures of data processing. This chapter may well be considered a digression (but a necessary one) in the presentation of statistical ideas, especially since a better understanding of the value of these procedures will come after the basic ideas of statistics are more thoroughly examined.

2. DATA REPRESENTATION

Example 2.1. A new process for making automobile tires was developed by the research staff of the Wearwell Rubber Company. They took a sample of 60 of these tires and tested them on a device which simulates the ordinary environment of automobile tires in use. When the tread was worn off, the mileage recorded for the tire was listed. (These numbers are rounded off to the nearest hundred miles, e.g., 43,689 miles and 43,743 miles are listed as 43.7

thousand miles.) The list is given in Table 2.1 and called the raw data since they are unprocessed (except for the rounding).

TABLE 2.1

RAW DATA: WEAR FOR SIXTY NEW WEARWELL TIRES (THOUSANDS OF MILES)

40.1	50.2	48.9	40.4
47.5	43.6	42.3	43.7
46.9	48.8	44.4	41.5
45.8	45.0	47.7	43.3
47.2	46.0	47.7	43.9
45.2	44.2	45.5	43.9
44.1	41.3	42.8	46.7
42.9	48.2	39.1	44.7
47.0	49.8	37.4	43.6
52.0	47.9	40.7	46.3
42.1	42.6	40.6	43.1
42.6	49.1	46.9	41.8
41.9	46.1	46.4	45.5
43.9	50.8	44.5	48.3
46.7	51.2	43.4	44.8

An array of 60 numbers such as this contains all the information afforded by the experiment, but in its present form it is a more or less meaningless jumble of numbers which the mind cannot grasp nor interpret. A useful way of organizing the data is to group them by size. Often about 10 groups are a satisfactory choice; so we note that the smallest rounded mileage among the 60 is 37.4 and the largest is 52.0. The total *range* is $14.6 = 52.0 - 37.4$. Since $14.6/10 = 1.46$ is near to 1.5, we will break up the range of values into separate *intervals* each of length 1.5, and every observation should then fall into one of ten intervals. In choosing just which possible set of intervals of length 1.5 to use, it is convenient to select them so that their *mid-points* are easy numbers to work with, such as 38.0, 39.5, 41.0, etc., rather than inconvenient ones such as 38.23, 39.73, 41.23, etc. Choice of interval length and interval mid-points establishes the interval boundaries since they must lie half way between mid-points. For the mid-points 38.0, 39.5, 41.0, etc., the boundaries are 37.25, 38.75, 40.25, etc.

We now write down a list of the intervals showing the boundaries (and also the mid-points); then we run through the data in Table 2.1 and, as we come to each observation, enter a tally mark for the interval in which the observation lies. The result is shown in Table 2.2.

TABLE 2.2

TALLY SHEET AND FREQUENCY DISTRIBUTION FOR WEAR DATA ON WEARWELL TIRES (THOUSAND OF MILES)

Mid-points	Intervals	Tallies	Number of Observations in Interval (frequency)	Proportion of Observations in Interval (relative frequency)
38.0	37.25–38.75	/	1	0.0167
39.5	38.75–40.25	//	2	0.0333
41.0	40.25–41.75	𝍸	5	0.0833
42.5	41.75–43.25	𝍸 ////	9	0.1500
44.0	43.25–44.75	𝍸 𝍸 ///	13	0.2167
45.5	44.75–46.25	𝍸 ///	8	0.1333
47.0	46.25–47.75	𝍸 𝍸 /	11	0.1833
48.5	47.75–49.25	𝍸 /	6	0.1000
50.0	49.25–50.75	//	2	0.0333
51.5	50.75–52.25	///	3	0.0500

This table was constructed by reading down the list of data in Table 2.1 and, upon reading 40.1, entering a tally in the interval 38.75–40.25, then upon reading 47.5, entering a tally in the interval 46.25–47.75, etc. The number of tallies for each interval can be seen visually, literally "giving a picture of the data." After the tally is completed, the last two columns may be filled in. The first of these lists the number of observations in each interval (frequency). Relative frequencies, shown in the last column, are obtained by dividing the *frequency* by the sample size (60 in this example).

In arriving at this convenient representation, or tabular summary of the 60 observations, we tacitly avoided one pitfall of which the student should take note. Reference to Table 2.2 shows that the mid-points chosen all have one figure after the decimal point, exactly as the rounded data themselves have, and that the boundaries of the intervals all have an extra decimal digit 5. This forces each rounded observation to lie *inside* one interval, and never on the boundary between two intervals. It also ensures that any rounded observation will be tallied in the same interval as it would have been had it not been rounded. For example, the fourth observation in column 3 of Table 2.1 is shown as 47.7, and it was tallied in the interval 46.25–47.75. Before that observation was rounded, it might have been 47.659 or 47.749 ; in either case it would be rounded to 47.7 and, in either case, it would properly belong to the interval into which it was finally put. But suppose that we had chosen a system of mid-points and an interval length so that one of the cell boundaries was 47.7. In this case, we would not know into which interval to tally 47.7. This pitfall (of having cell boundaries be possible values for the data) was avoided by choosing the mid-points to be possible values (i.e., have one figure after the decimal place in this example) and using an interval length which was an *odd* number (1.5) instead of an even number (such as 1.4).

Exercise 2.1. List appropriate mid-points and intervals for a tally sheet with eight instead of ten intervals for wear data on Wearwell tires.

Exercise 2.2. List appropriate mid-points for a tally sheet with eight intervals based on the weights of students in your class. The weights are to be rounded off to the nearest pound.

3. GRAPHICAL REPRESENTATIONS: HISTOGRAM

From the data as summarized in Table 2.2, we can derive a simple graphical representation which yields at a glance an effective picture of the data, and which sacrifices practically no information. This graph is called the histogram.

Graph paper consists of paper marked with horizontal and vertical lines. One of the horizontal lines is called the *horizontal axis*, and one of the vertical lines is called the *vertical axis*. The point where these two lines intersect is called the *origin*. Each point on the paper can be represented by a pair of numbers. These are called the *coordinates*. The first coordinate is also called the *abscissa* and represents the horizontal distance of the point from the vertical axis. The second coordinate is called the *ordinate* and represents the vertical distance of the point from the horizontal axis. For points below the horizontal axis the ordinate is negative, and for points to the left of the vertical axis the abscissa is negative. In Figure 2.1, the coordinates of the points P, Q, R, and S are $(2, 1)$, $(-2, 1)$, $(-2, -1)$, and $(2, -1)$ respectively. For point P, 2 is the abscissa and 1 is the ordinate.

For the data on the new Wearwell tires, the histogram is obtained as follows. First we refer to Table 2.2. In Figure 2.2, mark off,

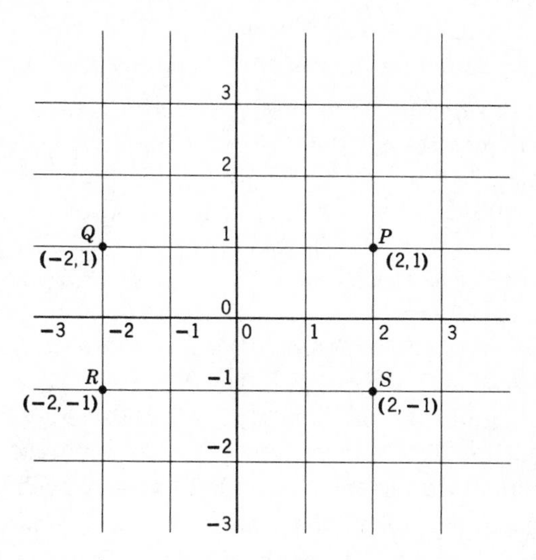

Figure 2.1. Graph paper.

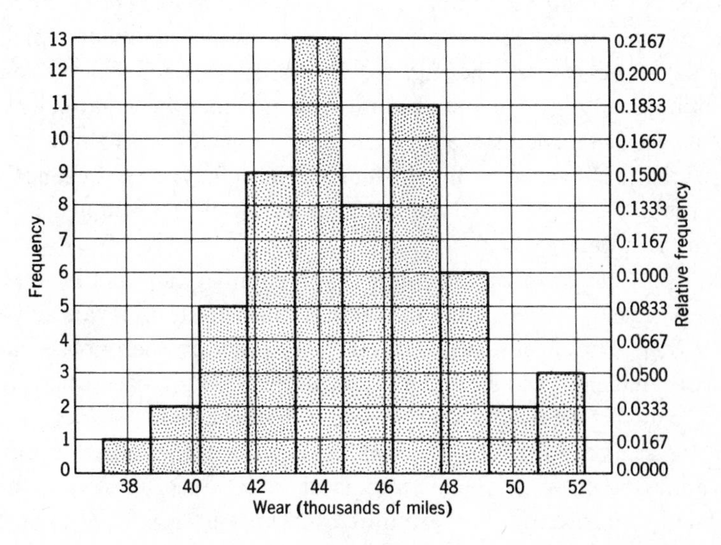

Figure 2.2. Histogram for wear data for sixty new Wearwell tires.

along the horizontal axis, intervals corresponding to those in Table 2.2. For each interval we draw a " bar " at a height equal to the number of observations in that interval.

The following characteristics of the histogram should be noted.

1. *All intervals are of the same length.* Because of this, the histogram gives a good idea of the " relative frequency " of getting various mileages. If the first three intervals had been bunched into one large interval, it would have had eight observations. A bar of size 8 from 37.25 to 41.75 would give the misleading impression that a wear of 38.0 is as common as one of 45.5. This is definitely not the case.

2. *The intervals have mid-points which are three-digit numbers just like the tabulated data.* It is convenient for the mid-points to have no more digits than the tabulated data because a good deal of computing which follows is then simplified. To select the intervals to have such mid-points, one need only select one such mid-point (such as 38.0) and make sure that the length of the interval has the same number of digits after the decimal point. For example, a length of 1.3 or 1.4 would have accomplished this aim.

3. *The end points of the intervals have one more digit after the decimal point than the tabulated data. This last digit is a* 5. When our intervals have this property, there is no doubt about which interval an observation belongs to. Furthermore, each observation is automatically inserted into the interval in which it would belong had it not been rounded off. To achieve intervals with this characteristic, it is necessary for the lengh to be odd. Hence the length of 1.4 is not adequate but 1.3, 1.5, or 1.7 would accomplish the desired results.

4. *About ten intervals are used.* This number can fluctuate for several reasons. One should make sure that most intervals have more than 4 or 5 observations. If there are very many observations, then it pays to select the intervals so as to increase both the number of intervals and the number of observations per interval. If there are very few observations, it does not pay to draw a histogram. One may achieve approximately a specified number of intervals by subtracting the smallest observation from the largest and dividing this by the desired number of intervals. The interval

length should be slightly larger than this number. In our case we wanted a length a bit larger than $(52.0 - 37.4)/10 = 1.46$.

In review, we want a fixed interval length which is odd (in the units in which the observations are tabulated) and of a size so that we have a reasonable number of intervals. A mid-point should be selected so that it is convenient, i.e., it has no more digits after the decimal than the tabulated observations.

In the following we shall use the terms "cell" and "cell frequency" to mean interval and the number of observations in the interval. The relative frequency is the proportion of the total number of observations that lie in a given cell. In the cell 41.75–43.25 there are nine observations which constitute $9/60 = 3/20 = 0.15$ (or 15%) of the total sample. Both frequencies and relative frequencies are often useful.

The only information in Table 2.1 which is not available in Figure 2.2 is the precise disposition of the observations within each cell. For most purposes, this loss of information is not serious and is readily compensated for by the advantages of the pictorial representation.

4. GRAPHICAL REPRESENTATIONS: CUMULATIVE FREQUENCY POLYGON

A second graphical representation which is useful is that of the cumulative frequency polygon. This representation is obtained as

TABLE 2.3

SUBDIVISION OF WEAR DATA FOR SIXTY NEW WEARWELL TIRES INTO INTERVALS WITH CUMULATIVE FREQUENCIES (THOUSAND OF MILES)

Cell Number	Cells	Cell Mid-points	Frequency	Relative Frequency	Cumulative Frequency	Cumulative Relative Frequency
1	37.25-38.75	38.0	1	0.0167	1	0.0167
2	38.75-40.25	39.5	2	0.0333	3	0.0500
3	40.25-41.75	41.0	5	0.0833	8	0.1333
4	41.75-43.25	42.5	9	0.1500	17	0.2833
5	43.25-44.75	44.0	13	0.2167	30	0.5000
6	44.75-46.25	45.5	8	0.1333	38	0.6333
7	46.25-47.75	47.0	11	0.1833	49	0.8167
8	47.75-49.25	48.5	6	0.1000	55	0.9167
9	49.25-50.75	50.0	2	0.0333	57	0.9500
10	50.75-52.25	51.5	3	0.0500	60	1.0000

follows. First, Table 2.2 is extended by inserting a column of cumulative frequencies. That is to say, a new column is added which has the number of observations falling in all cells to the left of or in the given cell. (See Table 2.3.)

The corresponding cumulative frequency polygon (Figure 2.3) is obtained as follows: For tire wear corresponding to cell boundary 37.25, put a dot at the horizontal axis (since zero observations were less than or equal to 37.25). For tire wear 38.75, a dot is put at a height of one unit since one observation was less than or equal to 38.75. For tire wear 40.25, a dot is put at a height of three units since three observations were no larger than 40.25. Continue in this fashion and connect successive dots by line segments.

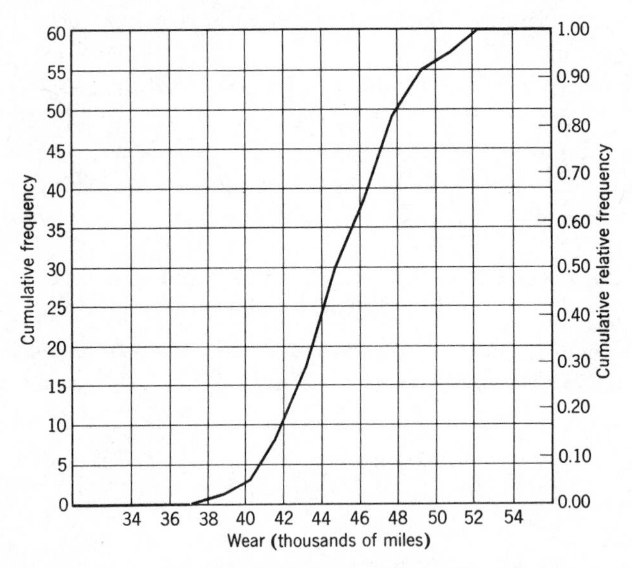

Figure 2.3. Cumulative frequency polygon for the wear of sixty new Wearwell tires.

The cumulative frequency polygon has the advantage that one can readily compute the number of observations lying between any two values. For example, the number of observations between 40.25 and 47.75 is the difference between the corresponding heights, i.e., $49 - 3 = 46$. This method works for the case where the given values correspond to points plotted. For other values, say 45, the height of 31.33 is an estimate of the number of observations not exceeding the given mileage. In fact this estimate, which is not even a whole number, is based on the tacit, although not necessarily correct, assumption that the observations are evenly spread throughout each cell.

Another advantage of this polygon is that it furnishes a convenient method of estimating the "percentiles" of the sample. A percentile, say the 20th percentile of the sample, is the number such that 20% of the observations are less than or equal to this number, and the rest are no smaller. Referring to our cumulative frequency polygon, we see that the 20th percentile of the sample is approximately 42.42. A particular percentile of especial interest is the 50th percentile of the sample, which is called the sample *median*. According to our cumulative polygon, the median is approximately 44.75. Suppose that the observations had been listed in ascending order (ordered observations). Then the median would

TABLE 2.4

ORDERED DATA: WEAR FOR SIXTY NEW WEARWELL TIRES ARRANGED IN ASCENDING ORDER (THOUSANDS OF MILES)

37.4	42.9	44.8	47.2
39.1	43.1	45.0	47.5
40.1	43.3	45.2	47.7
40.4	43.4	45.5	47.7
40.6	43.6	45.5	47.9
40.7	43.6	45.8	48.2
41,3	43.7	46.0	48.3
41.5	43.9	46.1	48.8
41.8	43.9	46.3	48.9
41.9	43.9	46.4	49.1
42.1	44.1	46.7	49.8
42.3	44.2	46.7	50.3
42.6	44.4	46.9	50.8
42.6	44.5	46.9	51.2
42.8	44.7	47.0	52.0

be the observation which was in the middle of the list. However, for our sample size 60, which is even, there is no middle of the list. Then the median is considered to be halfway between the 30th and 31st ordered observation. In this case the 30th and 31st ordered observations are 44.7 and 44.8, and the median is precisely 44.75, which, accidentally, coincides with the above approximation. The median has the property that as many observations exceed it as are exceeded by it. For an odd sample size, say 61, the median would be the 31st ordered observation.

Data are called "grouped" when the records or tables present only the numbers of observations in certain cells rather than the original observations. Hence, Tables 2.2 and 2.3 and Figures 2.2 and 2.3 describe grouped data and suffer from the slight loss of information described at the end of Section 3. In particular, the cumulative frequency polygon is the grouped version of the so-called cumulative frequency graph.

To prepare a cumulative frequency graph, it is necessary to arrange the data in increasing order of size. Table 2.4 shows the data so arranged.

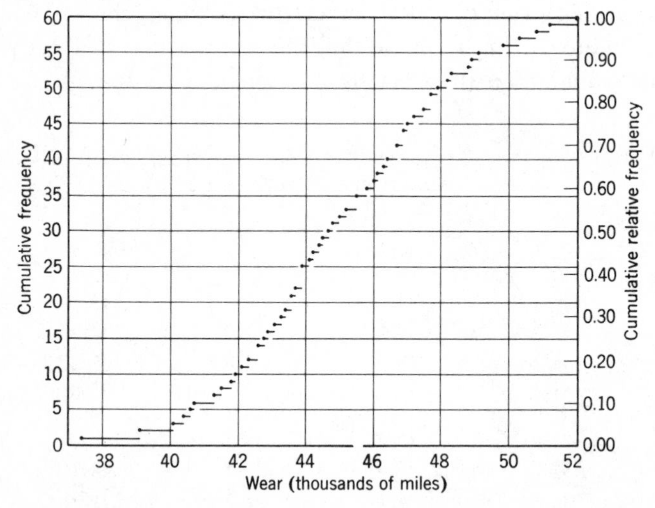

Figure 2.4. Cumulative frequency graph for the wear of sixty new Wearwell tires.

The cumulative frequency graph, presented in Figure 2.4, gives the exact number of observations less than or equal to a given value ; for example, the height of the graph at 43.9, 43.95, 44.0, 44.08 is 25 because, as can be seen from Table 2.4, there are just 25 observations which are no larger than 43.9, and the same 25 are also no larger than 43.95 or 44.0 or 44.08, but there are 26 observations no larger than 44.1, so the graph rises by one step at that point.

This graph is necessarily similar to the grouped representation,

Figure 2.3. If samples are very small, then the cumulative frequency graph is preferable to the cumulative frequency polygon. But if samples are large enough to justify grouping data, the polygon is to be preferred because it is much easier to draw, and can be constructed by cumulating the frequencies obtained from tallying the observations. In large samples, tallying is easier than preparing the list of ordered observations.

**Exercise 2.3.* Draw a histogram and cumulative frequency polygon for the 75 observations in Table 2.5 on annual rainfall on the MacDonald nylon farm in South Phiggins. South Phiggins has the following interesting geographic features. The soil is uniform throughout the region and will allow the growth of an amount of nylon proportional to the annual rainfall. No other crop can be economically grown on this soil. Secondly, the rainfall is affected considerably by small variations in location. Hence the value of a farm is dependent upon its "average" annual rainfall, and for 75 years the annual rainfall on each farm has been recorded by law by the county clerk.

TABLE 2.5

ANNUAL RAINFALL ON THE MACDONALD NYLON FARM 1881–1955 (INCHES)

Year	Rainfall	Year	Rainfall	Year	Rainfall	Year	Rainfall
1881	717	1901	887	1921	798	1941	651
1882	811	1902	805	1922	747	1942	939
1883	748	1903	773	1923	760	1943	881
1884	848	1904	766	1924	729	1944	791
1885	943	1905	752	1925	791	1945	737
1886	643	1906	619	1926	788	1946	703
1887	754	1907	824	1927	831	1947	754
1888	874	1908	635	1928	834	1948	853
1889	820	1909	673	1929	780	1949	723
1890	905	1910	824	1930	843	1950	685
1891	751	1911	987	1931	826	1951	844
1892	802	1912	931	1932	872	1952	881
1893	583	1913	816	1933	617	1953	756
1894	629	1914	754	1934	796	1954	820
1895	747	1915	861	1935	792	1955	801
1896	885	1916	733	1936	951		
1897	661	1917	739	1937	668		
1898	818	1918	638	1938	760		
1899	763	1919	920	1939	805		
1900	766	1920	808	1940	809		

Exercise 2.4. The cumulative frequency polygon is a useful tool for graphically comparing two sets of data ; for this purpose it can be much more suitable than the histogram. Below are two frequency distributions, presenting (in grouped form) the heights (in inches) of 50 men and 40 women enrolled in the same statistics course. (see Tables 2.6 and 2.7.)

(a) Complete the two tables by calculating and entering the relative frequencies and the cumulative relative frequencies.

TABLE 2.6

HEIGHTS OF FORTY WOMEN (INCHES)

Mid-point	Interval	Frequency	Relative Frequency	Cumulative Relative Frequency
62	61.5–62.5	2		
63	62.5–63.5	4		
64	63.5–64.5	7		
65	64.5–65.5	10		
66	65.5–66.5	8		
67	66.5–67.5	5		
68	67.5–68.5	3		
69	68.5–69.5	0		
70	69.5–70.5	1		

TABLE 2.7

HEIGHTS OF FIFTY MEN (INCHES)

Mid-point	Interval	Frequency	Relative Frequency	Cumulative Relative Frequency
63	62.5–63.5	1		
64	63.5–64.5	0		
65	64.5–65.5	3		
66	65.5–66.5	5		
67	66.5–67.5	4		
68	67.5–68.5	6		
69	68.5–69.5	4		
70	69.5–70.5	5		
71	70.5–71.5	7		
72	71.5–72.5	5		
73	72.5–73.5	4		
74	73.5–74.5	4		
75	74.5–75.5	0		
76	75.5–76.5	2		

(b) In *one* graph plot the two histograms using relative frequency for the ordinate. (If this is correctly carried out for these data, the two histograms will overlap each other.)

(c) Plot the cumulative frequency polygons for the two samples together on a second sheet of paper. Use relative frequency for the ordinate so that both polygons will rise to the same height.

Note how (c) is more useful than (b) for comparing the two samples.

5. DESCRIPTIVE MEASURES: SUMMATION

For the example of wear for 60 tires, there are a couple of numbers or measures which indicate much of the essential information contained in the entire *sample* of 60 observations. One of these measures is the *sample mean* or *average*. In the above example, the sample mean is

$$\frac{40.1 + 47.5 + 46.9 + \cdots + 44.8}{60} = 45.008.$$

In general, the sample mean or average is defined as the sum of all the observations divided by the number of observations in the sample. To deal with this measure, and at least one other important one, it is convenient to introduce the summation and subscript notation.

If a typical observation is labeled[1] $\mathbf{X}$, and it is desired to distinguish between successive observations, we can do so by labeling the first $\mathbf{X}_1$ (read $\mathbf{X}$ sub one), the second $\mathbf{X}_2$, the third $\mathbf{X}_3$, etc. In general the ith observation would be $\mathbf{X}_i$. If there are 60 observations, the last will be $\mathbf{X}_{60}$. If there were n observations, the last would be $\mathbf{X}_n$. We represent the sum of these observations by either $\sum_{i=1}^{n} \mathbf{X}_i$ (read aloud as "summation from one to n of $\mathbf{X}$ sub i") or

$$\mathbf{X}_1 + \mathbf{X}_2 + \mathbf{X}_3 + \cdots + \mathbf{X}_n$$

[1] Throughout this book we shall indicate observations or numbers which can be obtained from observations by boldfaced symbols. This use of a different kind of print enables and indeed requires one to distinguish random quantities. This is essential to clear thinking in statistics. Since boldfaced symbols cannot easily be written, we recommend for use on paper or at the blackboard that the boldfaced symbols be replaced by a corresponding symbol with an extra line through it. For example, we can use, X̸, Y̸, $, ꝑ̂ for $\mathbf{X}$, $\mathbf{Y}$, $\mathbf{s}$, $\hat{\mathbf{p}}$.

or

$$\mathbf{X}_1 + \mathbf{X}_2 + \mathbf{X}_3 + \cdots + \mathbf{X}_{n-1} + \mathbf{X}_n$$

or sometimes by slightly modified versons of the last two forms. The reader may note that for $n = 2$

$$\sum_{i=1}^{n} \mathbf{X}_i = \mathbf{X}_1 + \mathbf{X}_2$$

and it seems ridiculous, and even slightly misleading, to use the notation

$$\mathbf{X}_1 + \mathbf{X}_2 + \mathbf{X}_3 + \cdots + \mathbf{X}_n$$

in this case. Nevertheless, there are situations where we do not wish to specify the value of n and the latter representation will be very useful because of its suggestiveness.

In the notation $\sum_{i=1}^{n} \mathbf{X}_i$, the symbol i represents the possible values of the subscript. These are the integers from 1 to n. If for some reason we could not use the symbol i, we could represent these possible values by the subscript j or by some other symbol which has not previously been assigned some meaning. Then $\sum_{j=1}^{n} \mathbf{X}_j$ would indicate the same sum. The letter i has no special significance since

$$\sum_{i=1}^{4} \mathbf{X}_i = \sum_{j=1}^{4} \mathbf{X}_j = \mathbf{X}_1 + \mathbf{X}_2 + \mathbf{X}_3 + \mathbf{X}_4.$$

If for some reason it was decided to have only the sum of those $\mathbf{X}$'s with subscripts between 5 and n, we would write $\sum_{i=5}^{n} \mathbf{X}_i$. More generally, for the sum of those $\mathbf{X}$'s with subscripts between m and n inclusively, we have the representations

$$\sum_{i=m}^{n} \mathbf{X}_i = \mathbf{X}_m + \mathbf{X}_{m+1} + \mathbf{X}_{m+2} + \cdots + \mathbf{X}_n.$$

If there are two types of observations being taken simultaneously, we may label them $\mathbf{X}_1, \mathbf{Y}_1, \mathbf{X}_2, \mathbf{Y}_2, \cdots, \mathbf{X}_n, \mathbf{Y}_n$. Then we may apply our notation to represent the sum of all the $\mathbf{X} + \mathbf{Y}$'s as follows:

$$\sum_{i=1}^{n} (\mathbf{X}_i + \mathbf{Y}_i) = (\mathbf{X}_1 + \mathbf{Y}_1) + (\mathbf{X}_2 + \mathbf{Y}_2) + (\mathbf{X}_3 + \mathbf{Y}_3) + \cdots + (\mathbf{X}_n + \mathbf{Y}_n).$$

We shall also be interested in expressions such as

$$\sum_{i=1}^{n} X_i^2 = X_1^2 + X_2^2 + X_3^2 + \cdots + X_n^2$$

which represents the sum of the squares of the X's,

$$\sum_{i=1}^{n} aX_i = aX_1 + aX_2 + aX_3 + \cdots + aX_n$$

which represents the sum of aX_1, aX_2, aX_3, etc., and

$$\sum_{i=1}^{n} (X_i - a)^2 = (X_1 - a)^2 + (X_2 - a)^2 + (X_3 - a)^2 + \cdots + (X_n - a)^2.$$

Exercise 2.5. For a sample of heights in inches (X) and weights in pounds (Y) of $n = 7$ West Phiggindians, we obtain

$$\begin{array}{llll} X_1 = 48, & Y_1 = 207; & X_2 = 50, & Y_2 = 200; \\ X_3 = 52, & Y_3 = 210; & X_4 = 50, & Y_4 = 207; \\ X_5 = 48, & Y_5 = 200; & X_6 = 50, & Y_6 = 210; \\ X_7 = 52, & Y_7 = 200. & & \end{array}$$

Evaluate:

$$\sum_{i=1}^{n} X_i,\ \sum_{i=1}^{n} Y_i,\ \sum_{i=1}^{n} (X_i + Y_i),\ \sum_{i=1}^{n} 20X_i,\ \text{and} \sum_{i=1}^{n}(X_i - 50)^2.$$

The following equations are often useful in the applications of the summation symbol.

$$\sum_{i=1}^{n} (X_i + Y_i) = \sum_{i=1}^{n} X_i + \sum_{i=1}^{n} Y_i. \tag{2.1}$$

$$\sum_{i=1}^{n} aX_i = a\left(\sum_{i=1}^{n} X_i\right). \tag{2.2}$$

$$\sum_{i=1}^{n} a = na. \tag{2.3}$$

They may be explained in terms of the following simple interpretations. If the X's add up to 10 and the Y's to 20, the (X + Y)'s should add up to 30. If the X's add up to 10, tripling each X will triple the sum. Adding a number $a = 3$ to itself, $n = 10$ times will give $na = 30$. Please note that for Equations (2.2) and (2.3), the important respect in which a differs from the X's and Y's is that it does not change when i changes.

One may easily derive these equations. Equations (2.1) is developed thus:

$$(2.1)\quad \sum_{i=1}^{n}(X_i+Y_i)=(X_1+Y_1)+(X_2+Y_2)+(X_3+Y_3)+\cdots+(X_n+Y_n).$$

Rearranging the terms so that the X's are next to one another, we have

$$\begin{aligned}\sum_{i=1}^{n}(X_i+Y_i) &= (X_1+X_2+X_3+\cdots+X_n)\\ &\quad +(Y_1+Y_2+Y_3+\cdots+Y_n)\\ &= \sum_{i=1}^{n}X_i+\sum_{i=1}^{n}Y_i.\end{aligned}$$

Equation (2.2) is shown in the following way:

$$(2.2)\quad \begin{aligned}\sum_{i=1}^{n}aX_i &= aX_1+aX_2+aX_3+\cdots+aX_n\\ &= a(X_1+X_2+X_3+\cdots+X_n)=a\sum_{i=1}^{n}X_i\end{aligned}$$

Equation (2.3) is shown by:

$$(2.3)\quad \sum_{i=1}^{n}a=\underbrace{a+a+a+\cdots+a}_{n\text{ times}}=na.$$

**Exercise 2.6*. Use the summation symbol to abbreviate the following expressions:

(a) $X_3^2+X_4^2+X_5^2+X_6^2+X_7^2$

(b) $X_1+2X_2+3X_3+\cdots+nX_n$

(c) $1^2(x-1)+2^2(x-2)+\cdots+7^2(x-7)$.

**Exercise 2.7*. Given that $\sum_{i=1}^{7}X_i=17$ and $\sum_{i=1}^{7}X_i^2=53$, find

(a) $\sum_{i=1}^{7}(X_i-2)$ (b) $\sum_{j=1}^{7}(2X_j+1)$ (c) $\sum_{j=1}^{7}(X_j+3)^2$.

Exercise 2.8. Compute

(a) $\sum_{i=1}^{10}i$ (b) $\sum_{i=2}^{5}(i+1)^2$ (c) $\sum_{j=1}^{4}(j^2+4)$.

Exercise 2.9. Let $x_1=4$, $x_2=3$, $x_3=1$, $x_4=3$, $y_1=2$, $y_2=0$, $y_3=-2$, $y_4=-4$. Compute

(a) $\sum_{i=1}^{4} x_i$ (b) $\sum_{i=1}^{4} y_i$ (c) $\sum_{i=1}^{4} (x_i + y_i)$

(d) $\sum_{i=1}^{4} x_i^2$ (e) $\sum_{i=1}^{4} y_i^2$ (f) $\sum_{i=1}^{4} (x_i + y_i)^2$

(g) $\sum_{i=1}^{4} (x_i - y_i)^3$.

Exercise 2.10. Given only that $\sum_{i=1}^{7} \mathbf{X}_i = 17$ and $\sum_{i=1}^{7} \mathbf{X}_i^2 = 53$, which of the following expressions can be evaluated? Evaluate them.

(a) $\sum_{i=1}^{7} (\mathbf{X}_i + \mathbf{X}_i^2)$ (b) $\sum_{i=1}^{7} \mathbf{X}_i + 2\sum_{j=1}^{7} \mathbf{X}_j^2$ (c) $\sum_{i=1}^{7} i\mathbf{X}_i$

(d) $\sum_{j=1}^{7} (j + \mathbf{X}_j)$ (e) $\sum_{j=1}^{7} (j + \mathbf{X}_j)^2$ (f) $\sum_{i=1}^{7} (2 + \mathbf{X}_i)^3$.

**Exercise 2.11.* Either derive or illustrate with simple examples three of the following equations.

(2.4) $$\sum_{i=1}^{n} (\mathbf{X}_i + \mathbf{Y}_i + \mathbf{Z}_i) = \sum_{i=1}^{n} \mathbf{X}_i + \sum_{i=1}^{n} \mathbf{Y}_i + \sum_{i=1}^{n} \mathbf{Z}_i$$

(2.5) $$\sum_{i=1}^{n} (\mathbf{X}_i - \mathbf{Y}_i) = \sum_{i=1}^{n} \mathbf{X}_i - \sum_{i=1}^{n} \mathbf{Y}_i$$

(2.6) $$\sum_{i=1}^{n} (a\mathbf{X}_i + b\mathbf{Y}_i) = a\sum_{i=1}^{n} \mathbf{X}_i + b\sum_{i=1}^{n} \mathbf{Y}_i$$

(2.7) $$\sum_{i=1}^{n} (a + b\mathbf{X}_i) = na + b\sum_{i=1}^{n} \mathbf{X}_i$$

(2.8) $$\sum_{i=1}^{n} (\mathbf{X}_i - a)^2 = \sum_{i=1}^{n} \mathbf{X}_i^2 - 2a\sum_{i=1}^{n} \mathbf{X}_i + na^2.$$

6. DESCRIPTIVE MEASURES: SAMPLE MEAN AND VARIANCE

Previously we mentioned the sample mean as an important measure associated with the sample. Referring to the definition given (sum of observations divided by sample size), we see that we could denote the *sample mean* of a sample of n observations $\mathbf{X}_1, \mathbf{X}_2, \mathbf{X}_3, \cdots, \mathbf{X}_n$ by

(2.9) $$\overline{\mathbf{X}} = \frac{1}{n}\sum_{i=1}^{n} \mathbf{X}_i = \frac{1}{n}(\mathbf{X}_1 + \mathbf{X}_2 + \mathbf{X}_3 + \cdots + \mathbf{X}_n).$$

In the example of the wear of 60 Wearwell tires, the sample mean has a good deal of significance. Consider the manager of a large fleet of taxis who has to decide on what brand of tires to use. The sample mean divided by the cost per tire is a good indication of how many miles of use he can expect (on the average) per dollar invested in Wearwell tires. In this case, the sample mean is a very relevant measure of what constitutes a "typical" wear for Wearwell tires.

The problem which faces an individual motorist is somewhat different. On one hand, he would like to average a great deal of wear per tire. On the other hand, his costs are affected by the savings which he can make if he trades in five old tires simultaneously for five new ones. If it should happen that the tread on two of his tires wears out at 37,000 miles, while the tread on the others seems only two-thirds worn out, then he may have to buy his new tires individually. It is clear that he would be willing to accept some sacrifice in average mileage per tire if he could reduce the *variability* in the mileage. A measure of variability is of interest to him.

Variability in a sample may be interpreted visually as spread among the points along a line where each point represents the value of an observation. Thus Figure 2.5 depicts the following samples.

Sample A	15, 16, 17, 19, 20
B	13, 16, 17, 19, 22
C	13, 14, 17, 21, 22
D	15, 16, 19, 23, 24

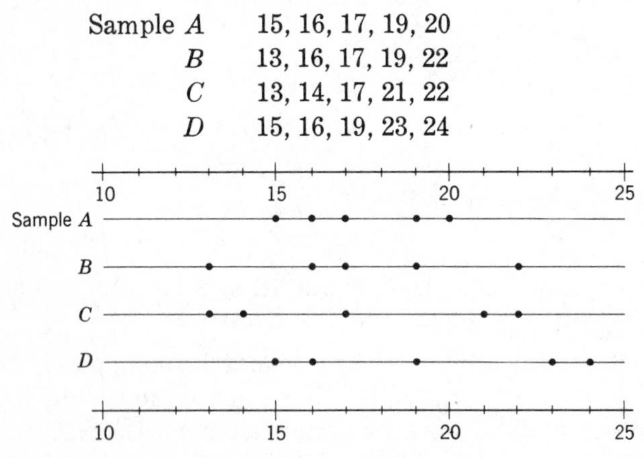

Figure 2.5. Graphical representation of four samples.

Study of the figure shows that sample A is the least spread out, samples C and D are equally spread out (one graph is exactly the same as the other except for being shifted two units). The spread of sample B is between that of A and the other two. "Spread" or "variability" or "dispersion" can be measured in *various* ways. One generally useful measure of variability is the *sample standard deviation* $\mathbf{d}_X$. This measure is obtained as follows. The *sample variance* (or average squared deviation from the mean) is given by

$$\mathbf{d}_X^2 = \frac{1}{n} \sum_{i=1}^{n} (\mathbf{X}_i - \overline{\mathbf{X}})^2. \tag{2.10}$$

The sample *standard deviation* $\mathbf{d}_X$ is the positive square root of the variance.[1] It is clear that $\mathbf{d}_X$ measures spread, because, if the observations are close together (little spread), they will be close to their mean $\overline{\mathbf{X}}$, and $\mathbf{d}_X^2$ will be the average of quantities which are small. On the other hand, if the observations are spread far apart, they will tend to be far from $\overline{\mathbf{X}}$, and $\mathbf{d}_X^2$ will be the average of quantities which are large. The four samples A, B, C, and D have standard deviations of 1.855, 3.007, 3.611, and 3.611 respectively. The Wearwell tire data yield $\mathbf{d}_X = 3.096$ (thousands of miles).

The following five results express basic and important properties of the sample mean and standard deviation.

$$\sum_{i=1}^{n} (\mathbf{X}_i - \overline{\mathbf{X}}) = 0 \tag{2.11}$$

$$\mathbf{d}_X^2 = \frac{1}{n} \sum_{i=1}^{n} (\mathbf{X}_i - \overline{\mathbf{X}})^2 = \left(\frac{1}{n} \sum_{i=1}^{n} \mathbf{X}_i^2\right) - \overline{\mathbf{X}}^2. \tag{2.12}$$

(2.13) If $\mathbf{Y}_i = a + \mathbf{X}_i$, then $\overline{\mathbf{Y}} = a + \overline{\mathbf{X}}$ and $\mathbf{d}_Y = \mathbf{d}_X$.

(2.14) If $\mathbf{U}_i = b\mathbf{X}_i$, where b is positive, $\overline{\mathbf{U}} = b\overline{\mathbf{X}}$ and $\mathbf{d}_U = b\mathbf{d}_X$.

(2.15) If $\mathbf{W}_i = a + b\mathbf{X}_i$, where b is positive, $\overline{\mathbf{W}} = a + b\overline{\mathbf{X}}$ and $\mathbf{d}_W = b\mathbf{d}_X$.

Equation (2.11) expresses the fact that the sum of the deviations from the sample mean is zero. This is easy to show since

[1] For technical reasons, most statisticians find it more convenient to define and use $\mathbf{s}_X$ and $\mathbf{s}_X^2$ for the sample standard deviation and variance where

$$\mathbf{s}_X^2 = \frac{1}{n-1} \sum_{i=1}^{n} (\mathbf{X}_i - \overline{\mathbf{X}})^2 = \frac{n}{n-1} \mathbf{d}_X^2.$$

For reasonably large sample size it makes very little difference whether $\mathbf{s}_X$ or $\mathbf{d}_X$ is used. Since most statisticians prefer to use $\mathbf{s}_X$, we shall mean $\mathbf{s}_X$ and $\mathbf{s}_X^2$ when we use the terms "standard deviation" and "variance" in later chapters. In this chapter, we shall continue to use the slightly simpler measures $\mathbf{d}_X$ and $\mathbf{d}_X^2$.

$$\sum_{i=1}^{n} (X_i - \overline{X}) = (X_1 - \overline{X}) + (X_2 - \overline{X}) + \cdots + (X_n - \overline{X})$$

$$= (X_1 + X_2 + \cdots + X_n) - n\overline{X} = n\overline{X} - n\overline{X} = 0.$$

Equation (2.12) is derived in Appendix E_1 and states that variance d_X^2 is the *average square minus the square of the average* and is very useful for computation. (On the other hand, it is seldom very helpful in theoretical work such as deriving the properties of Equations (2.13) through (2.15)). We illustrate the convenience of using Equation (2.12) by calculating in Table 2.8 the variance of a sample from both the definitional equation and the above result.

TABLE 2.8

VARIANCE

Use of Definitional Equation

i	X_i	$X_i - \overline{X}$	$(X_i - \overline{X})^2$
1	4	-1.167	1.362
2	3	-2.167	4.696
3	5	-0.167	0.028
4	8	2.833	8.026
5	6	0.833	0.694
6	5	-0.167	0.028
Sum	31		14.834
Average	5.167		2.472

$d_X^2 = 2.472$

Calculation Equation

i	X_i	X_i^2
1	4	16
2	3	9
3	5	25
4	8	64
5	6	36
6	5	25
Sum	31	175
Average	5.167	29.167

$$d_X^2 = \frac{175}{6} - \left(\frac{31}{6}\right)^2 = \frac{89}{36} = 2.472$$

For a long list of data, the task of computing and squaring the individual $X_i - \overline{X}$ in the first method is very tedious. The second method is especially well adapted for computation with a desk calculator with which both $\sum_{i=1}^{n} X_i$ and $\sum_{i=1}^{n} X_i^2$ are easy to obtain.

The results of Equations (2.13) through (2.15) state that if 10 (representing 10,000 miles) were added to the wear of each tire, the average wear would be increased by 10, but the standard deviation would be unaffected. This statement may also be interpreted graphically. Adding 10 to each observation merely shifts the histogram 10 units to the right. Thus the "center" or "typical value" is shifted 10 units but the spread is not affected. If the wear of each tire were tripled, both the average and the standard deviation would triple. (The variance would be multiplied by nine.) Equation (2.15) is a combination of the other two equations. In other words, if first we triple the wear $\mathbf{X}_i$ and then add 10, we get a wear which is $10 + 3\mathbf{X}_i$. It is clear that the new average is obtained by tripling the old average and then adding 10. The new standard deviation is obtained by merely tripling the old. These results may also be used to simplify computations. For example, subtracting a constant does not affect the standard deviation. Thus in samples A, B, C, and D we can subtract 17 or some other appropriate number before computing $\mathbf{d}_{\mathrm{X}}^2$. The new numbers so obtained are smaller in magnitude, easier to work with, and, incidentally, are less sensitive to rounding off errors.

Exercise 2.12. Either derive or use the sample 2, 4, 5, 6, 8 to illustrate:

(a) Equation (2.11);
(b) Equation (2.12);
(c) Equation (2.13).

Exercise 2.13. Compute $\mathbf{d}_{\mathrm{X}}^2$ for samples A, B, and D in Section 6.

Exercise 2.14. Compute the mean and standard deviation for the sample of height of seven West Phiggindians (see Exercise 2.5). A table of square roots may prove convenient. (See Appendix B_1.)

**Exercise 2.15.* Why would a prospective buyer of Mr. MacDonald's nylon farm in South Phiggins be especially interested in the mean and standard deviation of the annual rainfall? What sort of buyer would be relatively uninterested in the standard deviation? (See Exercise 2.3.)

Exercise 2.16. The lengths of five cat tails are 12, 14, 16, 10, and 8 inches respectively. For this sample, compute the mean, median, variance, and standard deviation.

Exercise 2.17. Compute the standard deviation of the five numbers 137, 139, 141, 140 and 142. *Hint:* It is convenient to subtract some common number from each of the above since doing so does not change the standard deviation, and smaller numbers are easier to work with.

Exercise 2.18. Show that:

$$\sum_{i=1}^{n} (X_i - a)^2 = \sum_{i=1}^{n} (X_i - \bar{X})^2 + n(\bar{X} - a)^2.$$

For what value of a will the above sum of squares be minimized? *Hint*: $(X_i - a) = (X_i - \bar{X}) + (\bar{X} - a)$.

† 7. SIMPLIFIED COMPUTATION SCHEME FOR SAMPLE MEAN AND STANDARD DEVIATION USING GROUPED DATA

In Table 2.9 we present a simplified scheme for computing the mean and standard deviation, using grouped data. Since the data are grouped we shall for computational simplicity assume that all the observations in a cell are at the mid-point of this cell. As long as the length of the cell interval is small, this assumption cannot lead to any serious error. Fundamentally the method is an application of Equations (2.12) and (2.15), but for our purposes it will suffice to illustrate the method, leaving the explanation of why it works to Appendix E_2.

In Table 2.9 appears a column labeled "associated values" which are designated by w_i. This column increases by increments of one and has the value zero for an arbitrarily selected cell, preferably near the middle. The frequenceis, f_i are displayed in the columns to the right of the w_i. The typical element of the next column is $f_i w_i$. The last column is obtained by multiplying $f_i w_i$ by w_i.

Compute $\sum_{i=1}^{k} f_i w_i$ and $\sum_{i=1}^{k} f_i w_i^2$, where k is the number of cells.

We compute

$$\bar{W} = \frac{1}{n} \sum_{i=1}^{k} f_i w_i$$

$$d_W^2 = \frac{1}{n} \sum_{i=1}^{k} f_i w_i^2 - \bar{W}^2$$

† Sections marked with a dagger may be bypassed at the discretion of the instructor. The method discussed in this section plays an important role in the practice of handling data but will not be called for again in this book. However, the presence of this section ought to be pointed out to prospective users of data.

TABLE 2.9

COMPUTATION OF MEAN AND STANDARD DEVIATION FOR A SAMPLE OF WEAR FOR SIXTY WEARWELL TIRES USING GROUPED DATA (THOUSANDS OF MILES)

Cells	Cell Mid-point x_i	Associated Values w_i	Frequency $\mathbf{f}_i$	$\mathbf{f}_i w_i$	$\mathbf{f}_i w_i^2$
37.25-38.75	38.0	-4	1	-4	16
38.75-40.25	39.5	-3	2	-6	18
40.25-41.75	41.0	-2	5	-10	20
41.75-43.25	42.5	-1	9	-9	9
43.25-44.75	44.0	0	13	0	0
44.75-46.25	45.5	1	8	8	8
46.25-47.75	47.0	2	11	22	44
47.75-49.25	48.5	3	6	18	54
49.25-50.75	50.0	4	2	8	32
50.75-52.25	51.5	5	3	15	75
Sums			$n = 60$	42	276

k = number of cells = 10

$$\overline{\mathbf{W}} = \frac{1}{n}\sum_{i=1}^{k} \mathbf{f}_i w_i = \frac{42}{60} = 0.7$$

$$\frac{1}{n}\sum_{i=1}^{k} \mathbf{f}_i w_i^2 = \frac{276}{60} = 4.6$$

$$\mathbf{d}_{\mathrm{W}}^2 = 4.6 - (0.7)^2 = 4.11$$

$$\mathbf{d}_{\mathrm{W}} = 2.027$$

n = number of observations = 60

$a = 44.0$

$b = 1.5$

$\overline{\mathbf{X}} = 44.0 + (1.5)(0.7) = 45.05$

$\mathbf{d}_{\mathrm{X}} = (1.5)(2.027) = 3.041$

and $\mathbf{d}_{\mathrm{W}}$ is the positive square root of $\mathbf{d}_{\mathrm{W}}^2$. Then a is the value of x_i for the cell where $w_i = 0$ and b is the length of the cell interval (the difference between any two successive mid-points).

$$\overline{\mathbf{X}} = a + b\overline{\mathbf{W}}$$

$$\mathbf{d}_{\mathrm{X}} = b\mathbf{d}_{\mathrm{W}}$$

which completes the computation.

For the example of tires, note how the results for the grouped data compare with those for the more precise ungrouped data. There

$$\overline{\mathbf{X}} = 45.008$$

$$\mathbf{d}_{\mathrm{X}} = 3.096.$$

Note that Table 2.9 contains all the information in Table 2.2.

Exercise 2.19. Using grouped data, compute the sample mean and standard deviation of annual rainfall on Mr. MacDonald's nylon farm in South Phiggins. (See Exercise 2.3.)

Exercise 2.20. Compute the sample mean and standard deviation of the heights of 40 women as given in Table 2.6.

Exercise 2.21. Compute the sample mean and standard deviation of the heights of 50 men as given in Table 2.7.

8. SUMMARY

The histogram and cumulative frequency polygon are convenient graphical methods of treating grouped data.

$$\sum_{i=1}^{n} \mathbf{X}_i = \mathbf{X}_1 + \mathbf{X}_2 + \mathbf{X}_3 + \cdots + \mathbf{X}_n$$

is defined as the sum of the n $\mathbf{X}$'s.

The sample mean, standard deviation, and variance are given by $\bar{\mathbf{X}}$, $\mathbf{d}_{\mathbf{X}}$, and $\mathbf{d}_{\mathbf{X}}^2$ where

$$\bar{\mathbf{X}} = \frac{1}{n} \sum_{i=1}^{n} \mathbf{X}_i = \frac{\mathbf{X}_1 + \mathbf{X}_2 + \cdots + \mathbf{X}_n}{n}$$

$$\mathbf{d}_{\mathbf{X}}^2 = \frac{1}{n} \sum_{i=1}^{n} (\mathbf{X}_i - \bar{\mathbf{X}})^2 = \left(\frac{1}{n} \sum_{i=1}^{n} \mathbf{X}_i^2\right) - \bar{\mathbf{X}}^2$$

and Table 2.9 gives a convenient method of computing these quantities for grouped data.

The sample mean is a measure of an average or typical value. The sample standard deviation is a measure of the variability of observations in the sample. The variance, which is the square of the standard deviation, is defined as the average squared deviation from the mean.

SUGGESTED READINGS

[1] Walker, H. M., and Joseph Lev, *Elementary Statistical Methods*, Henry Holt and Co., New York, 1958.

[2] Wallis, W. A., and H. V. Roberts, *Statistics: A New Approach*, The Free Press, Glencoe, Ill., 1956.

Walker and Lev give a thorough treatment of methods of summarizing data. Wallis and Roberts have many examples to illustrate the applicability of descriptive techniques in statistics.

CHAPTER 3

Introduction to Probability and Random Variables

1. INTRODUCTION

In the contractor example we indicated that the state of nature did not uniquely determine the observations. Thus, two customers who both need 15-amp wiring might give different responses to the contractor when asked how much they would need. In fact, we assumed that half of the customers who need 15-amp wiring would, when questioned, say that they use at most 10 amp, and half of these customers would respond 12 amp. In this case, the contractor could reasonably say that the probability is one-half that a customer who needs 15 amp will respond 10 amp. We made much use of this probability, which measured the randomness type of uncertainity. In this chapter, we discuss some of the ideas involved in the notion of probability and random variables with a view to applying them to problems in decision making.

2. TWO EXAMPLES

For a preliminary illustration, we shall discuss two examples which have as their main assets their simplicity and possible applications in certain gambling situations. They are useful as illustrations of the notions of probability and random variables though their applicability in statistics is limited.

Example 3.1. Mr. Sharp purchased a pair of dice from a firm which advertises novelties designed to "amuse and entertain your friends." One of these dice is green and the other red. Each die is a cube with six faces numbered one to six. Mr. Sharp plays craps with his friends. In this game, the dice are rolled and one adds the numbers showing on the two upward faces. The number seven plays a prominent role in this game, and Mr. Sharp has observed that with his new dice this number appears as the sum about 30% of the time. It is interesting that in this game the initial roll of a seven is desired by the thrower. However, if the first roll is not a seven, the subsequent roll of a seven is deplored. At first Mr. Sharp's friends are elated at the frequency with which their initial rolls give sevens. They usually end up having lost all their money.

In this example, let us focus our attention on the act of rolling the dice. We may consider this act as the performance of an experiment the outcome of which is not determined in advance. Associated with the outcome of this experiment is a number $\mathbf{X}$, namely, the sum of the numbers on the faces showing. In general a number $\mathbf{X}$ determined by the outcome of an experiment is called a "*random variable*."

Because of the interest in the game of craps, Mr. Sharp is especially interested in the frequency with which $\mathbf{X}$ is equal to seven when his experiment is repeated many times. Since this happens about 30% of the time, Mr. Sharp is tempted to claim that the probability that $\mathbf{X}$ is equal to seven is 0.3. We shall think of probability as follows.

If the proportion of times that $\mathbf{X}$ is equal to seven tends to get very close to 0.3 as the experiment is repeated many times under similar circumstances, then we shall say that the probability that $\mathbf{X}$ is equal to seven is 0.3 and we shall write

$$P\{\mathbf{X} = 7\} = 0.3.$$

**Exercise 3.1.*[1] In order to illustrate the notion of probability, it would be desirable to repeat the above experiment several million times and find the proportion of times that $\mathbf{X} = 7$. Instead we shall take fewer observations and save the expense of buying dice, by using coins. Toss a nickel and a penny, observe the results, count the number of heads, and repeat 100 times. Construct a chart as follows. The last column $\mathbf{m}_n/n$ represents the proportion of times $\mathbf{X} = 1$ in the first n trials. Draw a graph where the horizontal scale gives n and the vertical scale $\mathbf{m}_n/n$. Plot the points $(n, \mathbf{m}_n/n)$ for $n = 4, 8, 12, \cdots, 100$ and connect adjacent points by straight lines as in the diagram following the chart.

[1] Exercises 3.4 and 3.8 depend on the data obtained here. If these exercises are to be assigned, Exercise 3.1 should not be turned in until they are completed.

n (Trial)	Nickel	Penny	$\mathbf{X}$ = No. of Heads	$\mathbf{m}_n$ = Number of Times $\mathbf{X}=1$ in the First n Trials	$\mathbf{m}_n/n$
1	H	H	2	0	0
2	H	T	1	1	0.5
3	H	T	1	2	0.667
4	T	H	1	3	0.750
5	T	T	0	3	0.600
6	H	H	2	3	0.500
7	H	H	2	3	0.429
8	T	T	0	3	0.375
.	.	.	.	.	.
.	.	.	.	.	.
.	.	.	.	.	.

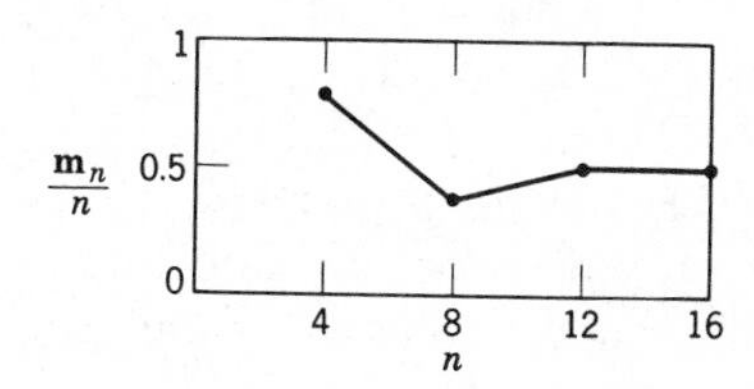

Does $\mathbf{m}_n/n$ seem to tend to be very close to some number as n gets large? Would this tendency be very strongly affected if the $\mathbf{X}$'s for the first ten repetitions of the experiment were arbitrarily replaced by 0? Do not cheat. For future reference, count the total number of times $\mathbf{X}$ is equal to zero and to two respectively.

It should be emphasized that there is a certain amount of simplification of reality in our treatment of probability. It would be impossible for Mr. Sharp to repeat indefinitely the experiment of dice rolling under similar circumstances. In fact, after 6 billion rolls, the dice would be so worn that the circumstances could no longer be considered similar. This minor difficulty does not prevent Mr. Sharp from using his new dice.

Frequently, dice manufacturers go to great effort and expense to obtain dice which are well balanced. They feel that, for a well-balanced die, one side should show up almost as frequently as another. Let us designate the outcome of the roll of two dice by two numbers, where the first corresponds to the green and the second to the red die. If Mr. Sharp's dice were well balanced, all the outcomes (1, 1), (1, 2), (1, 3), (1, 4), (1, 5), (1, 6), (2, 1), (2, 2), (2, 3), (2, 4), (2, 5), (2, 6), (3, 1), (3, 2), (3, 3), (3, 4), (3, 5), (3, 6), (4, 1), (4, 2), (4, 3), (4, 4), (4, 5), (4, 6), (5, 1), (5, 2), (5, 3), (5, 4), (5, 5), (5, 6), (6, 1), (6, 2), (6, 3), (6, 4), (6, 5), (6, 6) should occur with relatively equal frequencies. Of these 36 possible outcomes, the following six (1, 6), (2, 5), (3, 4), (4, 3), (5, 2), (6, 1) lead to $\mathbf{X} = 7$. Consequently, if the dice were well balanced, one would expect $\mathbf{X} = 7$ about $6/36 = 1/6$ of the time, i.e., 16.67% instead of 30%.

In fact, to get a die well-enough balanced to suit professional gamblers is quite difficult, and such dice are expensive. For such dice it may be assumed that $P\{\mathbf{X} = 7\}$ is approximately 1/6. In the future when we talk about "well-balanced dice," we shall approximate reality by assuming not only that the dice were carefully constructed but also that each of the above outcomes has probability exactly 1/36 and $P\{\mathbf{X} = 7\}$ is actually equal to 1/6. This kind of idealization is very useful in general, although occasionally someone has made a pile of money by detecting that a supposedly well-balanced gambling device was not really very well balanced.

Example 3.2. Mr. Sharp recently put his expert knowledge to work at the town fair by installing a dial at a gambling booth. The circumference of this dial is labeled with numbers from zero to one.

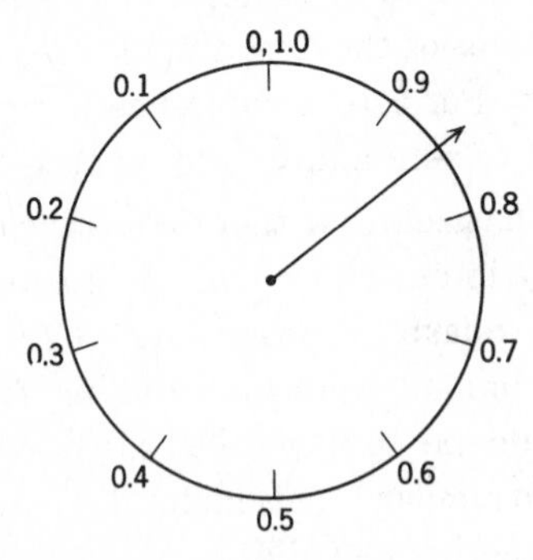

Figure 3.1. Mr. Sharp's dial.

These numbers are evenly spaced so that 0.5 is half way around the dial from zero. A long pointer is balanced at the center of the dial and, when spun, circles around and around the dial, finally coming to rest at some position. (See Figure 3.1.) If we treat the spin of the pointer as the experiment, a random variable of interest is **X** equal to the number corresponding to the position at which the pointer comes to rest.

Since Mr. Sharp will give away a teddy bear whenever **X** lies between 0.4 and 0.6, i.e., whenever $0.4 \leq \mathbf{X} \leq 0.6$, there is much interest in the frequency with which **X** satisfies this condition or restriction.[1] Mr. Sharp has managed to modify the pointer mechanism slightly so that this will happen only 7.3% of the time. That is, we may say that

$$P\{0.4 \leq \mathbf{X} \leq 0.6\} = 0.073.$$

Before Mr. Sharp's modification, the pointer was carefully constructed, and the makers felt that most professional gamblers would be satisfied that it was sufficiently well balanced. For them this would mean that in many trials (repetitions) of the experiment one would obtain $0.4 \leq \mathbf{X} \leq 0.6$, about 20% of the time. In fact, the probability that **X** would lie in any interval should be close to the length of the interval.

Exercise 3.2.*[2] For lack of well-balanced pointers and carefully marked dials, we shall simulate an ideal wheel and pointer by a table of random numbers. (See Appendix C_1.) The table was constructed as though an ideal ten-faced die with faces numbered 0, 1, 2, 3, 4, 5, 6, 7, 8, and 9 was rolled many times, and the results were recorded. Because of this construction, it can be shown that the first four digits may be assumed to be the first four digits of **X for an ideal dial. The next four digits will correspond to the **X** for the next trial, etc. Thus the first two simulated spins yield $0.0347 \cdots$ and $0.4373 \cdots$. Use the table of random numbers to simulate an ideal dial. As in Exercise 3.1, construct a chart tabulating the values of **X** in 100 trials. Let $\mathbf{m}_n$ be the number of times $0.4 \leq \mathbf{X} \leq 0.6$ in the first n trials. List and then plot $\mathbf{m}_n/n$ for $n = 4, 8, 12, \cdots, 100$.

The dial example represents one which differs from the dice example in an important respect. In the dice problem, the possible

[1] The symbols $\leq$ and $\geq$ are read "less than or equal to" and "greater than or equal to," respectively. If "less than" and "greater than" are desired, we use the symbols $<$ and $>$. Thus the inequalities $0.4 \leq \mathbf{X} \leq 0.6$ represent the condition that **X** is between 0.4 and 0.6 inclusive.

[2] Exercises 3.5 and 3.12 use the data obtained in this exercise.

values of **X** were 2, 3, 4, 5, 6, 7, 8, 9, 10, 11, and 12, and these are separated from one another, whereas in the dial problem the possible values of **X** were all the numbers between 0 and 1 and represented a "continuous" range. In the first case, **X** is called a *discrete* random variable, whereas in the second case it is called a *continuous* random variable. We shall say a bit more about this distinction later.

3. PROBABILITY DISTRIBUTIONS AND CUMULATIVE DISTRIBUTION FUNCTIONS

In the examples of the previous section, we were interested in whether **X** satisfied certain conditions or not. In the dice problem we were interested in whether $\mathbf{X} = 7$ or not, and in the dial problem whether $0.4 \leq \mathbf{X} \leq 0.6$ or not. If one were to play craps or to use dice in a variety of ways, one would be interested in other possibilities. For example, the craps player may be interested in the probability that $\mathbf{X} = 2$, the probability that $\mathbf{X} = 3$, the probability that $\mathbf{X} = 11$, the probability that $\mathbf{X} = 12$, and the probability that $\mathbf{X} = 4$ or 10. As for the dial problem, Mr. Sharp revived interest in his game by offering a toy electric train to anyone who obtained an **X** between 0.29 and 0.31 or between 0.79 and 0.81, and a new automobile to anyone who obtained $\mathbf{X} = 0.1$. Obviously he and his customers were interested in the probability that **X** would satisfy certain other conditions than $0.4 \leq \mathbf{X} \leq 0.6$. We shall be interested in the *probability distribution* of our random variables. The probability distribution of a random variable **X** is the rule which assigns a probability to each restriction on **X**. To illustrate, let us take the example where **X** is the number of heads in the toss of an ideal nickel and penny. The possible outcomes of the experiment can be denoted (H, H), (H, T), (T, H), and (T, T). For these possible outcomes, the corresponding values of **X** are 2, 1, 1, and 0 respectively, and hence we have

$$P\{\mathbf{X} = 2\} = 1/4$$
$$P\{\mathbf{X} = 1\} = 1/2$$
$$P\{\mathbf{X} = 0\} = 1/4$$
$$P\{\mathbf{X} = 1 \text{ or } 2\} = 3/4$$
$$P\{\mathbf{X} = 0 \text{ or } 2\} = 1/2$$

$$P\{\mathbf{X} = 0 \text{ or } 1\} = 3/4$$
$$P\{\mathbf{X} = 0,1, \text{ or } 2\} = 1.$$

In this experiment there are essentially no other possible restrictions on **X** and the above list represents the probability distribution of **X**.

For the ideal dice example, we list a few of the probabilities that characterize the probability distribution of **X**:

$$P\{\mathbf{X} = 2\} = 1/36$$
$$P\{\mathbf{X} = 3\} = 2/36$$
$$P\{\mathbf{X} = 12\} = 1/36$$
$$P\{\mathbf{X} = 2, 3, \text{ or } 12\} = 4/36$$
$$P\{\mathbf{X} = 7 \text{ or } 11\} = 8/36.$$

For the ideal dial example, we also list a few of the probabilities that characterize the probability distribution of **X**:

$$P\{0.4 \leq \mathbf{X} \leq 0.6\} = 0.2$$
$$P\{0.29 \leq \mathbf{X} \leq 0.31 \text{ or } 0.79 \leq \mathbf{X} \leq 0.81\} = 0.04$$
$$P\{0.09 \leq \mathbf{X} \leq 0.11\} = 0.02$$
$$P\{0.099 \leq \mathbf{X} \leq 0.101\} = 0.002$$
$$P\{0.0999 \leq \mathbf{X} \leq 0.1001\} = 0.0002$$
$$P\{\mathbf{X} = 0.1\} = 0.$$

In the dial example it is clearly impossible to list all conditions which can be imposed on **X** and the corresponding probabilities. On the other hand, it is quite clear how the listed values were obtained and how the method could be applied to many other restrictions.

It is an important convenience to realize that the probability distributions can be summarized concisely by the "*cumulative distribution function.*" The *cumulative distribution function* (cdf), F, gives the probabilities for restrictions of the form $\mathbf{X} \leq a$. For the coin example,

$$P\{\mathbf{X} \leq 0\} = 1/4$$
$$P\{\mathbf{X} \leq 1\} = 3/4$$
$$P\{\mathbf{X} \leq 2\} = 1.$$

In more detail:

$$
\begin{aligned}
P\{\mathbf{X} \leq a\} &= 0 && \text{for } a < 0 \\
P\{\mathbf{X} \leq a\} &= 1/4 && \text{for } 0 \leq a < 1 \\
P\{\mathbf{X} \leq a\} &= 3/4 && \text{for } 1 \leq a < 2 \\
P\{\mathbf{X} \leq a\} &= 1 && \text{for } 2 \leq a .
\end{aligned}
$$

This cdf is represented by the graph in Figure 3.2. For the coin problem the probability distribution is so simple that the cdf represents no gain in the way of conciseness. Note that, at $a = 0, 1$, and 2, the cdf jumps. The values of the cdf at $a = 0, 1$, and 2 are consequently marked by heavy dots.

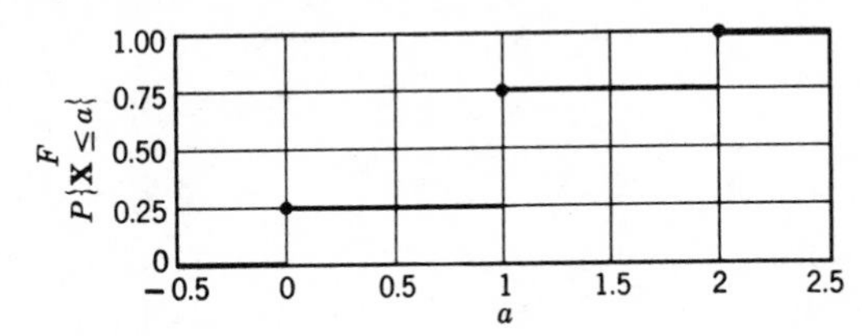

Figure 3.2. Cdf F for $\mathbf{X}$ equal to the number of heads in the toss of two ideal coins.

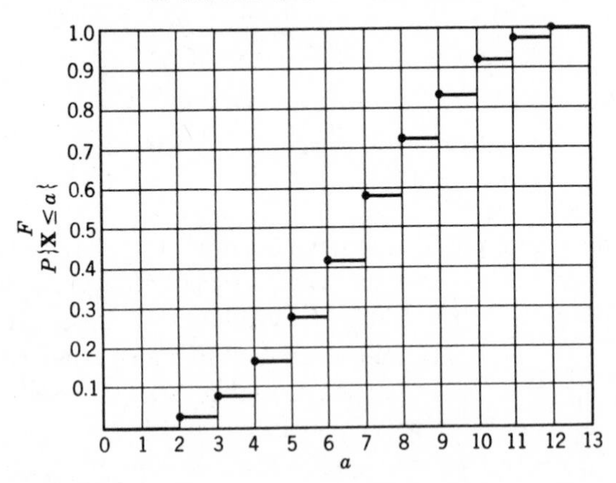

Figure 3.3. Cdf F for $\mathbf{X}$ equal to the sum of the two faces showing in the roll of two ideal dice.

The cdf for the ideal dice example is given in Figure 3.3. Here we have essentially gained in conciseness for it would be a rather long undertaking to list all probabilities. This conciseness would be of little advantage if it were not possible to recover the probability distribution from the cdf. In fact, it is possible to do so, for one may find $P\{\mathbf{X} = 2\}$ by taking the jump in the cdf at $a = 2$. Similarly, one may find $P\{\mathbf{X} = 3\}$, $P\{\mathbf{X} = 4\}$, etc. To find $P\{\mathbf{X} = 2, 3, \text{ or } 12\}$, one need only add $P\{\mathbf{X} = 2\} + P\{\mathbf{X} = 3\} + P\{\mathbf{X} = 12\}$.

In the continuous case, the use of the cdf is a great advantage. For the ideal dial problem, the cdf is given in Figure 3.4. The reader should check a few points on this graph. For example, it is obvious that

$$
\begin{aligned}
P\{\mathbf{X} \leq -0.1\} &= 0 \\
P\{\mathbf{X} \leq 0\} &= 0 \\
P\{\mathbf{X} \leq 0.3\} &= 0.3 \\
P\{\mathbf{X} \leq 0.9\} &= 0.9 \\
P\{\mathbf{X} \leq 1.0\} &= 1.0 \\
P\{\mathbf{X} \leq 1.7\} &= 1.0, \text{ etc.}
\end{aligned}
$$

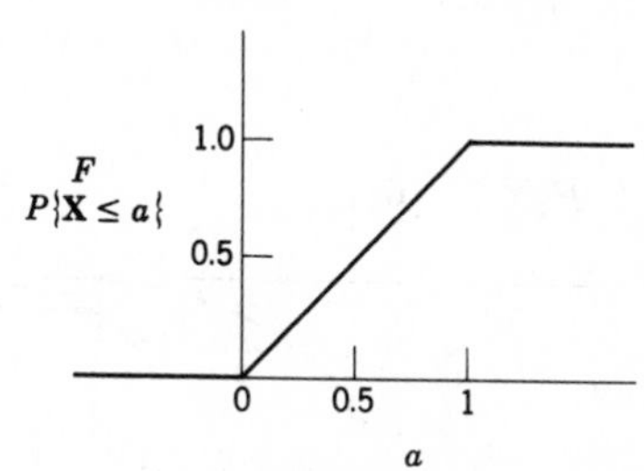

Figure 3.4. Cdf F for $\mathbf{X}$ equal to the outcome of the spin of an ideal dial.

How can the cdf be used to compute probabilities? This question is essentially answered by the following: "*A rise in the* cdf *corresponds to probability.*" Thus in the coin problem, $P\{\mathbf{X} = 1\} = 1/2$, because the cdf rises by 1/2 when the abscissa (horizontal position) is at one. In the dice problem, $P\{5.3 \leq \mathbf{X} \leq 7.2\} = 11/36 = 0.306$ because the cdf rises by 11/36 as the abscissa (horizontal position) moves from 5.3 to 7.2. In the dial problem, $P\{0.29 \leq \mathbf{X} \leq 0.31\} = 0.02$ because the cdf rises by 0.02 as the abscissa goes from 0.29 to 0.31. We get

$$P\{0.29 \leq \mathbf{X} \leq 0.31 \quad \text{or} \quad 0.79 \leq \mathbf{X} \leq 0.81\} = 0.04$$

by adding the probabilities that $\mathbf{X}$ will fall in these two nonoverlapping intervals.

Exercises 3.4 and 3.5 illustrate that the cumulative frequency graph for a large sample of observations of a random variable tends to resemble the cdf.

Exercise 3.3. Figure 3.5 shows the cumulative distributions of three random variables. Curve A is the cdf of **X** and shows $P\{\mathbf{X} \leq a\}$. Curve B is the cdf of **Y** and shows $P\{\mathbf{Y} \leq a\}$, and $P\{\mathbf{Z} \leq a\}$ is given by the straight-line sections. From the figure, estimate the following:

1. $P\{\mathbf{X} \leq 14\}$
2. $P\{9 < \mathbf{X} \leq 14\}$
3. $P\{12 \leq \mathbf{Y} \leq 14\}$
4. A value of t so that $P\{\mathbf{X} > t\} = P\{\mathbf{Y} > t\}$
5. $P\{\mathbf{X} \leq 13.5\} - P\{\mathbf{Y} \leq 13.5\}$
6. $P\{\mathbf{Z} \leq 14\}$
7. $P\{\mathbf{Z} < 14\}$
8. $P\{\mathbf{Z} = 14\}$
9. $P\{\mathbf{Z} \leq 16\} - P\{\mathbf{Y} \leq 16\}$.

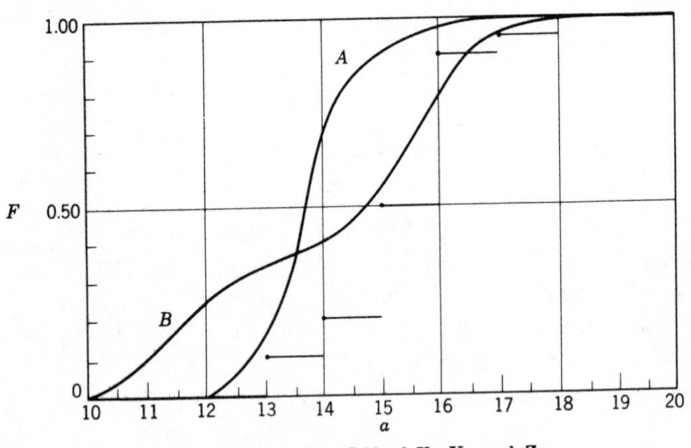

Figure 3.5. Cdf of **X**, **Y**, and **Z**.

Exercise 3.4. Refer to the data of Exercise 3.1, and graph **F** the cumulative (relative) frequency function. That is, plot the points $(a, \mathbf{F})$ where $\mathbf{F} = (1/100)$ (number of times $\mathbf{X} \leq a$). Thus,

$$\mathbf{F} = 0 \qquad \text{if } a < 0$$

$$\mathbf{F} = \frac{\text{No. of times } \mathbf{X} = 0}{100} \qquad \text{if } 0 \leq a < 1$$

$$\mathbf{F} = \frac{\text{No. of times } \mathbf{X} = 0 \text{ or } 1}{100} \qquad \text{if } 1 \leq a < 2$$

$$\mathbf{F} = 1 \qquad \text{if } 2 \leq a.$$

Exercise 3.5. For the experiment of Exercise 3.2, it would be too laborious to compute **F** for all abscissas. Instead, plot the points $(a, \mathbf{F})$ for $a = 0$, 0.1, 0.2, 0.3, 0.4, 0.5, 0.6, 0.7, 0.8, 0.9, 0.95, and 1.00. Then draw the polygon connecting these points. Compare this polygon with the cdf F by overlapping both the polygon and F on one graph.

An essential difference between the discrete and continuous cases is illustrated in Figures 3.2, 3.3, and 3.4. In the discrete case, the cdf rises in jumps and is flat elsewhere. A jump of 0.25 at 2 represents the fact that $P\{\mathbf{X} = 2\} = 0.25$. On the other hand, for the continuous case, the cdf has no jumps but rises smoothly. Because the cdf has no jumps for a continuous random variable, it follows that, if **X** is a continuous random variable as in the dial problem and a is any specified number,

$$P\{\mathbf{X} = a\} = 0.$$

In particular, $P\{\mathbf{X} = 0.1\} = 0$. This fact does not imply that it is impossible for **X** to be equal to 0.1. No matter what the outcome of the experiment, it will have to be some number such that the probability of obtaining that particular number is zero. A probability of zero may mean that the corresponding value of **X** is possible but, if the experiment is repeated many times, the relative frequency of occurrence of that particular value becomes very small. Hence **X** may equal $0.346000\cdots$ on the first spin of the dial, but it may never again equal $0.346000\cdots$.

In connection with this property of random variables, it may be pointed out that Mr. Sharp was sued by an irate customer for a car. This customer, Mr. Cox, has 20 witnesses who all agree that **X** was equal to 0.1 on his spin. Mr. Sharp claims that $P\{\mathbf{X} = 0.1\}$ is zero, hence $\mathbf{X} = 0.1$ is impossible and that the witnesses were not reading the dial accurately enough when they claimed that **X** had been equal to 0.1. The judge ruled against Mr. Sharp on the following grounds. First of all, a probability of zero does not imply impossibility. However, the judge granted that such a small probability was stronger evidence than that of 20 honest witnesses with ordinary eyesight. Therefore, the judge granted that **X** had not in fact been equal to 0.1. On the other hand, when Mr. Sharp enticed customers by offering a car, he either had fraud in mind, for which he should be prosecuted, or he meant to offer the car to anyone who obtained an **X** *reasonably close* to 0.1. Assuming the latter, and since reasonably close had not been defined by Mr. Sharp, the judge felt that he could define **X** to be reasonably close to 0.1 if 20 honest witnesses could not distinguish between the actual value of **X** and 0.1. The judge ordered Mr. Sharp to deliver the car to Mr. Cox.

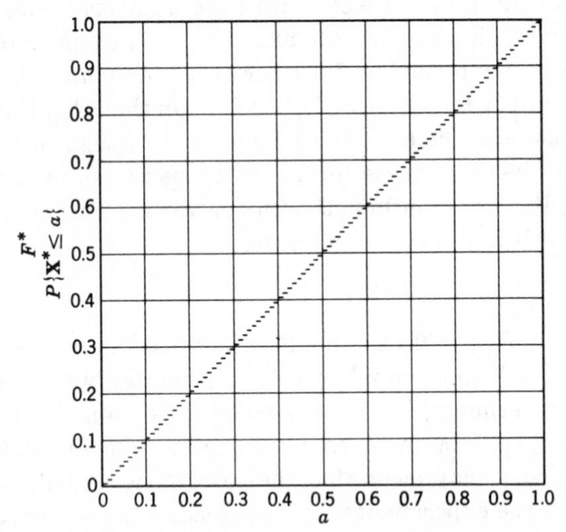

Figure 3.6. Cdf F^* for $\mathbf{X}^*$ where $\mathbf{X}^*$ is the outcome of the spin of an ideal dial rounded off to the nearest hundreth.

Note: $F^*(0) = 0.005$, $F^*(0.01) = 0.015$, $F^*(0.02) = 0.025, \cdots F^*(0.98) = 0.985$, $F^*(0.99) = 0.995$, $F^*(1.00) = 1.00$.

If you were to look at some measuring device such as a ruler, you would see that there is no way of measuring a distance exactly. However, you could approximate a distance with a certain amount of precision depending on how the ruler is marked and on your eyesight. Suppose that an observer is capable of measuring **X** in the dial problem to the nearest hundredth. That is to say, if

X were equal to $0.3763 \cdots$, the witness would estimate **X** by 0.38. As far as this witness is concerned he is observing a discrete random variable $\mathbf{X}^*$ whose possible values are $0.00, 0.01, 0.02, \cdots, 0.99, 1.00$, where the values $0.00, 0.01, 0.02, \cdots, 0.98, 0.99, 1.00$ have probabilities $0.005, 0.01, 0.01, \cdots, 0.01, 0.01, 0.005$, for an ideal dial. The cdf of this discrete random variable appears in Figure 3.6. It is clear that for many practical purposes the continuous distribution of **X** may be approximated by the discrete distribution of $\mathbf{X}^*$. *In general, if* **X** *is any continuous random variable, its distribution can be approximated by a discrete distribution* by a rounding-off process. That is, **X** is replaced by $\mathbf{X}^*$, which is obtained from **X** by rounding after a certain decimal position. The more decimal points that are kept before rounding off, the closer will $\mathbf{X}^*$ tend to be **X** and the finer will be the approximation of the cdf of $\mathbf{X}^*$ to the cdf of **X**. This remark has a good deal of theoretical importance. There arise many theoretical situations where discrete random variables are easier to treat than continuous ones. The possibility of approximating a continuous random variable arbitrarily well by a discrete one is often applied.

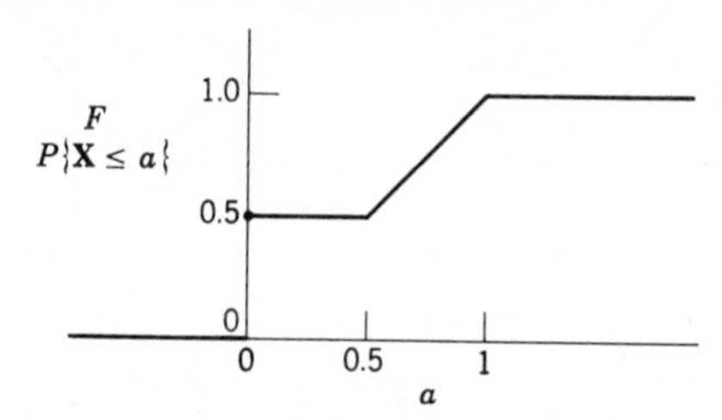

Figure 3.7. Cdf F for **Y** where **X** is the outcome of the spin of an ideal dial, $\mathbf{Y} = \mathbf{X}$ if $\mathbf{X} \geq 0.5$ and $\mathbf{Y} = 0$ otherwise.

It should be pointed out that not all random variables are continuous or discrete. For example, one may have combinations of these. An example is the following. Suppose Mr. Sharp has decided to offer **X** dollars if someone spins an **X** between 0.5 and 1.0. If, however, **X** is less than 0.5, the spinner receives nothing. The amount the spinner receives is a random variable **Y** which has the cdf given in Figure 3.7. Note that the cdf has a jump at $\mathbf{Y} = 0$ but its rise is "smooth" elsewhere. Although most random variables that we come across are either discrete or continuous, there are exceptions. For these exceptions, it is nice to have a tool such as the cdf which is applicable to all kinds of distributions.

4. PROBABILITY DENSITY FUNCTION—DISCRETE CASE

In the preceding section we have discussed the cdf as a tool for summarizing the probability distribution concisely. We have also pointed out that the cdf is similar to the cumulative frequency polygon. When we have a discrete random variable such as the number of heads in the toss of two ideal coins, it is just as concise and often more convenient and descriptive to summarize the probability distribution by the values of $P\{\mathbf{X} = 0\}$, $P\{\mathbf{X} = 1\}$, and $P\{\mathbf{X} = 2\}$. That the cdf can be recovered from these values is clear because

$$\begin{aligned}
P\{\mathbf{X} \le 0\} &= P\{\mathbf{X} = 0\} = 1/4 \\
P\{\mathbf{X} \le 1\} &= P\{\mathbf{X} = 0\} + P\{\mathbf{X} = 1\} = 1/4 + 1/2 = 3/4 \\
P\{\mathbf{X} \le 2\} &= P\{\mathbf{X} = 0\} + P\{\mathbf{X} = 1\} + P\{\mathbf{X} = 2\} \\
&= 1/4 + 1/2 + 1/4 = 1.
\end{aligned}$$

In general, if $\mathbf{X}$ is a discrete random variable, the possible values it may have can be labeled in some order $x_1, x_2, x_3, \cdots$. The discrete probability density function which is sometimes called the *discrete density* is defined by

$$P\{\mathbf{X} = x_i\}.$$

As the cdf is related to the cumulative polygon, the discrete density is related to the histogram. In Figures 3.8 and 3.9 we give the discrete densities for the ideal coins and the ideal dice together with the cdf's. Note that the value of the discrete density is nothing but the jump in the cdf.

The discrete density has the important property that the sum of its values is one. From a probability point of view, this means merely that the probability of observing some one of the possible values is one. This fact can be written as:

$$\sum_i P\{\mathbf{X} = x_i\} = P\{\mathbf{X} = x_1\} + P\{\mathbf{X} = x_2\} + \cdots = 1.$$

An example where there are infinitely many possible values of a discrete random variable is the following. An experiment consists of tossing an ideal coin until the appearance of heads. Let $\mathbf{X}$

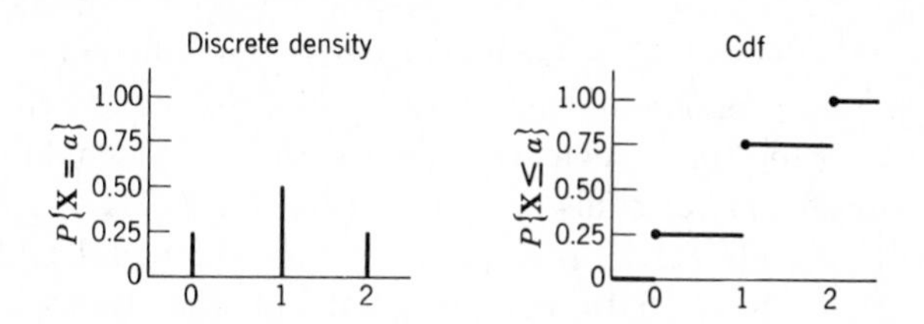

Figure 3.8. Discrete density and cdf for **X** equal to the number of heads in the toss of two ideal coins.

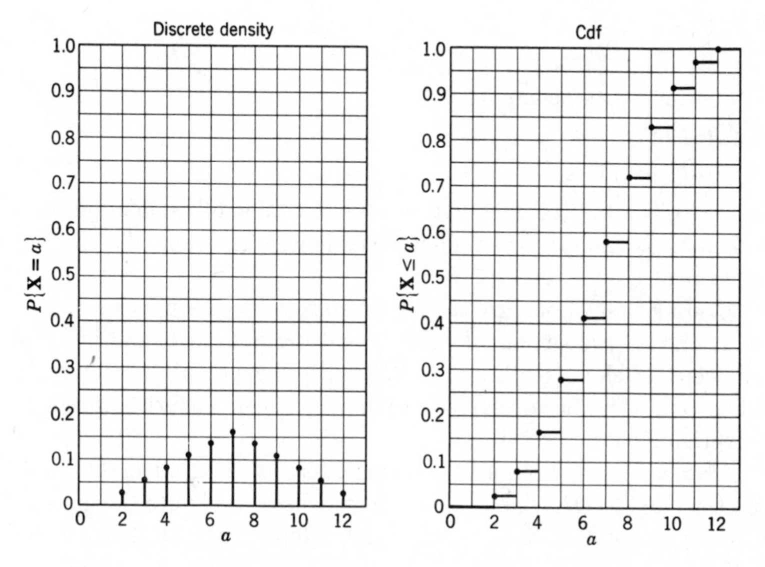

Figure 3.9 Discrete density and cdf for **X** equal to the sum of the two faces showing in the toss of two ideal dice.

be the number of times the coin has been tossed. It is clear that the probability of heads on the first toss is 1/2. That means $P\{\mathbf{X} = 1\} = 1/2$. We will have $\mathbf{X} = 2$ if we get tails followed by heads. We believe that most readers will grant that this should occur in 1/4 of the experiments. That is to say, $P\{\mathbf{X} = 2\} = 1/4$. Similarly, we shall have $\mathbf{X} = 3$ if we obtain tails twice followed by heads once, which should occur in 1/8 of our experiments. Hence, $P\{\mathbf{X}=1\} = 1/2$, $P\{\mathbf{X}=2\} = 1/4$, $P\{\mathbf{X}=3\} = 1/8$, $P\{\mathbf{X}=4\} = 1/16$, etc. It is clear that **X** may be any positive integer although it is unlikely that it will be a very large integer. Note also that $1/2 + 1/4 + 1/8 + 1/16 + \cdots$ represents the sum of a simple geometric progression and is, in fact, equal to one. If the reader is unacquainted with geometric progressions, he should consider the so-called partial sums, $1/2$, $1/2 + 1/4$, $1/2 + 1/4 + 1/8$, $1/2 + 1/4 + 1/8 + 1/16$, etc. These partial sums are equal to $1 - 1/2$, $1 - 1/4$, $1 - 1/8$, $1 - 1/16$, etc., respectively, and clearly approach one.

Exercise 3.6. A very unbalanced die might have the following distribution for $\mathbf{Y}$, the face falling uppermost when it is cast:

$$P\{\mathbf{Y} = 1\} = 1/2, \quad P\{\mathbf{Y} = 2\} = P\{\mathbf{Y} = 3\} = P\{\mathbf{Y} = 4\} = P\{\mathbf{Y} = 5\} = P\{\mathbf{Y} = 6\} = 1/10.$$

For such a die, compute

(a) $P\{\mathbf{Y} \leq 3\}$
(b) $P\{1 < \mathbf{Y} \leq 4\}$.

Exercise 3.7. Tell how you could use a table of random digits to take observations on a random variable with the same distribution as that of $\mathbf{Y}$ in Exercise 3.6.

Exercise 3.8. Compare the observed relative frequencies of $\mathbf{X} = 0$, 1, and 2, (proportion of trials resulting in 0, in 1, and in 2) for the coin problem (see Exercise 3.1) with the probabilities for the ideal coin. Do these relative frequencies seem surprisingly close, surprisingly far, or neither?

Exercise 3.9. Three ideal coins are tossed. List all eight possibilities, (*HHH*, *HHT*, etc.) and use this list to compute the discrete density and cdf of $\mathbf{X}$ = no. of heads. Represent the density and cdf by graphs.

Exercise 3.10. What is the probability that, in the toss of two ideal coins, the two coins will match?

Exercise 3.11. For the toss of a pair of ideal dice, compute

(a) $P\{\mathbf{X} \text{ is odd}\}$
(b) $P\{\mathbf{X} \text{ is a multiple of } 3\}$.

5. PROBABILITY DENSITY FUNCTION—CONTINUOUS CASE

The continuous random variable is characterized by the existence of a probability density function with the following properties. This *probability density function* is represented by a curve on a graph. The curve never lies below the horizontal axis. The total area between the curve and the horizontal axis is one. *Area between the curve and the horizontal axis represents probability.* Thus $P\{-0.5 < \mathbf{X} \leq 1.0\}$ is the area between the curve and the horizontal axis as the abscissa ranges over the interval from -0.5 to 1.0. This probability is also represented by the rise in the cdf as the abscissa ranges from -0.5 to 1.0. See Figure 3.10. Note that the height of the density at a point indicates how rapidly the cdf is rising at that point.

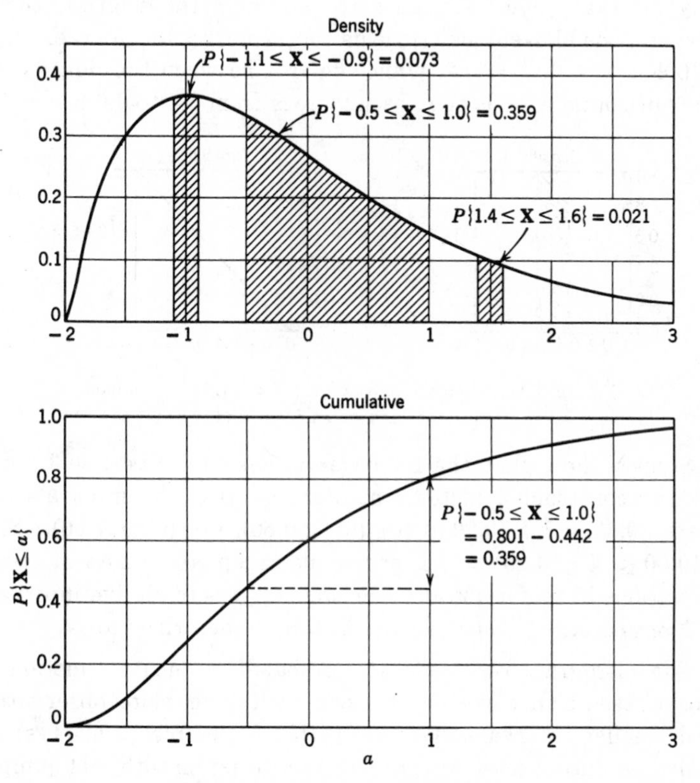

Figure 3.10. The probability density function and cdf of the random variable $\mathbf{X}$ = the change in height (measured in feet) in MacDonald's rain barrel over a one-year period.

If two possible values of the random variable $\mathbf{X}$ are specified, then $\mathbf{X}$ is more likely to be near the value for which the density is larger than near the other. Thus, for Figure 3.10, the probability that $\mathbf{X}$ will fall within 0.1 of 1.5 is approximately $(0.2)(0.105) = (0.021)$, whereas the probability that $\mathbf{X}$ will be within 0.1 of -1.0 is much larger.

In the ideal dial problem, $\mathbf{X}$ is just as likely to fall in one interval between 0 and 1 as in any other such interval of the same size. There the density is equal to one for abscissas between 0 and 1, and is equal to zero for other abscissas; see Figure 3.11. This can be further checked by noting that the probability of falling within any interval lying between zero and one is the length of the interval. The above density yields the same result. For example, $P\{0.4 \leq \mathbf{X} \leq 0.6\} = 0.2$, while the area between the density and the horizontal axis as the abscissa goes from 0.4 to 0.6 is also 0.2.

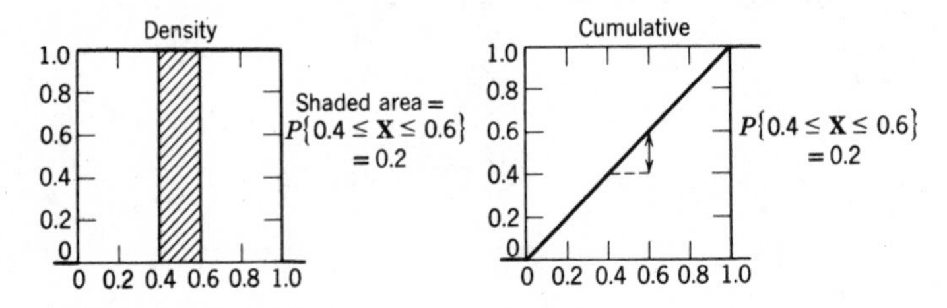

Figure 3.11. Density and cdf for $\mathbf{X}$ equal to the outcome of the spin of an ideal dial.

Suppose now that the 100 observations from Exercise 3.2 were used to construct a histogram. Suppose that the intervals used were (0.00, 0.20), (0.20, 0.40), (0.40, 0.60), $\cdots$, (0.80, 1.00). Since $P\{0.00 \leq \mathbf{X} \leq 0.20\} = 0.2$, one would except about 20% of the observations in this interval. The same applies to all five intervals.

Exercise 3.12. Construct the histogram described above.

The histogram resembles the probability density function. A closer resemblance can be obtained by taking more observations and smaller intervals. Generally (but by no means always), the densities encountered in practice are rather smooth, not jumping abruptly. Then, for reasonably large samples and fine enough intervals, the histogram gives a very good idea of the density of the corresponding random variable.

Exercise 3.13. Use the density of Figure 3.10 to estimate $P\{-1.0 < \mathbf{X} \leq -0.5\}$, $P\{-0.5 < \mathbf{X} \leq 0\}$, $P\{0 < \mathbf{X} \leq 0.5\}$, and

$P\{0.5 < \mathbf{X} \leq 1.0\}$. Use the cdf to obtain these quantities and compare them with the estimates.

Exercise 3.14. Figure 3.12 shows five probability densities labeled A, B, C, D, and E. Each of them gives rise to a cdf. Answer

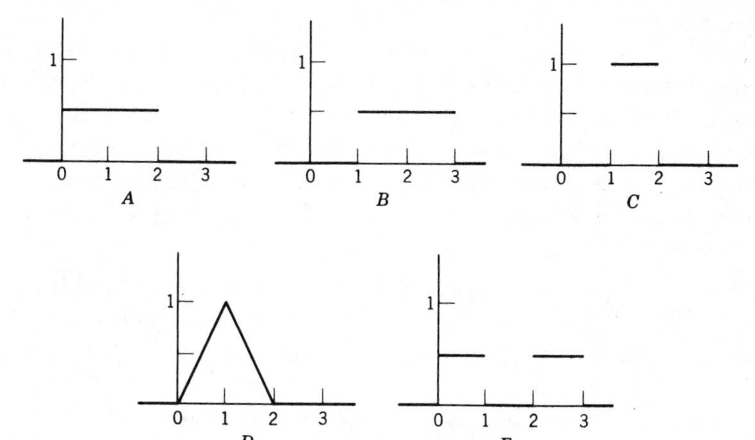

Figure 3.12. Probability densities.

the following questions about their associated cdf's. (Rough sketches may help, but are not a required part of the exercise.)

(a) Which is rising most rapidly at $a = 1.5$?

(b) Which is rising least rapidly at $a = 1.5$?

(c) Which is not polygonal in form?

Exercise 3.15. Suppose that the bearings in our dial develop friction so that the pointer is twice as likely to stop in any interval between 0.5 and 1.0 as in any interval of the same length between 0 ond 0.5. Draw the density and cdf.

†6. POPULATION AND RANDOM SAMPLES

North Phiggins is an unusual community in that, when an inhabitant is asked a question, he will always answer truthfully. This characteristic has led to the enrichment of a local statistician who has convinced a marketing research company that, in all other respects, this community is typical of the country.

Recently a clothing manufacturer requested the marketing research company for information about waist measurements of married women. The company promptly asked the statistician to carry out a local survey. He was asked to take a *random sample* of 100 waist measurements from the *population* of waist measurements of the 643 married women in North Phiggins and to deliver a histogram of the 100 waist measurements together with the mean and standard deviation.

Although a sample of 100 waist measurements can give much more information about waists than a sample of two measurements, it does not give too much more information about the principles of sampling. For simplicity then, we shall discuss three methods of obtaining a sample of two waist measurements. For each of these methods, we label the married women of North Phiggins by numbers from 1 to 643.

Method 1. Roll a well-balanced die with 643 sides numbered from one to 643. Take the waist measurement of the woman whose number appears face down. Repeat this experiment, obtaining a second measurement.

Method 2. This method is the same as the above except, if the second number is the same as the first, ignore it and repeat the roll until a different number appears.

Method 3. Obtain the first measurement as in the other methods. Let the second measurement be that of the woman labeled $\mathbf{N} + 1$, where $\mathbf{N}$ is the first number rolled. (If $\mathbf{N} = 643$, let the second measurement be that of the woman labeled 1.)

Method 1 is called *random sampling with replacement.* Method 2 is called *random sampling without replacement.* We shall not name Method 3. The above terminology is derived from the fact that Method 1 is equivalent to numbering 643 tickets mixed in a box, picking one out, *replacing* it, picking a ticket out again, and taking the measurements corresponding to the two numbers. Method 2 is equivalent to the same procedure except that the first ticket is not replaced.

When sampling with *replacement* we have the following properties satisfied.

1. *Each woman is equally likely to be sampled.*
2. *The first choice does not influence the second.*

When sampling without replacement we still have:

1. *Each woman is equally likely to be sampled.*

However, Property 2 is replaced by:

2a. *Each distinct pair of women has an equal chance of being sampled.*

Sampling with replacement has one disadvantage and one advantage over the other method. First, if the population is small, *replacement tends to give less information* since there is a chance that the same woman will be picked twice. Despite this disadvantage, the replacement method is often used because *the replacement method is easier to deal with mathematically.*

When the population size is large compared with the sample size, these two methods tend to have almost the same properties. Thus, for large populations, where sampling has been carried out without replacement, mathematicians will frequently make the approximation of computing as though there had been replacement.

Interestingly enough, Method 3 also has Property 1. But where Method 2 barely failed to have Property 2 for the sake of more information, Method 3 can fail to have Property 2 with disastrous consequences. Suppose that women were labeled in alphabetical order. Then Method 3 is liable to get women who are related, and we may obtain both measurements large or both measurements small. The data will tend to be less informative than under the alternative methods and, what is worse, it will not be known to what extent the first choice influenced the second measurement.

From the above example we see that taking a random sample of two observations with replacement is equivalent to repeating an experiment two times where the outcome is a random variable with a probability distribution determined by the population. This statement is also approximately true for a sample without replacement from a "very large" population. If we repeated an experiment, it would be suggestive to call the results a "sample" from a population. Thus, the number on the face showing in the roll of a well-balanced die can be treated as an observation obtained by random sampling from the population consisting of the six numbers one through six. It can also be treated as an observation obtained by random sampling from the population consisting of 600 numbers, 1/6 of which are one, 1/6 of which are two, etc.

It is convenient to speak of this experiment as sampling from an *infinitely large population* where 1/6 of the members are one,

1/6 are two, etc. This usage is particularly convenient if we have an experiment with an outcome which has a continuous probability distribution. Then we speak of this experiment as sampling from an *infinitely large population* where the proportion of elements between a and b is equal to $P\{a \leq \mathbf{X} \leq b\}$.

Because of the relation of the population and the associated *probability distribution*, we shall use these terms interchangeably. When we talk about observations from a population, we may mean (a) random variables with a specified probability distribution or (b) actual observations taken from a population by random sampling with replacement.

Exercise 3.16. Take two random samples of 15 observations each from the population of numbers 1 to 27. The first sample should be without replacement and the second should be with replacement. To avoid using a jar or a 27-sided die which is nonexistent, use a table of random numbers. Indicate how you used the table of random numbers.

Exercise 3.17. In Section 2, Chapter 1, assume θ_1 is the state of nature. Select an observation by using the table of random digits with the appropriate table of response frequencies. Repeat ten times.

Exercise 3.18. Follow the directions of Exercise 3.17 in connection with your example in Exercise 1.3.

Exercise 3.19. Follow the directions of Exercise 3.17 in connection with the example of Exercise 1.5.

Exercise 3.20. There are several ways to use a table of random digits to select an observation from the population of numbers 1 to 27. Some are more efficient than others in that they tend to use fewer random digits. Present two methods and indicate which seems more efficient.

7. THE NORMAL POPULATION

A special family of populations or distributions of great importance in statistics is the family of *normal distributions.* For theoretical reasons which will be discussed later, these distributions arise frequently.

One example where a normal distribution occurs in a rather artificial way[1] is the scores on the graduate record examination, which

[1] Exercise 3.24 involves an explanation of why the normal distribution occurs artificially in the graduate record examination.

is given to a large number of students who have graduated from college. Hence, if a student is selected at random from the population of students who have taken the examination, the corresponding grade **X** has the density and cdf given in Figure 3.13. For each

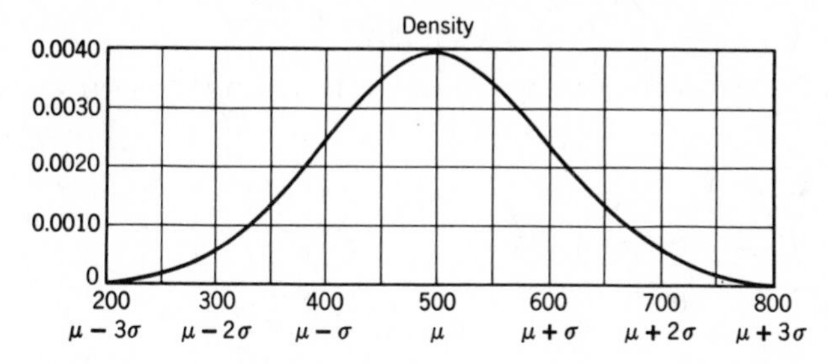

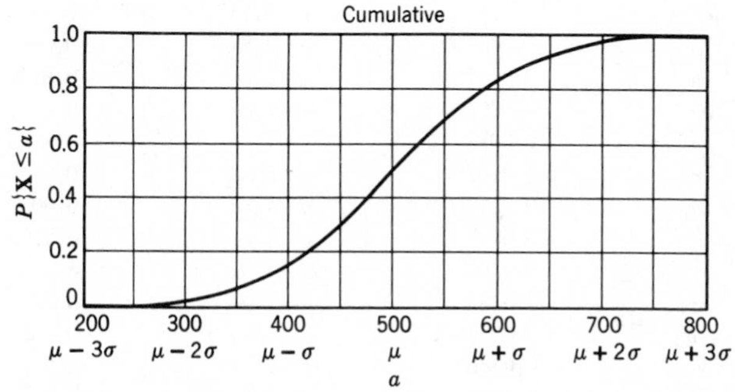

Figure 3.13. Probability density function and cdf for the grade **X** on the graduate record examination. The random variable **X** has the normal distribution with $\mu = 500$ and $\sigma = 100$.

number μ and each positive number σ, there is a normal density with the characteristic bell-shaped appearance. The number μ represents the location of the "center" of the distribution. Here the density reaches its highest value. Between μ and $\mu + \sigma$ there is 34% of the area. That is, $P\{\mu \leq \mathbf{X} \leq \mu + \sigma\} = 0.34$. If μ is increased, the density distribution merely shifts to the right. If σ is increased, then the curve flattens out; see Figure 3.14. The value of σ is a measure of the variability of observations from the population (for large samples the sample standard deviation will tend to be close to σ). When σ is two, observations will tend to be scattered twice as far from the center as they would be if σ were one. For the graduate record examination, $\mu = 500$ and $\sigma = 100$. Hence

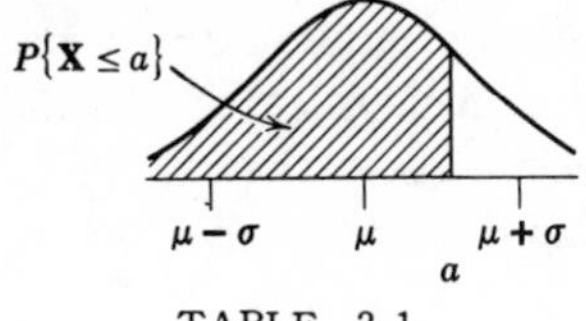

TABLE 3.1.

CUMULATIVE DISTRIBUTION FOR THE NORMALLY DISTRIBUTED RANDOM VARIABLE*

a	$P\{\mathbf{X} \leq a\}$	a	$P\{\mathbf{X} \leq a\}$
$\mu-3.5\sigma$	0.0002	$\mu+0.0\sigma$	0.5000
$\mu-3.4\sigma$	0.0003	$\mu+0.1\sigma$	0.5398
$\mu-3.3\sigma$	0.0005	$\mu+0.2\sigma$	0.5793
$\mu-3.2\sigma$	0.0007	$\mu+0.3\sigma$	0.6179
$\mu-3.1\sigma$	0.0010	$\mu+0.4\sigma$	0.6554
$\mu-3.0\sigma$	0.0013	$\mu+0.5\sigma$	0.6915
$\mu-2.9\sigma$	0.0019	$\mu+0.6\sigma$	0.7258
$\mu-2.8\sigma$	0.0026	$\mu+0.7\sigma$	0.7580
$\mu-2.7\sigma$	0.0035	$\mu+0.8\sigma$	0.7881
$\mu-2.6\sigma$	0.0047	$\mu+0.9\sigma$	0.8159
$\mu-2.5\sigma$	0.0062	$\mu+1.0\sigma$	0.8413
$\mu-2.4\sigma$	0.0082	$\mu+1.1\sigma$	0.8643
$\mu-2.3\sigma$	0.0107	$\mu+1.2\sigma$	0.8849
$\mu-2.2\sigma$	0.0139	$\mu+1.3\sigma$	0.9032
$\mu-2.1\sigma$	0.0179	$\mu+1.4\sigma$	0.9192
$\mu-2.0\sigma$	0.0227	$\mu+1.5\sigma$	0.9332
$\mu-1.9\sigma$	0.0287	$\mu+1.6\sigma$	0.9452
$\mu-1.8\sigma$	0.0359	$\mu+1.7\sigma$	0.9554
$\mu-1.7\sigma$	0.0446	$\mu+1.8\sigma$	0.9641
$\mu-1.6\sigma$	0.0548	$\mu+1.9\sigma$	0.9713
$\mu-1.5\sigma$	0.0668	$\mu+2.0\sigma$	0.9773
$\mu-1.4\sigma$	0.0808	$\mu+2.1\sigma$	0.9821
$\mu-1.3\sigma$	0.0968	$\mu+2.2\sigma$	0.9861
$\mu-1.2\sigma$	0.1151	$\mu+2.3\sigma$	0.9893
$\mu-1.1\sigma$	0.1357	$\mu+2.4\sigma$	0.9918
$\mu-1.0\sigma$	0.1587	$\mu+2.5\sigma$	0.9938
$\mu-0.9\sigma$	0.1841	$\mu+2.6\sigma$	0.9953
$\mu-0.8\sigma$	0.2119	$\mu+2.7\sigma$	0.9965
$\mu-0.7\sigma$	0.2420	$\mu+2.8\sigma$	0.9974
$\mu-0.6\sigma$	0.2742	$\mu+2.9\sigma$	0.9981
$\mu-0.5\sigma$	0.3085	$\mu+3.0\sigma$	0.9987
$\mu-0.4\sigma$	0.3446	$\mu+3.1\sigma$	0.9990
$\mu-0.3\sigma$	0.3811	$\mu+3.2\sigma$	0.9993
$\mu-0.2\sigma$	0.4207	$\mu+3.3\sigma$	0.9995
$\mu-0.1\sigma$	0.4602	$\mu+3.4\sigma$	0.9997
		$\mu+3.5\sigma$	0.9998

*For a more detailed table see Appendix D_1.

34% of the population have grades between 500 and 600, 84% have grades less than or equal to 600. Table 3.1 (a tabular representation of the cdf) enables one to compute the probability of a normal random variable falling in some interval.

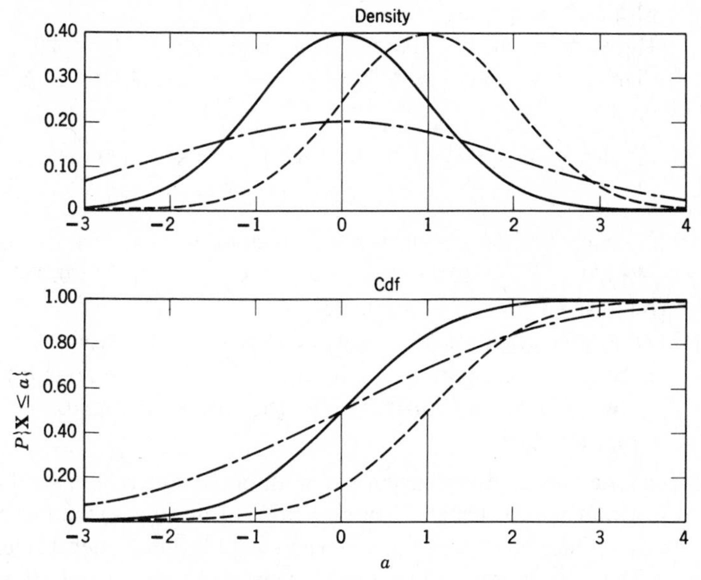

Figure 3.14. The densities and Cdf's of three normal distributions.
Normal with $\mu = 0$, $\sigma = 1$ ————————
Normal with $\mu = 0$, $\sigma = 2$ —·—·—·—·—·—·
Normal with $\mu = 1$, $\sigma = 1$ — — — — — — — —

Example 3.3. Suppose it is desired to compute the proportion of the population achieving grades on the graduate record examination:

1. Less than or equal to 800.
2. Over 800.
3. Between 300 and 600.
4. At least 200 away from 500.

1. $P\{\mathbf{X} \leq 800\}$

The number 800 is 300 units away from $\mu = 500$. Since $\sigma = 100$, 800 is 3σ from μ. Thus

$$P\{\mathbf{X} \leq 800\} = P\{\mathbf{X} \leq \mu + 3\sigma\} = 0.9987 .$$

2. $P\{\mathbf{X} > 800\}$

Those grades which are not less than or equal to 800 are over 800. Thus,

$$P\{\mathbf{X} > 800\} = 1 - P\{\mathbf{X} \leq 800\} = 0.0013.$$

3. $P\{300 \leq \mathbf{X} \leq 600\}$

Here $600 = 500 + 100 = \mu + \sigma$ and $300 = 500 - 200 = \mu - 2\sigma$. Thus, $P\{\mathbf{X} \leq 600\} = P\{\mathbf{X} \leq \mu + \sigma\} = 0.8413$, and $P\{\mathbf{X} < 300\} = 0.0227$. Then,

$$\begin{aligned} P\{300 \leq \mathbf{X} \leq 600\} &= P\{\mathbf{X} \leq 600\} - P\{\mathbf{X} < 300\} \\ &= 0.8413 - 0.0227 = 0.8186. \end{aligned}$$

4. The proportion of students whose grades are within 200 of 500 is $P\{300 \leq \mathbf{X} \leq 700\} = P\{\mu - 2\sigma \leq \mathbf{X} \leq \mu + 2\sigma\} = 0.9546$. Hence the desired proportion is 0.0454.

In these computations, when we want $P\{\mathbf{X} \leq a\}$, we have to express a as a certain number of σ's from μ. The distance from a to μ is $a - \mu$ and, thus, as illustrated in the above examples, a is $(a - \mu)/\sigma$ sigmas from μ.

Because the normal distribution plays an important role in statistics, it is sometimes desirable to have a table of numbers which corresponds to observations from a normal population. Such tables exist, and one is given in Appendix C_2. On the other hand, it is possible to construct such a table from that of random digits. A method that could also be used is as follows. Suppose that it is desired to obtain a normal random variable with mean 500 and standard deviation 100. We could select a student at random from the population of students taking the graduate record examination and take his score for $\mathbf{X}$.

The following alternative takes less time. Take a random number from the rectangular population between zero and one. This can be obtained approximately as before by putting four random digits after a decimal point. Find the location of this number in the second column of the cdf table (Table 3.1) for the normal distribution. The corresponding a is the desired normal random variable. For example, the first four digits in our table of random numbers in Appendix C_1 give us 0.0347. This corresponds to an a close to $\mu - (1.817)\sigma$ which is approximately $500 - (1.817)100 = 318.3$. How do we know that this procedure yields a random variable with the appropriate cumulative distribution function? Let $\mathbf{X}$ be the random number between 0 and 1. Let $\mathbf{Y}$ be the corresponding number yielded by this procedure. We have $\mathbf{Y} \leq \mu - 2\sigma$ if and only if $\mathbf{X} \leq 0.0227$. Hence, $P\{\mathbf{Y} \leq \mu - 2\sigma\} = P\{\mathbf{X} \leq 0.0227\} = 0.0227$. From the argument used for this special value of $a = \mu - 2\sigma$, it is clear that the cdf of $\mathbf{Y}$ is exactly the desired normal cdf.

Exercise 3.21. Suppose the heights of navy recruits are normally distributed with $\mu = 69$ and $\sigma = 3$. Let $\mathbf{X}$ be the height of a recruit selected at random.

Evaluate:

$P\{69 \leq \mathbf{X}\}$

$P\{69 \leq \mathbf{X} \leq 72\}$

$P\{66 \leq \mathbf{X} \leq 78\}$

$P\{42 \leq \mathbf{X}\}$

Find such heights A and B that:

$P\{\mathbf{X} \leq A\} = 0.50$

$P\{\mathbf{X} \leq B\} = 0.05$

Exercise 3.22. Apply the table of random normal deviates (Appendix C_2) to obtain two samples of ten observations each from a normal population with $\mu = 30$ and $\sigma = 10$. Compute the sample mean for each of the samples.

Exercise 3.23. In East Phiggins, the men love to engage in cooperative sports. The most favored game is Tug of War. There are two versions. In one of these, 50 men pull against 50 others. In the second, their cooperative spirit is manifested by all 100 men pulling against a given load. It takes 200,000 lb of pulling force to move the load. The pulling force of the 100 men is the sum of the individual pulling forces. A consequence of an elegant theorem in probability theory is that the sum of many observations on a random variable $\mathbf{X}$ is approximately normally distributed. We shall assume then that, for 100 men selected at random in East Phiggins, the total pulling load is normally distributed with $\mu = 160{,}000$ lb and $\sigma = 25{,}000$ lb. What proportion of the time do 100 men, selected at random, succeed in pulling the load?

Exercise 3.24. Mr. Evans, Principal of East Phiggins High School, is a great believer in normal distributions. Having once read a book on statistics, he has the fixed notion that examination grades should be normally distributed with $\mu = 50$, and $\sigma = 10$. Although the students of East Phiggins High School have adjusted to receiving low grades consistently, Mr. Evans' teachers are unhappy because even in large classes the histograms usually look quite different from the bell-shaped normal density. In fact, they resemble a two-humped camel more than a bell. Under the compulsion of being consistent with "statistics", the teachers have taken to *modifying* the grades. When they are finished, the distribution of grades in each class is approximately normal, with $\mu = 50$ and $\sigma = 10$. This modification is not carelessly done. If Bill Jones had a lower grade than Tom Smith before the modification, he will have a lower grade after the modification. How would you accomplish this end? Illustrate by indicating the final grades you would give to the students who ranked 300th, 500th, and 650th (from the bottom) in a class of 1000 students. Incidentally, this is exactly what is done with graduate record examinations, and explains why the distribution there is normal, although artificially so. The technical mathematical advantages of working with a normal distribution are such that it is often worth while to use the above techniques for large populations so long as one is under no illusions about where the normality comes from.

Exercise 3.25. In North Phiggins, the banks estimate the number of kopeks in a pile by weighing them. The estimate, based on two kopeks to the ounce, is twice the weight, in ounces, of the pile of kopeks rounded off to the nearest integer. Thus the pile will be estimated as having 100 kopeks if its weight is between 49.75 and 50.25 oz. Compute the probability that 100 kopeks will weigh between 49.75 and 50.25 oz if the weight of 100 kopeks is considered to be normally distributed with $\mu = 50$ and $\sigma = 0.125$. If the weight does not fall between 49.75 and 50.25 oz, the size of the pile of 100 kopeks will be incorrectly estimated.

8. SETS AND FUNCTION

Thus far in this chapter, and also to a certain extent in the preceding chapters, we have been handicapped in our language. A reluctance to introduce the notions of set and function has involved us in a great deal of circumlocution. In this section, we shall discuss these notions and, using them in the next section, we shall summarize concisely many of the remarks made up to now about probability.

The notions of set and function appear in many places, and, like prose, are used by many persons who are unaware of what they are using. In many cases, to point out that these notions are being used does not add any great amount of understanding to a problem. On the other hand, in many cases, including probability theory, the use of these notions not only makes for conciseness but often seems necessary for the understanding of all but the most trivial questions.

Any collection, such as the possible outcomes of an experiment in which two coins are tossed, is called a *set*. The items in the collection are called the *elements* of the set. If one set consists exclusively of some of the elements of a second set, then the first set is called a *subset* of the second. For example, $\{(H, H), (H, T)\}$ is a subset of $\{(H, H), (H, T), (T, H)\}$. Notice that a set is frequently represented by indicating its elements within a pair of braces. It is customary to distinguish between an element and the set consisting of that element. Thus (H, H) is an element of the rather trivial set $\{(H, H)\}$. Two sets are said to be *nonoverlapping* if they have no elements in common. For example, the sets $\{(H, H), (H, T)\}$ and $\{(T, H)\}$ are nonoverlapping.

Some examples where we previously dealt with sets are the following. In the introduction we were concerned with:

1. The set $\{\theta_1, \theta_2, \theta_3\}$ of possible states of nature;
2. the set $\{a_1, a_2, a_3\}$ of available actions;
3. the set $\{s_1, s_2, s_3, s_4, s_5\}$ of strategies; and
4. the set $\{z_1, z_2, z_3, z_4\}$ of possible observations (outcomes of the experiment).

Frequently we have occasion to deal with:

5. The set of all numbers; and
6. the set of positive integers $\{1, 2, 3, \cdots\}$.

The abscissa and ordinate representing a point on graph paper are two numbers and the set of points on graph paper can be represented by:

7. *The set of pairs of numbers* where the first number is the abscissa and the second is the ordinate. This set is sometimes called the *plane*.

Certain notations and abbreviations for sets are often convenient to use. We use $\{x: x \text{ has a specified property}\}$ to represent the set of all elements which have the specified property. For example, $\{x: x \text{ is a horse}\}$ is the set of horses and $\{x: 3 \leq x \leq 4\}$ is the set of numbers between 3 and 4 including 3 and 4. The symbol x is used here to denote a typical element of the set but has no special meaning. If we replaced x by y in the above, we would still have the same set. Thus $\{y: y \text{ is a horse}\}$ is still the set of horses. The set $\{x: 3 \leq x \leq 4\}$ has a simple geometric representation. Consider the horizontal axis on graph paper. Each point on this line corresponds to a number. The set $\{x: 3 \leq x \leq 4\}$ corresponds to a segment or interval of the axis (including the end points). Hence, we shall frequently call sets of real numbers, and sets of pairs of real numbers, point sets. A set $\{x: a \leq x \leq b\}$ will be called an interval.

In the experiment of tossing a nickel and a penny, we were interested in the number of heads obtained in the two tosses. For each of the possible outcomes of the experiment, there is a number which represents the corresponding number of heads obtained. *A correspondence or rule which associates with each element of one set an element of a second set is called a function.* More explicitly, it is sometimes called a function on the first set to the second set. The element of the second set which corresponds to the given element of the first set is called the *corresponding value* of the function. In some cases a function can be represented by a table. For example, Table 3.2 represents the above "number of heads" function.

TABLE 3.2.

f_1 = NUMBER OF HEADS IN THE TOSS OF A NICKEL AND A PENNY

Possible outcomes of experiment	(H, H)	(H, T)	(T, H)	(T, T)
Corresponding values of f_1	2	1	1	0

It is clear that this particular function f_1 represents the random variable we have discussed previously. This function f_1 is a function on the set of possible outcomes to the set of numbers.

Some other functions we have come across before are represented in the Tables 3.3 and 3.4.

Some functions cannot be represented quite so easily. For example, consider the function f_4 which makes correspond to each number between 0 and 1 the square of the given number.

TABLE 3.3.

f_2 = STRATEGY s_1 (SEE TABLE 1.3)

Possible outcomes of the experiment	z_1	z_2	z_3	z_4
Corresponding values of s_1 (actions assigned by s_1)	a_1	a_1	a_2	a_3

TABLE 3.4.

f_3 = AVERAGE LOSS ASSOCIATED WITH STRATEGY s_1 (SEE TABLE 1.4)

Possible states of nature	θ_1	θ_2	θ_3
Corresponding values of f_3 (average losses for strategy s_1)	1	3.5	4

There are infinitely many elements in the set of all numbers between 0 and 1, and a tabular representation of f_4 would never terminate. One may partially represent this function by an incomplete table which would give the squares of the numbers 0.00, 0.01, 0.02, 0.03, etc. One may approximately represent this function by a graph where a typical point has coordinates (x, y), where x is a number between zero and one and $y = x^2$ (see Figure 3.15).

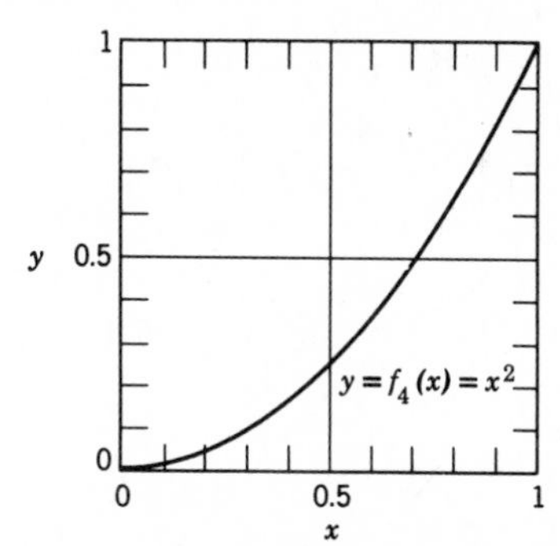

Figure 3.15. Graphical representation of the function f_4 where $f_4(x) = x^2$.

Again this is only a partial representation because of limitations in (1) drawing points and (2) the number of points one can actually draw. In fact, this partial representation is quite serviceable because f_4 is such a continuous and well-behaved function.

A *brief* representation which is very convenient consists of describing f_4 by

$$f_4(x) = x^2 \qquad \text{for } 0 \leq x \leq 1.$$

Here $f_4(x)$ denotes the *value* of f_4 corresponding to an arbitrary number x between 0 and 1. In general, if f is a function on A to

B, $f(x)$ represents the *value* of f, i.e., the element of B, corresponding to an element x of A. Thus $f_1((H, H)) = 2$ $f_2(z_2) = a_1$, $f_3(\theta_3) = 4$, and $f_4(0.5) = 0.25$.

Some other functions are described as follows:

$f_5(x) = 2 + 3x$ for an arbitrary number x

$f_6(x) = 4$ for an arbitrary number x

$f_7(x) = x$ for an arbitrary number x

$f_8(x) =$ mother of x for an arbitrary person x.

Sometimes the description of a function is not quite so simple. For example, consider f_9 represented by

$f_9(x) = x$ for $0 \leq x \leq 1$

$f_9(x) = 2 - x$ for $1 \leq x \leq 2$

$f_9(x) = 0$ for all numbers not between 0 and 2.

See Figure 3.16 for a graphical representation of f_9.

x	$f_9(x)$
−1.0	0
−0.5	0
0	0
0.5	0.5
1.0	1.0
1.5	0.5
2.0	0
2.5	0

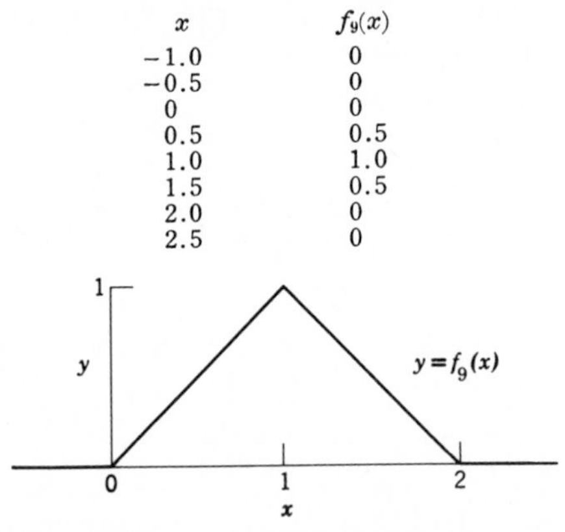

Figure 3.16. Graphical and tabular representation of f_9 defined by:

$f_9(x) = x$ for $0 \leq x < 1$

$= 2 - x$ for $1 \leq x < 2$

$= 0$ otherwise (i.e., for all other numbers).

**Exercise 3.26*. Consider the experiment of tossing three coins (a penny, a nickel, and a dime). List the set of all possible outcomes where heads or tails is recorded for each coin. Tabulate $\mathbf{X}$, the number of heads corresponding to each outcome. If the coins were well balanced, what would be the probability of exactly two heads?

*_Exercise 3.27_. Plot and describe in geometric terms the following sets. Are they overlapping?

(a) $\{(x, y)\colon 2x + y = 4\}$
(b) $\{(x, y)\colon x - 2y = 1\}$
(c) $\{(x, y)\colon y = x^2\}$

Exercise 3.28. Plot and describe in geometric terms the following sets. Are they overlapping?

(a) $\{(x, y)\colon x^2 + y^2 = 4\}$
(b) $\{(x, y)\colon x^2 + y^2 \leq 4\}$
(c) $\{(x, y)\colon (x - 2)^2 + (y - 1)^2 < 1\}$

Exercise 3.29. There are many possible normal distributions, one for each pair of values (μ, σ), where $-\infty < \mu < \infty$ and $\sigma > 0$. Each such possible pair (μ, σ) can be represented as a point having two coordinates. Describe the set of all such points which can thus represent normal distributions.

Exercise 3.30. Present a graphical and tabular representation of the function f_{10} described by $f_{10}(x) = 1/(1 + x^2)$.

Exercise 3.31. Present a graphical and tabular representation of the function f_{11} described by

$$f_{11}(x) = 0 \qquad \text{for } x \leq 0$$

$$f_{11}(x) = \frac{x}{1 + x^2} \qquad \text{for } x > 0.$$

9. REVIEW OF PROBABILITY

We shall be interested in certain subsets of the set $\mathscr{X}$ of all possible outcomes of an experiment. For example, in the experiment of tossing a nickel and a penny, we may be interested in $E_1 = \{(H, T), (T, H)\}$, which is the set of all those outcomes for which heads appeared exactly once.

Let E be a subset of $\mathscr{X}$. As the experiment is repeated many times under similar circumstances, the proportion of times in which the outcome of the experiment is an element of E tends to be close to some number. This number will be called the probability of E and written $P\{E\}$.

This correspondence or function denoted by P is called the *probability distribution for the experiment*. Note that P is a function on the set of subsets E of $\mathscr{X}$ to the real numbers between 0 and 1.

A possible probability distribution corresponding to some rather bent coins is represented in Table 3.5.

A random variable is represented by a rule or function g which assigns a number to each possible outcome of the experiment. Label the outcome of the experiment **Z**. The value of the random variable is $\mathbf{X} = g(\mathbf{Z})$. Thus, if the above experiment, when carried out,

TABLE 3.5.

PROBABILITY DISTRIBUTION FOR AN EXPERIMENT OF TOSSING TWO RATHER BENT COINS

E	$P\{E\}$
$\{(H, H)\}$	6/24
$\{(H, T)\}$	10/24
$\{(T, H)\}$	3/24
$\{(T, T)\}$	5/24
$\{(H, H), (H, T)\}$	16/24
$\{(H, H), (T, H)\}$	9/24
$\{(H, H), (T, T)\}$	11/24
$\{(H, T), (T, H)\}$	13/24
$\{(H, T), (T, T)\}$	15/24
$\{(T, H), (T, T)\}$	8/24
$\{(H, H), (H, T), (T, H)\}$	19/24
$\{(H, H), (H, T), (T, T)\}$	21/24
$\{(H, H), (T, H), (T, T)\}$	14/24
$\{(H, T), (T, H), (T, T)\}$	18/24
$\{(H, H), (H, T), (T, H), (T, T)\}$	24/24

TABLE 3.6.

PROBABILITY DISTRIBUTION OF **X**=NUMBER OF HEADS IN THE EXPERIMENT OF TOSSING TWO RATHER BENT COINS (REFER TO TABLE 3.5)

E	$P\{E\}$
$\{\mathbf{X}=0\}=\{(T, T)\}$	5/24
$\{\mathbf{X}=1\}=\{(H, T), (T, H)\}$	13/24
$\{\mathbf{X}=2\}=\{(H, H)\}$	6/24
$\{\mathbf{X}=0 \text{ or } 1\} =\{(T, T), (H, T), (T, H)\}$	18/24
$\{\mathbf{X}=0 \text{ or } 2\} =\{(T, T), (H, H)\}$	11/24
$\{\mathbf{X}=1 \text{ or } 2\} =\{(H, T), (T, H), (H, H)\}$	19/24
$\{\mathbf{X}=0 \text{ or } 1 \text{ or } 2\}=\{(T, T), (T, H), (H, T), (H, H)\}$	24/24

Note: The notation {**X** satisfies a certain property} is an abbreviation for the set of *outcomes* such that the corresponding value of **X** satisfies this property, i.e., {**Z**: **Z** is an outcome such that **X** satisfies the property}. Note that the only outcome for which $\mathbf{X}=0$ is (T, T) and, hence, $\{\mathbf{X}=0\}=\{(T, T)\}$.

led to $\mathbf{Z} = (H, T)$, the random variable f_1 (number of heads as in Section 8) would have the value $\mathbf{X} = f_1(\mathbf{Z}) = 1$. When there is no fear of real confusion, it is common to call $\mathbf{X}$ the random variable, although, strictly speaking, $\mathbf{X}$ is the *value* of the random variable and not the random variable. For example, we talk of the *probability distribution of the random variable* $\mathbf{X}$; this is defined as the rule which assigns probabilities to sets of outcomes which can be expressed in terms of restrictions on $\mathbf{X}$. *The probability distribution of the random variable* is generally an abbreviated version of the *probability distribution for the experiment*; see Table 3.6.

The probability distribution of a random variable $\mathbf{X}$ is summarized concisely by the *cumulative distribution function* F which is defined by

$$F(a) = P\{\mathbf{X} \leq a\} = P\{\mathbf{Z}: \mathbf{Z} \text{ is an outcome such that } \mathbf{X} = g(\mathbf{Z}) \leq a\}$$

for all numbers a. In the discrete case, where the possible values of $\mathbf{X}$ are separated from one another, the probability distribution is adequately summarized by the *discrete probability density function* f defined by

$$f(x) = P\{\mathbf{X} = x\}.$$

The continuous case is characterized by the existence of a *probability density function* f which has the properties:

1. $f(x) \geq 0$ for all numbers x; and

2. the area between the horizontal axis and the curve whose typical points are $(x, f(x))$ is one.

This density corresponds to the random variable $\mathbf{X}$ if

3. $P\{a \leq \mathbf{X} \leq b\}$ = area between the horizontal axis and graph corresponding to abscissas between a and b.

In the discrete case, the values of the discrete density correspond to the jumps of the cdf. In the continuous case, the cdf has no jumps. There the value of the cdf for a given abscissa a is the area to the left of a between the graph of the density and the horizontal axis.

The "rectangular" density function for the ideal dial problem is f_{12} defined by

$$\begin{aligned} f_{12}(x) &= 1 \qquad \text{for } 0 \leq x \leq 1 \\ f_{12}(x) &= 0 \qquad \text{for all other numbers.} \end{aligned}$$

The *probability density function* f for the normal distribution is defined by

$$f(x) = \frac{1}{\sqrt{2\pi\sigma^2}} e^{-(x-\mu)^2/2\sigma^2} \quad \text{for all numbers } x.$$[1]

When $\mu = 0$ and $\sigma = 1$, the above expression becomes

$$\frac{1}{\sqrt{2\pi}} e^{-x^2/2}.$$

Exercise 3.32. Show that f_9 of Section 8 is a probability density function. (This is the density for the random variable which is the sum of the outcomes of two spins of a well-balanced dial.)

Exercise 3.33. Tabulate the discrete density for $\mathbf{X}$ = number of heads obtained in the toss of four well-balanced coins.

Exercise 3.34. What is the probability of throwing a 7, 8, 9, or 10 with a pair of dice?

Exercise 3.35. A poker hand contains four spades and one heart. The heart is discarded and another card is drawn. What is the probability that the new card will be a spade?

Exercise 3.36. A poker hand contains a 3, 4, 5, 6, and 10. The 10 is discarded. What is the probability of drawing a 2 or 7 if the 10 is discarded?

Exercise 3.37. A poker hand contains a 5, 6, 7, 9, and Jack. What is the probability of drawing an 8 if the Jack is discarded?

Exercise 3.38. A poker hand contains two aces, two 7's and a 9. What is the probability of drawing an ace or a 7 if the 9 is discarded?

Exercise 3.39. Let $\mathbf{Y} = \mathbf{X}^2$, where $\mathbf{X}$ is the outcome of the ideal dial experiment. Find the cdf of $\mathbf{Y}$.

Exercise 3.40. Let $\mathbf{X}$ equal the number of people in a sample of 100 voters from East Phiggins who favor prohibition. If the proportion of all voters who favor prohibition is p, $\mathbf{X}$ is approximately normally distributed with $\mu = 100p$ and $\sigma = 10\sqrt{p(1-p)}$. We call the data misleading if $\mathbf{X} > 50$ when $p < 1/2$ or if $\mathbf{X} < 50$ when $p > 1/2$. Find the probability that the data will be misleading if $p = 0.40$.

Exercise 3.41. Given that $\mathbf{X}$ has cdf F defined by

[1] The number $e = 2.71828 \cdots$ occurs in mathematics so often that it is honored with the special symbol e. Another example of such a number is $\pi = 3.14159 \cdots$. For later convenience in typography, we shall often represent e^x by $\exp(x)$.

$$\begin{aligned} F(a) &= 0 && \text{for } a \leq -1 \\ &= (a+1)^2/4 && \text{for } -1 < a \leq 1 \\ &= 1 && \text{for } a > 1 . \end{aligned}$$

Compute

(*a*) $P\{-0.5 < \mathbf{X} \leq 0.5\}$ (*b*) $P\{\mathbf{X} > 0.5\}$ (*c*) $P\{0 < \mathbf{X} \leq 0.6\}$.

Exercise 3.42. A point is selected at random from the interior of a circle of radius one. By this we mean that the probability that the point lies in a given subset of the circle is proportional to the area of the subset. Graph the cdf of **R** where **R** is the distance of the point to the center of the circle.

°*Exercise 3.43.* What is the probability that two socks, selected at random (without replacement) from a drawer containing six red and three brown socks, will match?

Exercise 3.44. What is the probability that three socks, selected at random (without replacement) from a drawer containing six red and three brown socks, will contain a matching pair?

°*Exercise 3.45.* Socks are taken (without replacement) from a drawer containing six red and three brown socks until a red sock is drawn. Describe the discrete density of **X** equal to the number of socks drawn.

°*Exercise 3.46.* A poker hand contains three spades, one heart, and one club. What is the probability of drawing two spades if the heart and club are discarded?

°*Exercise 3.47.* Find the density of **Y** in Exercise 3.39.

°*Exercise 3.48.* Find the probability density function for **X** in exercise 3.41.

°*Exercise 3.49.* A point (**X**, **Y**) is selected at random on the circumference of the circle $x^2 + y^2 = 1$. (The probability of falling on an arc segment is proportional to the length of the segment.) Find the cdf and density of **X**.

°A convenient method of doing this problem involves the idea of combinations which are not covered in this text. Problems involving the use of mathematical ideas not covered in this text nor in high school mathematics courses will be marked with a°.

SUGGESTED READING

[1] Cramér, Harald, *The Elements of Probability Theory and Some of its Applications*, John Wiley and Sons, New York, 1955; Almqvist and Wiksell, Stockholm, 1955.

[2] Feller, William, *An Introduction to Probability Theory and its Applications*, Vol. 1, John Wiley and Sons, New York, first edition, 1950, second edition, 1957.

[3] Parzen, Emanuel, *An Introduction to Modern Probability Theory*, John Wiley and Sons, New York (to be published).

Each book presents probability theory from a rather elementary point of view. However, calculus becomes a prerequisite for going deeply into any of these books.

CHAPTER 4

Utility and Descriptive Statistics

1. INTRODUCTION

As was previously pointed out, there are two types of uncertainty. One is due to randomness and the second is due to ignorance of which law of randomness (state of nature) applies. Suppose that the state of nature is known and randomness is the only type of uncertainty left. What principles does one use to make decisions then? Certainly this problem is simpler than the one where the state of nature is unknown.

Under certain assumptions we shall see that the prospect facing a person has a numerical value called *utility*, and that he should make those decisions for which his prospects would have as large a utility as possible. Generally speaking, this is easier said than done, but, in many applications, certain simple properties of utility make it feasible for one to make intelligent decisions.

2. UTILITY

Put yourself in the following hypothetical positions where the state of nature is known and see whether you would accept the following bets (each bet is offered only once).

1. You receive $2 if a (well-balanced) coin falls heads and you pay $1 if it falls tails.
2. Your entire fortune has a cash value of $10,000,000. You receive $20,000,000 extra if the coin falls heads and you lose your fortune otherwise.
3. You intend to spend all your cash on beer to drink this evening. You have $3 which will buy a good deal of beer. You receive $3 extra if the coin falls heads and you lose your $3 otherwise.
4. You are desperate to see the " big game." You have $3 but a ticket costs $5. You receive $3 extra if the coin falls heads and you lose your $3 otherwise.

The authors' reactions to these situations would be: (1) bet; (2) do not bet; (3) do not bet; and (4) bet. The situations (1) and (2) were similar in that the favorable 2-to-1 odds were offered when "even" odds (1-to-1) seemed appropriate. Still, our responses were different. What is the essence of the difference between the two situations? Vaguely speaking, the authors feel that in situation 1, where the probability of winning is one half, the amount to be gained is greater than that to be lost. In situation 2, it is also true that there is more money to be gained than lost, but the winning of $20,000,000 would increase our happiness very little while the loss of our $10,000,000 would lead to considerable misery. Thus in situation 2, there is more "happiness" to be lost than to be gained.

In situations 3 and 4, we have the similarity that the bets were at even odds when even odds seemed called for, but the responses were different. Once again, the essence of the difference seems to be in the value of the money gained or lost. The pleasure gained by the beer drinker out of drinking $6 worth of beer instead of $3 worth is small compared to the displeasure of being deprived of the original $3 worth of beer. On the other hand, if one is desperate to go to the big game, $3 is almost as useless as nothing, whereas $6 makes all the difference in the world.

In other words, the usual use of "fair" odds in money bets is not always appropriate because the *value* of money to the owner does not always seem to be proportional to the *amount* of money. (Thus $30,000,000 is not three times as valuable as $10,000,000 to the millionaire but $6 is many times more valuable than $3 to the big-game enthusiast.)

It would be convenient to have some measure of value which did not have the above shortcoming. Furthermore, it would be desirable that this measure apply to valuable considerations, other than money, such as leisure, reputation, etc.

Under assumptions to be listed, we shall see that there is a measure of value, namely, *utility*, which can be used to measure situations or prospects[1] in such a way that the choice of actions is sometimes relatively easy to make. In fact, it has been proved that if an individual has tastes which satisfy the four assumptions discussed in Section 2.1, then *there is a utility function u on the set of*

[1] The term "prospect" will be explained in more detail shortly. Until then, one may regard it in a nontechnical sense.

prospects to the set of numbers. That is, to each prospect P, there corresponds a number $u(P)$ which is called the utility of the prospect P. This function has the following properties, called the utility function properties:

UTILITY FUNCTION PROPERTY 1. $u(P_1) > u(P_2)$ *if and only if the individual prefers* P_1 *to* P_2.

UTILITY FUNCTION PROPERTY 2. *If P is the prospect where, with probability p, the individual faces P_1 and with probability $1 - p$ he faces P_2, then*

$$u(P) = pu(P_1) + (1 - p)u(P_2).$$

The first property states that utility increases when the prospect improves and the second states that utility, *unlike money*, can *always* be computed with according to ordinary odds. We shall discuss this type of computation in more detail in Section 3. We shall see that, if an individual feels that $u(P_1) = 1$, $u(P_2) = 2$, and $u(P_3) = 1.6$, he would prefer P_3 to the prospect of obtaining P_1 if a well-balanced coin falls heads and P_2 if it falls tails.

It must be pointed out that different people have different tastes and will have different utility functions (provided, of course, that their tastes satisfy the assumptions and that, therefore, they do have utility functions).

†2.1. The Assumptions Behind Utility

Let H denote the entire future history of a given individual, including all his joys and sorrows. Suppose that he were given a choice between two possible histories H_1 and H_2. In practice, many people may find it difficult to select between two different histories, not that they are often given such a clear-cut choice. However, let us assume that they can always decide which they prefer or whether they like each history equally well.

Now let us suppose that the individual were told that his future history would be selected according to some random rule (probability distribution) which he is given. This situation may be called a prospect P. An example of a prospect is the following. A young man rushes across the street at the sight of a pretty young lady. With probability 0.8 he crosses the street, meets her, marries her and, to summarize briefly, lives happily for 60 additional years.

With probability 0.2, a car hits him and he dies.[1]

We discuss four assumptions:

ASSUMPTION 1. *With sufficient calculation an individual faced with two prospects P_1 and P_2 will be able to decide whether he prefers prospect P_1 to P_2, whether he likes each equally well, or whether he prefers P_2 to P_1.*

ASSUMPTION 2. *If P_1 is regarded at least as well as P_2, and P_2 at least as well as P_3, then P_1 is regarded at least as well as P_3.*

Suppose that the individual is given a choice between P_2 on one hand and gambling between P_1 and P_3 on the other hand. A prospect P which consists of applying some random device to select one of several prospects is called a *mixture* of these prospects. Thus the prospect of facing P_1 with probability 0.8 and P_3 with probability 0.2 is a mixture of P_1 and P_3. (See the example of the young man crossing the street for a rather simple version of a mixture of two prospects.) Ordinarily if P_1 is preferred to P_3, it would seem reasonable to assume that among mixtures of P_1 and P_3 the ones which give higher probabilities for P_1 are to be preferred. If the probability of facing P_1 should be close to one, is the mixture almost as good as P_1? If the probability of facing P_1 is close to zero, is the mixture almost as poor as P_3? Assumption 3 will essentially answer "yes."

ASSUMPTION 3. *If P_1 is preferred to P_2 which is preferred to P_3, then there is a mixture of P_1 and P_3 which is preferred to P_2, and there is a mixture of P_1 and P_3 over which P_2 is preferred.*

ASSUMPTION 4. *Suppose the individual prefers P_1 to P_2 and P_3 is another prospect. Then we assume that the individual will prefer a mixture of P_1 and P_3 to the same mixture of P_2 and P_3.*

The reader would do well to think about whether he believes that these assumptions apply to his method of choosing between prospects. For example, the third assumption stated is quite important and well worth considering. Suppose P_1 involves living a very happy life, P_2 involves living a miserable life, and P_3 involves the possibility of hell. Could it be that you would prefer the miserable life to any risk of P_3? That is, would you prefer certain

[1] In this case the prospect P involves two possible histories. Any history H can itself be considered as a somewhat trivial prospect where the outcome is H itself with probability 1.

misery to gambling between almost certain happiness and the slightest possibility of hell? Our assumption is that, if the probability of P_3 is small enough, you would gamble between P_1 and P_3 instead of facing P_2. The issues raised here are essentially the following. Are there prospects so terrible or so wonderful that the slightest possibility of facing them is also incomparably worse or better than ordinary prospects? This issue can be raised even if one refuses to accept the existence of hell. For example, the immediate loss of life is evidently not considered incomparably bad since people are always crossing dangerous intersections and risking immediate loss of life rather than face the inconvenience of waiting several hours for traffic to slow down.

Even if the existence of such terrible or wonderful prospects is accepted, our assumption may still apply to a limited extent. It may apply for comparisons among all prospects in which one ends up in heaven. Then, so long as one's decisions do not prejudice the possibility of ending up in heaven, one may apply the ensuing results.

The existence of a *utility* described in Section 2 has been derived on the basis of these four assumptions.[1]

From now on, we shall assume the existence of utility functions for individuals. Granting their existence, how does one compute them? What role does utility play in the problem of decision making ? To answer these questions, let us illustrate some characteristics of utility with a specific example. For our specific example we consider the young man who is faced with a mixture of P_0, the prospect of being killed in his attempt to cross the street, and P_1, the prospect of successfully crossing the street to meet the young lady, etc. Suppose that the value of the utility function is known for these two prospects. In fact, suppose $u(P_0) = -3$, and $u(P_1) = 5$. (Evidently the prospect of married life attracts the young man.) What would be the utility of the mixed prospect if the probability of successfully crossing the street is 1/2? It would be $(1/2)(-3) + (1/2)(5) = 1$. If we were to plot the utilities of P_0, P_1, and the above mixture on the horizontal axis of graph paper, we would have two points and one (P_2) halfway between them; see Figure 4.1.

[1] A careful and detailed derivation appears in reference [2]. For reference purposes (for mathematically sophisticated readers), a skeleton outline of a derivation is presented in Appendix F_2.

Suppose now that the probability of successfully crossing the street is 4/5. The utility would be $(1/5)(-3) + (4/5)(5) = 17/5 = 3.4$. This point P_3 is clearly between the points represented by P_0 and P_1 and,

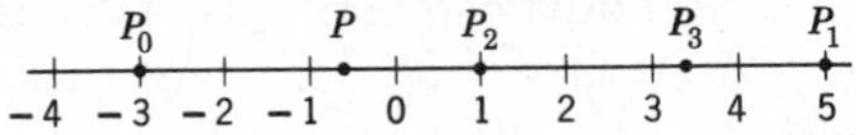

Figure 4.1. Utilities of various prospects.
P_2 = mixture of P_1 and P_0 each with probability 1/2.
P_3 = mixture of P_1 and P_0 with probabilities 4/5 and 1/5 respectively.

in fact, is 4/5 of the way from the first to the second. It is clear that, as the probability of successfully crossing the street increases from zero to one, the value of the utility of the mixed prospect moves from $u(P_0)$ to $u(P_1)$. Each point between those represented by P_0 and P_1 corresponds to a mixture of P_0 and P_1, with an appropriate probability of facing P_1.

Now consider the prospect P of not crossing the street and living a rather ordinary life. If P_1 is preferred to P which, in turn, is preferred to P_0, $u(P)$ should be between $u(P_0)$ and $u(P_1)$. But then the value of the utility for P is the same as for a certain mixture of P_0 and P_1. This means that there is a mixture of P_0 and P_1 such that the young man is indifferent between this mixture and P. If the young man were to sit and meditate sufficiently long, he would probably be able to determine that probability of successfully crossing the street for which he would just as soon not cross as cross. (If his calculations take too long, he may be too late to face P_1. For the sake of simplicity, we shall assume that he is a very rapid calculator.) Suppose that after sufficient introspection he decides that, if the probability of successfully crossing the street is only 0.3, he would just as soon not cross as cross. Then the utility of P is the same as that for the mixed prospect with probability (0.3) of successfully crossing the street. This is clearly $(0.7)(-3) + (0.3)(5) = -0.6$. In this way he can evaluate $u(P)$ for any prospect " between " P_0 and P_1.

Suppose that after some thought he had decided that death would be preferable to drab, ordinary life filled only with painful thoughts of what might have been. In other words, both P_1 and P_0 are preferred to P. Then how can he evaluate $u(P)$? In this case, P_0 will have the same utility as some mixture of P and P_1. Suppose that he is indifferent between P_0 and a mixture of P and

P_1, where P_1 has probability 0.4. Then,

$$u(P_0) = 0.6u(P) + 0.4u(P_1)$$
$$-3 = 0.6u(P) + (0.4)(5)$$
$$u(P) = -8.33.$$

A similar technique would apply if he had decided that, after all, he preferred drab bachelorhood (P) to a married life even with this gorgeous creature (P_1).

We have shown how to evaluate $u(P)$ for an arbitrary P once we know the value of u for any two special prospects P_0 and P_1 (which have different utilities). How do we find $u(P_0)$ and $u(P_1)$? The fact is that we can arbitrarily fix these two utilities at any two values, say 0 and 1, or -3 and 5, keeping in mind that the larger value goes to the preferred prospect. Instead of indicating why, we point out that a similar process applies to the measurement of temperatures. The effect of temperature is indicated by the length of a column of mercury in a small tube. The hotter the temperature the higher the column. A ruler is put down next to the tube and a scale is marked on the tube. Two points on this scale are arbitrarily taken. The point corresponding to the temperature at which water freezes is arbitrarily labeled 0 and the point corresponding to the temperature at which water boils is labeled 100. A temperature corresponding to a position halfway between would be called 50. This measurement of temperature is in the so-called Centigrade scale. Another scale more commonly used in English speaking countries is the Fahrenheit scale. Here 0 and 100 were selected for two different points. Zero and 100 in the Centigrade scale correspond to 32 and 212 in the Fahrenheit scale. It clearly does not matter which scale is used as long as it is identified.

This discussion of the assumptions behind utility has been intended to illuminate their meaning. If they hold exactly for an individual, then it is a logical consequence that he has a utility function, and expected utility is for him the sole guide to choice. To demonstrate that these assumptions actually do hold, even for one person, would be a task of unmanageable size. A more direct and limited approach has led to experiments which show that many people (not all) act, in certain carefully controlled experimental setups, as if their choices satisfied the utility assumptions. Any such experiment concerned with bets involving, say, books or records or cash, can at most demonstrate the validity of the theory for that person in that narrow context. Lacking solid evidence for the practical validity of the assumptions, must we conclude that the use of utility is a precarious business? Not at all. Even if a person may be unable to decide which of two crippling diseases he " prefers " (see Assumption 1), even if sometimes his preferences are circular (see Assumption 2), etc., it may still be true that in many practical choice situations he will, and should, act much as though he had a utility function. This is especially likely to be true where the alternatives involved are not too dissimilar in character. Examples which illustrate the usefulness of utility in limited choice situations occur in the remainder of the book.

Exercise 4.1. If $u(P_0) = 7$ and $u(P_1) = 19$, find $u(P)$ in the following three cases where you are indifferent between:

(a) P, and the mixture of P_0 and P_1 where you face P_1 with probability 1/6;

(b) P_0, and the mixture of P and P_1, where you face P_1 with probability 1/5;

(c) P_1, and the mixture of P and P_0, where you face P_0 with probability 2/3.

Exercise 4.2. Suppose $u(P_0) = -3$ and $u(P_1) = 5$, and we find it convenient to transform to a new utility function v such that $v(P_0) = 0$ and $v(P_1) = 1$. Derive the relationship between $u(P)$ and $v(P)$ for values of $v(P)$ between 0 and 1.

Exercise 4.3. Express the relationship between the utility functions w and u if $w(P_0) = 4$, $w(P_1) = 8$, $u(P_0) = -3$, and $u(P_1) = 5$.

2.2 Application of Utility

What use do we make of utility in decision making when the laws of randomness (state of nature) are known? *If a and a^* are two available actions and these actions lead to prospects P and P^* which have utilities $u(P)$ and $u(P^*)$, then of these two actions we should take the one for which the corresponding utility is larger.*[1]

Frequently, relevant prospects are mixtures of prospects with

[1] Utility was shown to be derived from preferences among prospects. Hence there seems to be circularity here in deriving preferences from utilities. This is true, but utility offers computational convenience. The fact is that utility property 2 enables us to compute many relevant utilities once a few others are known. Thus simple computations may replace difficult and time-consuming introspection.

known utilities. Then the utility of the mixture may be readily computed. For example, as we shall see later, if P is a mixture which presents you with P_1 with probability p_1, P_2 with probability p_2, P_3 with probability p_3, and P_4 otherwise (with probability p_4), then

$$u(P) = p_1 u(P_1) + p_2 u(P_2) + p_3 u(P_3) + p_4 u(P_4).$$

This simple computation may save hours of introspection.

An important consideration is the following. No matter how carefully you think or introspect, it is really rather difficult to attach utilities to histories (histories of the whole life including all the joys and sorrows). In practice, one usually confines one's attention to a small aspect of the history which is relevant to the particular problem and, with some rough ideas of how utility is affected by certain changes, one is put in a position of using the properties of utility to make very reasonable decisions. We illustrate with the following example about which more will be said later.

Example 4.1. Mr. Campbell is rather well established in a position he likes and which he does not expect to leave. This position yields him a steady income of $90 per month. Having industriously saved money, he has about $8000 in liquid assets. The presence of this money is not essential to his continuing in his present way of life. On the other hand, the more he has the more he feels free to spend for some of the finer things of life. Furthermore, if he lost all his money, he would, considering his conservative nature, feel compelled to forego certain expenditures till he had saved a bit, even though these expenditures could lead to savings in the long run. (For example, he would keep his old car even though it is falling apart and uneconomical to run.) Suppose that for the sake of convenience he were to set his present utility at 1 and his utility in the event that he had no money at 0, his utility function could be expressed approximately as a function of money (assuming all other things remain more or less equal). Suppose that after a bit of introspection, Mr. Campbell expresses his utility function by the following graph, Figure 4.2. Should Mr. Campbell accept a bet in which he makes $2000 with probability 1/2 and loses $1000 with probability 1/2? His present utility is 1. If he takes the bet, he has a probability 0.5 of having $7000 and a utility of 0.86, and a probability 0.5 of having \$10,000 and a utility of 1.18. Hence the utility associated with the bet is $(0.86 + 1.18)/2 = 1.02$, and he should take it.

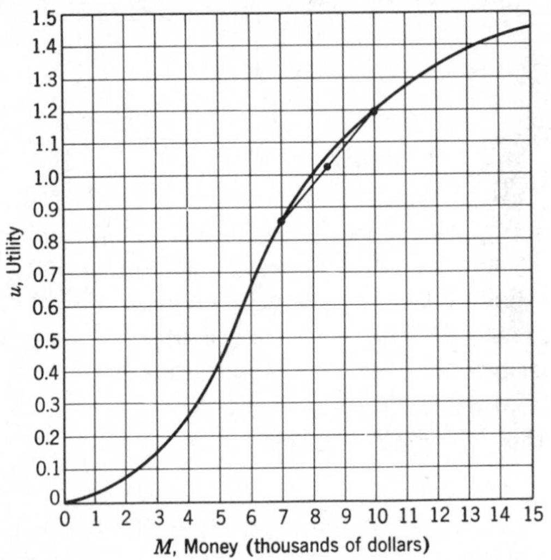

Figure 4.2. Mr. Campbell's utility for money.

Notice how Mr. Campbell's problem has been settled by reference to the utility of *money only*, assuming that "all other things remain more or less equal." Indeed only his utility for money over the range \$0 to \$15,000 has come under consideration. We may be fairly confident that a reasonable treatment of his problem has resulted, even though he may lack a full utility function embracing all possible prospects.

Exercise 4.4. At 2-to-1 odds on the flip of an ideal coin, how big a bet should Mr. Campbell be willing to take? That is, at what size of bet should he decide that he is indifferent between betting and not betting? (Try various sizes of bets and use Figure 4.2.)

**Exercise 4.5.* Suppose Mr. Campbell has only \$1000 and therefore utility 0.03. Should he invest \$1000 in an enterprise which will take his money with probability 5/6 and yield him \$5000 (this includes his investment) with probability 1/6?

**Exercise 4.6.* Note that in Figure 4.2 a chord is drawn connecting the points on the graph corresponding to the two possible outcomes of the bet in Example 4.1. The ordinates (heights) represent the utilities and, since either outcome is equally likely, the resulting utility is halfway between the two ordinates. This utility is the ordinate corresponding to the point of the chord halfway between the two other points. Present a similar geometric interpretation of your result for Exercise 4.5.

Exercise 4.7. Mr. Campbell has \$4000 and utility 0.25. He is offered a bet which yields him \$6000 with probability p and \$2000 with probability $1 - p$. For what value of p is he indifferent about accepting the bet?

Exercise 4.8. The prospect P is a mixture of P_1, P_2, and P_3 with probabilities 0.2, 0.5, and 0.3. If P_1, P_2, and P_3 have utilities 4, 8, and 16, find $u(P)$.

Exercise 4.9. Draw reasonable utility functions for Albert, Bertram, and Charles for money gifts up to \$10. Albert and Bertram have no cash. Albert would spend all he received on beer to drink tonight. Bertram would love to go to the big game (entrance fee \$5). Charles owns a bank and could always use more money.

3. PROBABILITY AND EXPECTATION

It is now expedient to review and expand on some aspects of the theory of probability. Probability refers to the long-run frequency in many repetitions of an experiment under similar circumstances. To be more precise, suppose $\mathscr{X}$ is the set of all possible outcomes of the experiment. To each set A of possible outcomes is associated a probability $P\{A\}$. This is the proportion of times, in many trials, that the outcome of the experiment will be an element of A. The rule or function P which associates to each set A of possible outcomes, the number $P\{A\}$, is called the probability distribution for the experiment. The probability distribution satisfies certain basic properties in terms of which the theory of probability is developed. To discuss these properties, we first present some set theory notation, some of which has been discussed previously. To illustrate this notation, we shall use as a special example the case of tossing a nickel and a penny, where the set $\mathscr{X}$ of possible outcomes is given by

$$\mathscr{X} = \{(H, H), (H, T), (T, H), (T, T)\}.$$

1. *Set A is a subset of B if B contains all the elements of A.* For example, the set $\{(H, H)\}$ is a subset of $\{(H, H), (T, H)\}$. Also

$\{(H, H), (T, T)\}$ is a subset of $\{(H, H), (T, T)\}$ although in a rather trivial way.

Exercise 4.10. For the sake of concreteness and to develop practice in set language, it is good exercise to label the various sets. Thus;

$A_1 = \{(H, H)\}$ corresponds to both heads, that is, $P\{A_1\}$ is the probability that both coins fall heads;

$A_2 = \{(H, H), (H, T)\}$ corresponds to "the nickel falls heads";

$A_3 = \{(H, T), (T, H)\}$ corresponds to "exactly one of the coins falls heads," or equivalently, "the two coins do not match."

List two of the other sets and label them.

2. *The complement $\tilde{A}$ of a set A is the set of all elements (under consideration) which are not in A.* Thus,

$$\widetilde{\{(H, H)\}} = \{(H, T), (T, H), (T, T)\}.$$

That is, the complement of "both heads" is "not both heads." Similarly, $\widetilde{\{(H, T), (T, H)\}} = \{(H, H), (T, T)\}$. In talking about the complement of a set, we must have decided previously on what elements are under consideration. Otherwise the complement of $\{(H, H)\}$ would have included, among other items, all the horses in the world and some mighty poor novels. In probability applications, the basic set under consideration is, of course, $\mathscr{X}$, the set of all possible outcomes. It is easy to see that the complement of $\tilde{A}$ is A. For example, $\widetilde{\{(H, H)\}} = \{(H, T), (T, H), (T, T)\}$ and $\widetilde{\{(H, T), (T, H), (T, T)\}} = \{(H, H)\}$. Generally speaking, if a set A can be given a name in English, the complement can be given the negative of that name. Thus, "the two coins match" and "the two coins do not match" are the complementary sets $\{(H, H), (T, T)\}$ and $\{(H, T), (T, H)\}$ respectively.[1]

3. One may ask what is the complement of $\mathscr{X}$. Since there are no elements under consideration which are not in $\mathscr{X}$, this set has no complement. It is convenient to eliminate this unusual characteristic by introducing a special set, the *null set* designated by ϕ which may be called the set consisting of no elements. Then we

[1] Because of the nature of the English translations of sets, we often use the two terms "the outcome is an element of A" and "A occurs" as equivalent. Thus $P\{A\}$ is often called the probability that A occurs and $P\{\tilde{A}\}$ is called the probabilitity that A does not occur.

may write

$$\tilde{\mathscr{X}} = \phi, \qquad \tilde{\phi} = \mathscr{X}.$$

4. *The union of several sets* $A_1, A_2, \cdots, A_n$ *is the set* A *of elements which are in at least one of the sets* $A_1, A_2, \cdots$. The union is denoted by

$$A = A_1 \cup A_2 \cup \cdots \cup A_n.$$

For example,

$$\{(H, H)\} \cup \{(H, H), (H, T)\} \cup \{(H, T), (T, H)\}$$
$$= \{(H, H), (H, T), (T, H)\}.$$

In general, if $A_1, A_2, \cdots, A_n$ are called by their names, the union can be and often is written in terms of *or*. Thus we shall usually write the union as

$$A = A_1 \text{ or } A_2 \text{ or } \cdots \text{ or } A_n.$$

(The everyday English usage of the word "or" is subject to some ambiguity. When we say A_1 or A_2, we shall mean A_1 or A_2 *or both*.)

Exercise 4.11. Take the union of two or more sets and interpret this union in terms of "or."

Exercise 4.12. Let $\mathscr{X}$ be the set of points in the plane. That is, $\mathscr{X} = \{(x, y) : x \text{ and } y \text{ are numbers}\}$. Represent the sets $A_1 = \{(x, y) : x^2 + y^2 \leq 4\}$, $A_2 = \{(x, y) : (x - 1)^2 + (y - 2)^2 \leq 1\}$, and their union geometrically.

Exercise 4.13. Consider various subsets of a deck of bridge cards. Let A be the spade suit, B the face cards (Jacks, Queens, and Kings), C the Jack of clubs, and D the red cards. (These sets contain 13, 12, 1, and 26 cards respectively.) How many elements are there in the following sets?

1. $A \cup B$
2. $\tilde{B}$
3. $A \cup C \cup B$
4. $(\widetilde{C \cup D})$
5. $(D \cup B)$
6. $(\widetilde{A \cup B \cup C \cup D})$.

Exercise 4.14. If 30% of the population like ice cream but not beer, 20% like beer but not ice cream, and 10% like both, what percent like ice cream or beer (or both)?

Remark: The following simple diagram (Figure 4.3), called a Venn diagram, is a convenient device to help avoid confusion.

For this problem, A represents the set of people who like ice cream, B the set who like beer. The overlapping portion represents the set of people who both like beer and like ice cream, the entire shaded portion represents A or B. The rectangle represents the entire population.

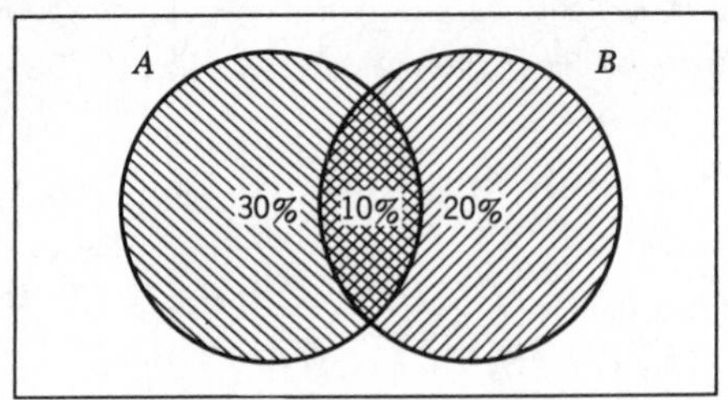

Figure 4.3. Venn diagram.

Exercise 4.15. (a) If 50% of the population like baseball, 60% like movies, and 30% like both, what percent like movies or baseball (or both)? (b) What percent like neither baseball nor movies?

5. *Two sets are called nonoverlapping if they have no elements in common.* Thus $\{(H, H)\}$ and $\{(H, T), (T, H)\}$ are nonoverlapping. Also $\{(x, y) : x^2 + y^2 \leq 4\}$ and $\{(x, y) : (x-4)^2 + (y-4)^2 \leq 1\}$ are nonoverlapping.[1] The two sets $\{(x, y) : x^2 + y^2 < 1\}$ and $\{(x, y) : x^2 + y^2 = 1\}$ are quite close to one another but they are nonoverlapping too. Three or more sets will be called nonoverlapping if every pair is nonoverlapping, that is, if no two of these sets have any element in common.

The basic properties of probability distributions are the following:

PROBABILITY PROPERTY 1. $0 \leq P\{A\} \leq 1$ for every set A of possible outcomes.

PROBABILITY PROPERTY 2. $P\{\mathscr{X}\} = 1, P\{\phi\} = 0.$

PROBABILITY PROPERTY 3. $P\{A\} + P\{\tilde{A}\} = 1.$

PROBABILITY PROPERTY 4. $P\{A_1 \text{ or } A_2 \text{ or } \cdots\} = P\{A_1\} + P\{A_2\} + \cdots$ if $A_1, A_2, \cdots$ are nonoverlapping sets.

PROBABILITY PROPERTY 5. $P\{A\} \leqq P\{B\}$ if A is a subset of B.

These properties are easily explained and form the core from which the theory of probability is developed. The first property merely states that a probability is always between zero and one. The second property merely states that the outcome of every trial

[1] These sets are circles (with their interiors). The first has radius 2 and center at the origin. The second has radius 1 and center at (4,4).

of the experiment is in $\mathscr{X}$. We sometimes call $\mathscr{X}$ the certain set since it is certain that the outcome of the experiment is in $\mathscr{X}$. Similarly, we sometimes call the null set ϕ the impossible set since it is impossible that the outcome of the experiment be in ϕ. Property 2 states that the certain set has probability one and the impossible set has probability zero.

Property 3 states that the probabilities of complementary sets add up to one. In English, the probability of something taking place plus the probability of that something not taking place is one. This is clear when we observe that, if something occurred in say 40% of the trials, it failed to occur in the remaining 60% of the trials.

Property 4 is related to Property 3. It states that, if the outcome is in A_1 30% of the time, and in A_2 25% of the time, and A_1 and A_2 are nonoverlapping, then in 55% of the experiments the outcome will be in A_1 or in A_2. If A_1 and A_2 overlap, this may not be true, for then it might be that 10% of the time the outcome is in both A_1 and A_2. Then the times that it is in A_1 or A_2 could be broken into three (nonoverlapping) parts :

In A_1 but not in A_2	20% of the time;
in A_2 but not in A_1	15% of the time; and
in A_1 and A_2	10% of the time; and thus
in A_1 or A_2	45% of the time.

Property 5 states merely that, if B occurs whenever A does, then $P\{B\}$ is greater than or equal to $P\{A\}$. Thus the probability of two heads is less than or equal to the probability of at least one head. That is, $P\{(H, H)\} \leqq P\{(H, H), (H, T), (T, H)\}$.

These properties are not only basic to the study of probability but we have used them implicitly many times. This is merely the first time they have been presented in a formal fashion but not the first time they have been considered or used.

Thus, we have always assumed that probabilities are between 0 and 1 (Property 1) and that, if the probability of heads on the toss of a coin is 0.4, then the probability of tails is 0.6 (Property 3). Properties 4 and 2 were assumed implicitly when we stated that the sum of the values of a discrete density was one. For there we really used the fact that $\{\mathbf{Z} : \mathbf{X} = x_1\}$, $\{\mathbf{Z} : \mathbf{X} = x_2\}$, $\cdots$ are all nonoverlapping sets, and that they exhaust all possibilities, i.e., their union is $\mathscr{X}$. Hence, the sum of their probabilities is the probability of $\mathscr{X}$ (Property 4) which is equal to 1 (Property 2). Property 5 is used so often by the man on the street that it hardly needs elaboration. For example, everyone knows that the probability of throwing a two, three, or twelve in craps exceeds the probability of throwing a three.

Finally, these properties are somewhat redundant in the sense that they can be proved by assuming only a few of them. In Appendix E_3 we present a proof of these properties, assuming only that: (a) $P\{A\} \geq 0$; (b) $P\{\mathscr{X}\} = 1$; and (c) Property 4.

Example 4.2. Mr. Sharp has taken up gambling as a business. Recently he investigated an interesting game which yields the customer a gain of \$8 if a certain bent coin falls heads and a loss of \$2 otherwise. Let $\mathbf{X} =$ Mr. Sharp's gain in dollars. Then $\mathbf{X}$ has -8 and 2 as possible values. After many observations, Mr. Sharp has decided that the probability of heads with this coin is about 0.3. He is interested in whether it would pay for him to attempt to entice customers to play this game. He estimates that, if customers play this game 1,000,000 times, he will lose \$8 about 300,000 times and gain \$2 about 700,000 times. Thus, in all, he will be ahead about

$$300{,}000(-8) + 700{,}000(2) = -1{,}000{,}000.$$

He finds that he expects to lose about \$1,000,000 which is an average gain of $-\$1$ per game. If Mr. Sharp's customers had played the game and the successive values of $\mathbf{X}$ were labeled

$$\mathbf{X}_1, \mathbf{X}_2, \cdots, \mathbf{X}_{1,000,000},$$

then $\overline{\mathbf{X}}$, the average value of these "observations," is what we have just estimated to be close to $-\$1$. Mr. Sharp does not care for this game. Instead, he develops a variation where, depending on the roll of a four-sided die, Mr. Sharp will win either \$2 with probability 0.4, \$1 with probabilty 0.3, \$0 with probability 0.2, or $-\$5$ with probability 0.1. Here he estimates that, if he has customers play a million times, giving him $\mathbf{X}_1, \mathbf{X}_2, \cdots, \mathbf{X}_{1,000,000}$, he will win \$2 about 400,000 times, \$1 about 300,000 times, \$0 about 200,000 times, and $-\$5$ about 100,000 times. That is, in all he will win

$$\mathbf{X}_1 + \mathbf{X}_2 + \cdots + \mathbf{X}_{1,000,000}$$

which is about equal (in dollars) to

$$(400{,}000)2 + (300{,}000)1 + (200{,}000)0 - (100{,}000)5 = 600{,}000$$

which gives an average value, $\bar{\mathbf{X}}$, which is about

$$\left(\frac{400{,}000}{1{,}000{,}000}\right)(2) + \left(\frac{300{,}000}{1{,}000{,}000}\right)(1) + \left(\frac{200{,}000}{1{,}000{,}000}\right)(0)$$
$$+ \left(\frac{100{,}000}{1{,}000{,}000}\right)(-5) = \left(\frac{600{,}000}{1{,}000{,}000}\right) = 0.6$$

or an average of 60 cents per game.

This "*long-run expected average of* **X**" is called the *expectation* of **X** and is denoted by $E(\mathbf{X})$. Mr. Sharp is well advised to entice customers to play this game. Even though the game never results in an exact gain of 60 cents, and even though he cannot be sure that in 100 games he will come out ahead, he can feel quite certain that, if enough customers play, his average winnings will tend to be close to 60 cents per game and, if enough customers play, he may become wealthy.

The technique used to compute $E(\mathbf{X})$ is subject to generalization in an obvious way. *That is, if* **X** *is a discrete random variable which can take on the values* $x_1, x_2 \cdots$ *with probabilities* $p_1, p_2, \cdots,$ *then*

$$E(\mathbf{X}) = p_1x_1 + p_2x_2 + \cdots. \tag{4.1}$$

The reader should compare this equation with the above examples to check that the same reasoning will yield this result.

Exercise 4.16*. Let **X be the number of heads obtained in the toss of four well-balanced coins (see Exercise 3.33). Compute $E(\mathbf{X})$. Suppose a gambler receives $\mathbf{Y} = \mathbf{X}^2$ dollars after the toss. What is $E(\mathbf{Y})$?

Exercise 4.17. In Example 1.1, suppose loss is measured in utility. If the state of nature is θ_3, compute the expected loss of utility involved in applying strategy s_1.

Exercise 4.18. If you receive one cent for every dot that appears in throwing two well-balanced dice, what is the expected gain (in cents)?

Exercise 4.19. The probability that a man aged 60 will live another year is 0.95. If the insurance company pays $1000 upon his death, what is the expected amount the insurance company will pay out for the man that year?

We have previously defined the expectation of a discrete random variable only. The notion extends to continuous random variables, and a discussion appears in Appendix E, after Consequence 6. For example, if Mr. Sharp were to use an ideal dial and receive from customers in gold dust the amount shown, e.g., 0.345 oz if the dial showed 0.345 etc., then it seems clear that, in the long run, he would average 1/2 oz per customer. Thus a reasonable definition of expectation would yield $E(\mathbf{X}) = 1/2$ for the outcome of the spin of an ideal dial. The fact that we are not prepared to discuss the extension of the definition of expectation in more detail need not trouble us. Whether a random variable is discrete or continuous or otherwise, it obeys the following properties which we shall find as useful as the general definition.

EXPECTATION PROPERTY 1. $E(\mathbf{X} + \mathbf{Y}) = E(\mathbf{X}) + E(\mathbf{Y})$.
EXPECTATION PROPERTY 2. $E(c\mathbf{X}) = cE(\mathbf{X})$.
EXPECTATION PROPERTY 3. $E(1) = 1$.
EXPECTATION PROPERTY 4a. $E(\mathbf{X}) > E(\mathbf{Y})$, if $\mathbf{X} > \mathbf{Y}$.
EXPECTATION PROPERTY 4b. $E(\mathbf{X}) \geq E(\mathbf{Y})$, if $\mathbf{X} \geq \mathbf{Y}$.

These four properties can be expressed as follows: Suppose Mr. Sharp has two different bets with two different friends or customers. Both bets depend on the outcome of the same game. Suppose that one friend pays him $\mathbf{X}$ dollars and the other pays him $\mathbf{Y}$ dollars at the end of the game, where $\mathbf{X}$ and $\mathbf{Y}$ are random variables, of course. The first property states that, if he averages \$4 a game from the first friend and \$2 a game from the second, on the whole he will average \$6 a game.

The second property states that, if all bets are doubled, he will average twice as much. The third property states that, if he collects \$1 no matter what happens, he will average \$1 per game. (This situation corresponds to the case where the house charges people \$1 per game to gamble but does not engage in the game otherwise.) Finally, the fourth property states merely that, if he always gets more money from one friend than from a second, he will average more from the first friend than from the second.

Because of the relation between expectation and average (exectation is a long-run average), it is not surprising that these properties correspond to similar properties of averages and the summation symbol [see Equations (2.1) through (2.3)].

Properties 1 and 2 can be easily combined and extended to yield

EXPECTATION PROPERTY 1a.

$$E(a\mathbf{X} + b\mathbf{Y} + c\mathbf{Z}) = aE(\mathbf{X}) + bE(\mathbf{Y}) + cE(\mathbf{Z}).$$

To illustrate, suppose Mr. Sharp bets on the outcome of the game with three different customers simultaneously. These wagers yield him **X**, **Y**, and **Z**. If he averages $1, $2, and $4 per game from these customers and they decide to double, triple, and quadruple their bets respectively, then he will average (in dollars)

$$2 \times 1 + 3 \times 2 + 4 \times 4 = 24.$$

A game is called a "*fair*" game if the gambler's winnings **X** are such that $E(\mathbf{X}) = 0$. Since Mr. Sharp has taken up gambling as a business and not as a recreation, he does not intend to run a "fair" game. On the other hand, if he can entice many more customers by making the game almost fair, he may decide to do so. His decision will be based on the principle that a larger volume of business may compensate for a decreased earning per customer.

When does it pay for a customer to gamble? This is not an entirely flippant question. In a sense, insurance companies are gambling establishments and the decision to take out or not to take out insurance corresponds roughly to the decisions not to gamble and to gamble respectively. Notice the peculiar order here. When you do not take out insurance, you gamble, and the insurance company does not. We will say more about this later.

If a customer is to gamble a relatively small sum at each play of the game and to do this many, many times, then he should gamble only if the expectation of his money winnings (per game) is positive, i.e., the game is "favorable." For then he is almost certain to come out winning money and his utility will be increased. If the expectation of his winnings (per game) is negative, he should stay away. If the game is a fair game, then after many plays his average winning or loss per game will be small. However, this small amount per play of the game may add up to a considerable sum if the game is played 1,000,000 times. In any case, if he were to play a fair game many, many times, eventually it would be unfavorable for him in the sense that the time wasted in playing would outweigh in value his winnings or losses. In fact, a gambling house must be certain of having some minimal expected winnings in order to be sure of being able to pay for the salaries of the gambling-device operators and other overhead costs.

Although the above discussion may satisfactorily treat the case of the gambler who plays his game many, many times, it does not say much about the person who has only one opportunity to play. For example, Mr. Campbell certainly will not be given hundreds of opportunities to win $2000 on investing $1000 on the toss of a coin. Even if he were given hundreds of opportunities to bet such a large amount of money, a *run of bad luck* might bankrupt him and stop the gambling before he had played often enough to take advantage of the long-run aspects of expectation. Previously we discussed how Mr. Campbell should decide about accepting a bet (assuming it was offered only once). His procedure consisted of comparing his utility when he does not bet $u(\$8000) = 1$ with $(1/2)u(\$7000) + (1/2)u(\$10{,}000) = (1/2)(0.86) + (1/2)(1.18) = 1.02$. In so doing he is applying the second utility function property which states that, if P is a prospect which leads to facing P_1 with probability p and P_2 with probability $1 - p$, then

$$u(P) = p\,u(P_1) + (1 - p)\,u(P_2).$$

In this case, P is a prospect which results in eventually facing a *random prospect* $\mathbf{P}$ with corresponding random utility $u(\mathbf{P})$. Here the possible values of the random prospect are P_1 and P_2 and the corresponding values of the random utility $u(P_1)$ and $u(P_2)$. Then $p\,u(P_1) + (1 - p)\,u(P_2)$ is precisely $E(u(\mathbf{P}))$.

It is possible to prove in general that, if P is a prospect which results in eventually facing a random prospect $\mathbf{P}$ (P is a mixed prospect consisting of a mixture of the prospects which are the possible values of $\mathbf{P}$), then

$$u(P) = E(u(\mathbf{P})).$$

From here on we shall call this property the *second utility function property*.

A major point of the preceding sections is the following. If a gamble is "favorable" from the point of view of expectation of money and you have the choice of repeating it many times, then it is wise to do so. For eventually, your amount of money and, consequently, your utility are bound to increase (assuming that utility increases if money increases). If you can gamble only once, then the issue is not expectation of money but expectation of utility. If engaging in a gamble increases the expected utility, then it should be engaged in. *In general, the expectation of the utility arising from a course of action is the one criterion for measuring how good that course of action is.*

Exercise 4.20. Mr. Campbell engages in a complicated game. When the game is over he will have either \$2000, \$4000, \$8000, or \$10,000, with probabilities 0.3, 0.2, 0.4, and 0.1. Compute his utility. (Use Figure 4.4.)

†4. APPLICATION OF UTILITY TO FAIR BETS

Let us consider a few problems that may conceivably face Mr. Campbell. Suppose that, before he decides to make his bet, the man who offered it notices that the odds are a bit out of line compared with the probabilities of winning and losing. He changes his offer to the following. A well-balanced, three-sided die is rolled. If side 1 turns up, Mr. Campbell collects \$2000; otherwise he pays \$1000. Because Mr. Campbell has not done so well by delaying, he now faces a less favorable gamble as an alternative to not gambling. This new gamble is "fair" in the sense that the expectation of profit is given by

$$(1/3)(\$2000) + (2/3)(-\$1000) = 0.$$

In Figure 4.4, we again present Mr. Campbell's utility function for money. The utility corresponding to the gamble is

$$u = (1/3)u(\$10{,}000) + (2/3)u(\$7000)$$
$$u = (1/3)(1.18) + (2/3)(0.86) = 0.97.$$

It clearly does not pay for Mr. Campbell to gamble. Geometrically, this fact is represented by considering the chord connecting the two points on the utility function corresponding to \$7000 and \$10,000. The point one-third of the way from the first point to the second has for its abscissa the expected money resulting from this gamble $(2/3)(7000) + (1/3)(10{,}000) = 8000$, and for its ordinate the expected utility 0.97. Since the gamble is fair, we know that the abscissa will be the same as the one before the gamble. Hence, to obtain the utility graphically, we just take the point on the chord with the same abscissa as before. It should be emphasized that this argument applies only to fair bets.

Why is the bet unfavorable to Mr. Campbell? There are several alternative ways to express this. One way that is standard in

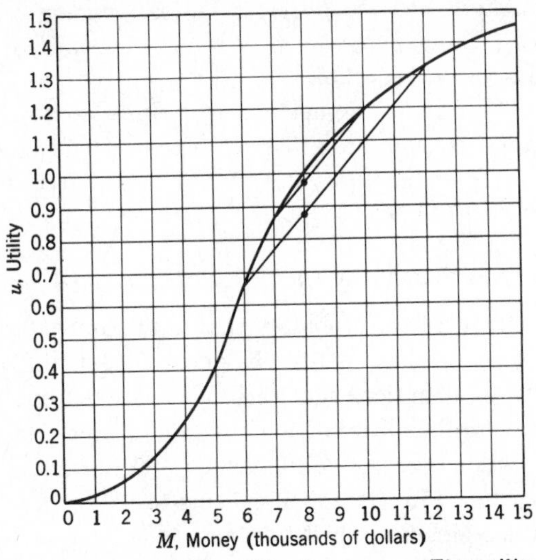

Figure 4.4. Mr. Campbell's utility for money. The utility for two fair bets at odds of two to one.

economic theory is that the marginal utility of money for Mr. Campbell is decreasing between \$7000 and \$10,000. This essentially means that, as more money is obtained, each extra dollar does less than the preceding to increase utility. In other words, the rate at which the curve goes upward is decreasing. A second way to express it is that the curve is concave between the relevant points. This merely means that all chords connecting points with abscissas between \$7000 and \$10,000 lie below the curve. There is a section of the curve which is convex (chords lie above the curve). If Mr. Campbell had only \$1000 and would therefore be in a convex section of the curve, it would pay for him to take the above-mentioned fair bet.

If the utility function were linear, i.e., represented by a straight line, then it would do no harm and no good to gamble on fair bets. This has the following application. It is well known that, if a powerful microscope is applied in the neighborhood of a point of a "smooth" curve, the curve will look like a straight line. This is analogous to the fact that driving a car along the circumference of a very large circle is like driving along a straight road. Hence, for " fair " gambles which involve very small stakes, the graph of the utility function behaves like a straight line and it makes *very* little difference in utility whether or not you gamble.

Now Mr. Campbell is an enterprising man and knows a bookie who will, for a small fee, offer him any fair bet. Mr. Campbell wants to know whether there is any fair bet for which he can improve his utility. To investigate this, all he has to do is to consider all chords starting on one side of $8000 and ending on the other. If any of these chords is above the utility graph at $8000, then it pays for him to make the corresponding gamble.

Suppose that Mr. Campbell had $4000. What would be his best bet and the corresponding utility? These can be obtained by drawing all chords about the point with abscissa $4000. On the other hand, we can save some fuss and bother in the constant re-evaluation of the utility for the best bet as Mr. Campbell's fortune varies. To do this we draw the *convex set generated by* the graph of the utility function ; this set automatically contains all the chords. It pays for him to make the gamble corresponding to that chord which gives the highest ordinate at $4000.

A *convex set* is a set such that if two points are in the set, all points on the line segment connecting them are also in the set. Hence, the interior of a circle is a convex set. The circumference of a circle is not convex. The interior and circumference of a circle together form a convex set. The points on a line form a convex set. The points in a triangle also form a convex set. Some, but not all quadrilaterals, are convex sets; see Figure 4.5.

The *convex set generated by* A is the smallest convex set containing all the points of A ; see Figures 4.6 and 4.7.

If Mr. Campbell has $4000, his utility is 0.25. If he decides to gamble appropriately, he can by the proper gamble raise his utility to 0.50. In fact, it pays for him to take even a slightly unfavorable bet to make the gamble. On the other hand, the bookie earns his living by taking bets which are slightly unfavorable to his customer. It is interesting that two people can both raise their utilities by engaging in a gamble in which one man's loss is another's gain (in money). It is this fact which keeps insurance companies in business.

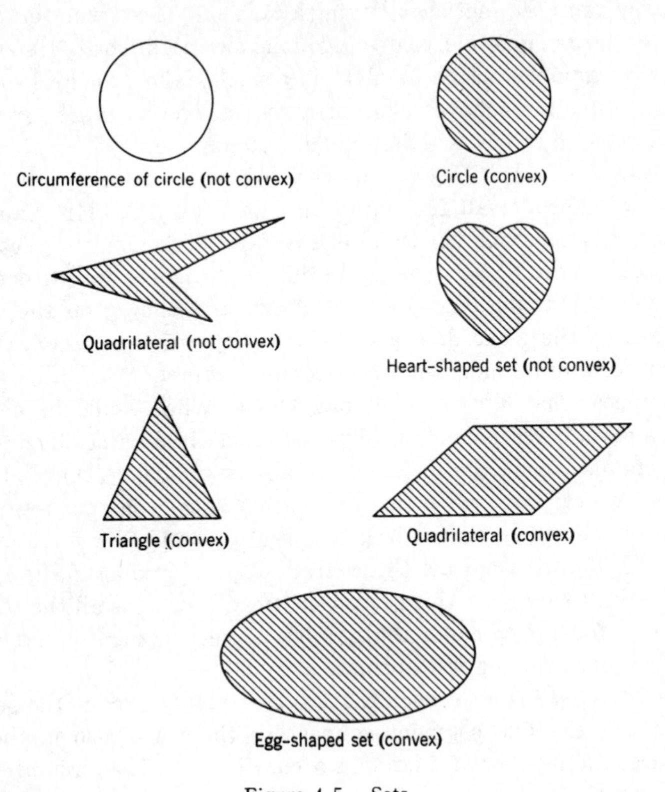

Figure 4.5. Sets.

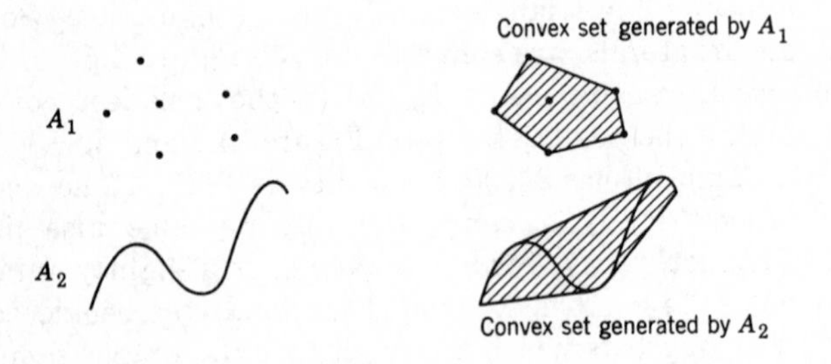

Figure 4.6. Convex sets generated by A_1 and A_2.

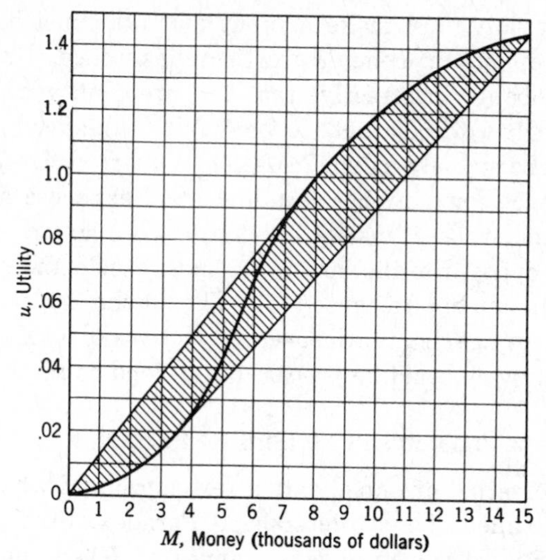

Figure 4.7. The set of utilities u available to Mr. Campbell through fair bets if he has M thousands of dollars. The set is obtained by taking the convex set generated by Mr. Campbell's utility curve for money.

Exercise 4.21. Some large companies do not carry insurance against certain kinds of losses, e.g., small fires; most householders do. Might both be acting in a reasonable way? Why?

Exercise 4.22. What is the best fair bet for Mr. Campbell to make if he has $4000? If he has $500?

Exercise 4.23. Suppose he has $4000 and is offered a fair bet yielding a profit of $2000 with probability 1/3 and a loss of $1000 with probability 2/3. Should he accept the bet? Suppose he is offered a compound bet where win or lose he has to repeat the above bet a second time. Should he accept the bet? (It is necessary to calculate the utilities and probabilities for all final outcomes.)

**Exercise 4.24.* What is the convex set generated by two points? What is the convex set generated by three points? Place five points haphazardly on graph paper and construct the convex set generated by these points.

Exercise 4.25. Mrs. Steele escorts her husband to the airport. For 25 cents she can get $6250 worth of insurance. Mr. Steele ridicules the idea. He does not expect the plane to crash. Besides, he once had a course in probability theory and explains that the insurance company must make a profit and this must be done by not giving people their money's worth in insurance. In fact, he estimates (probably incorrectly) that the probability of his wife collecting is only 1/62,500 and, to be "fair," the insurance company should charge only a dime instead of a quarter. Unfortunately, the insurance company has certain clerical expenses and is out to make a profit, thus, it insists on charging a quarter. Although Mr. Steele is opposed to the insurance, Mrs. Steele knows that by crying she can get him to agree to it. Should she invest? Give an argument for or against the insurance, and make use of certain utilities which you should be prepared to defend as reasonable.

†5. THE ST. PETERSBURG PARADOX

A very interesting example, only a few aspects of which will be discussed, is called the St. Petersburg Paradox. Suppose that a man tosses an ideal coin until heads appears. Then the game is over. He receives $\mathbf{X} = 2^{\mathbf{N}}$ cents, where $\mathbf{N}$ is the number of tosses of the coin. Thus, if heads appeared for the first time on the third toss, he collects 8 cents. It is to his advantage to get a very long run of tails before heads appears. With probability one, heads will eventually appear and he will collect some money. What is the expectation of the amount of money he will receive? In other words, how much should he pay for the privilege of playing this game to make it a "fair" game. In the discussion of Section 4, Chapter 3, we indicated that

$$P\{\mathbf{N} = i\} = 1/2^i.$$

For example, $P\{\mathbf{N} = 1\} = 1/2$, $P\{\mathbf{N} = 2\} = 1/4$, $P\{\mathbf{N} = 3\} = 1/8$, etc. Hence, $P\{\mathbf{X} = 2\} = 1/2$, $P\{\mathbf{X} = 4\} = 1/4$, $P\{\mathbf{X} = 8\} = 1/8$, etc. Then,

$$\begin{aligned} E(\mathbf{X}) &= 2P\{\mathbf{X} = 2\} + 4P\{\mathbf{X} = 4\} + 8P\{\mathbf{X} = 8\} + \cdots \\ &= 2(1/2) + 4(1/4) + 8(1/8) + \cdots \\ &= 1 + 1 + 1 + \cdots = \infty \text{ (symbol for infinity)}. \end{aligned}$$

Three questions arise from this example.

1. *What do we mean by the statement* $E(\mathbf{X}) = 1 + 1 + 1 + \cdots = \infty$? The equation $1 + 1 + 1 + \cdots = \infty$ expresses the following fact. No matter how large a number you take, say 7345, eventually, as

we keep adding terms, the sum will exceed that number. *The symbol* ∞ *is not to be construed as a number in the ordinary sense.* The statement $E(\mathbf{X}) = \infty$ means the following. No matter how large a number you take, say 7345, if you play this game many, many times, eventually you will average more than that amount per game. (It is a peculiarity of this problem that an immense number of games must be played before you can be reasonably sure of averaging more than 20 cents per game.) If we were capable of and interested in playing this game indefinitely, then there would be no amount of money that we could offer to make this a "fair" game. Paradoxically, no one seems willing to offer a mere $5 to play this favorable game once.

2. No one can offer us $7 billion if we should have a very long run of tails. Suppose we take into consideration the limited bankroll of the bookie in deciding the expectation of the winnings. *Then how much should we pay to make the limited game a "fair" game?* Suppose the bookie is rather poor and has only 2^{25} cents or $ 335,544.32. Then as long as $\mathbf{X} < 2^{25}$, we collect $\mathbf{X}$ cents. If $\mathbf{X} \geq 2^{25}$, that is, if we get 24 tails in a row and, hence, $\mathbf{N} \geq 25$, we collect his entire bankroll, a mere $335,544.32, and no more. Suppose we call the winnings of this modified game $\mathbf{Y}$. What is $E(\mathbf{Y})$? To compute this, we note that

$$P\{\mathbf{Y} = 2\} = P\{\mathbf{X} = 2\} = P\{\mathbf{N} = 1\} = 1/2$$
$$P\{\mathbf{Y} = 4\} = P\{\mathbf{X} = 4\} = P\{\mathbf{N} = 2\} = 1/4$$
$$\cdot$$
$$\cdot$$
$$\cdot$$
$$P\{\mathbf{Y} = 2^{24}\} = P\{\mathbf{X} = 2^{24}\} = P\{\mathbf{N} = 24\} = 1/2^{24}$$
$$P\{\mathbf{Y} = 2^{25}\} = P\{\mathbf{X} \geqq 2^{25}\} = P\{\mathbf{N} \geqq 25\}$$
$$= 1/2^{25} + 1/2^{26} + \cdots = 1/2^{24}.$$ [1]

Hence

$$E(\mathbf{Y}) = (1/2)2 + (1/4)4 + (1/8)8 + \cdots + (1/2^{24})2^{24} + (1/2^{24})2^{25}$$
$$E(\mathbf{Y}) = \underbrace{1 + 1 + 1 + \cdots + 1}_{24 \text{ times}} + 2$$
$$E(\mathbf{Y}) = 26.$$

[1] This result is based on the fact that $1/2^{25} + 1/2^{26} + \cdots = (1/2^{24})\{1/2 + 1/4 + 1/8 + \cdots\}$ and $1/2 + 1/4 + 1/8 + \cdots = 1$.

Thus, when playing against this poor bookie with less than a half million dollars in his bankroll, the game is worth only 26 cents from the point of view of expectation of money.

3. *How much should we be willing to pay for the privilege of playing either game (the unlimited one or the one with the poor bookie) once?* The authors, after some introspection, have decided that they would be willing to pay about 10 cents for either game. This amount is less than the expectation. The reason for this involves the shapes of our utility functions for money. If we were interested enough to want a more precise estimate of what the game was worth, we would try to draw a graph of our utility function for money. Then a simple computation would give us an approximate answer.

Another issue of some importance arises from this "paradox." Suppose that Mr. Campbell wonders whether his utility function is unbounded. Given any number, say 7345, can he be sure of finding a prospect P so wonderful that $u(P) > 7345$? *The answer is "No."* For, suppose his utility were unbounded, then there are prospects $P_1, P_2, \cdots$ such that $u(P_1) \geq 2$, $u(P_2) \geq 2^2$, $u(P_3) \geq 2^3$, etc. But then the mixed prospect P which yields him P_i with probability $1/2^i$ would have utility

$$u(P) \geq (1/2)2 + (1/2^2)2^2 + (1/2^3)2^3 + \cdots = \infty.$$

But this is impossible since P is a prospect and its utility is a (finite) number. Hence the utility function must be bounded.

In this argument we have used the St. Petersburg game to demonstrate that *the utility function is bounded.* Hereafter this fact will be called the *third utility function property.*

6. DESCRIPTIVE PARAMETERS[1]

In the preceding sections we considered several examples where Mr. Campbell had a choice of several actions. These were whether to gamble or not to gamble and, if he decided to gamble, to make a choice among various bets. In each case the action he took determined a probability distribution (which we assumed to be known) on the possible outcomes and a corresponding utility. His object

[1] It is suggested that this section be covered only partially until the course reaches estimation.

was to select the action which led to the greatest utility.

In many examples of statistical interest, where the relevant probability distributions are assumed to be known (because there is a great deal of data available) the utility can be expressed in terms of a simple property or characteristic of the probability distribution. Hence, knowing this property or characteristic of the probability distribution is of fundamental importance in deciding whether to take the corresponding action. A property of a probability distribution is called a *parameter*. If this property is relevant in that utility depends greatly on it, it will be called a *descriptive parameter*. Parameters are usually denoted by Greek letters and exceptions to this rule (which are due to ancient custom) will be clearly pointed out. We illustrate the notion of descriptive parameters with some examples.

Example 4.3. Target Shooting. Among Mr. Sheppard's large collection of rifles, there is one he generally prefers to use. Even though this rifle tends to point in a direction slightly different from the one in which it is aimed, Mr. Sheppard compensates for this " bias " and does very well at target shooting. We shall explain his principle for the simple case where we are interested only in how far to the right or left of the target we hit. For simplicity we ignore the possibility of misses due to vertical errors. This assumption is quite reasonable for bowling or croquet. Suppose that the rifle, when aimed directly at the target, hits a point $\mathbf{X}$ units to the right of the target. Negative values of $\mathbf{X}$ represent distances to the left of the target. Because the rifle is not a perfect device, and Mr. Sheppard's position tends to vary slightly from one shot to another, and sometimes winds blow and the amount of powder in a cartridge tends to vary slightly, etc., $\mathbf{X}$ is a random variable. Mr. Sheppard has studied the probability distribution of $\mathbf{X}$ very carefully and claims to know it well. Suppose Mr. Sheppard aims at a point a units to the right of the target. Then, instead of hitting $\mathbf{X}$, which he would if he aimed at the target, he will hit

$$\mathbf{Y} = \mathbf{X} + a.$$

Mr. Sheppard feels that if he hits $\mathbf{Y}$ instead of 0 he will lose an amount of utility equal to

$$l(\mathbf{Y}) = 2.3\mathbf{Y}^2.$$

If Mr. Sheppard is correct, then in selecting an aiming point a, he is selecting a probability distribution for $\mathbf{Y}$ and a corresponding expected loss of utility which is given by

$$L(a) = 2.3E(\mathbf{Y}^2) = 2.3E[(\mathbf{X} + a)^2].$$

Mr. Sheppard should aim at that point a for which $L(a)$ is a minimum. In Appendix E_4 we prove that

$$L(a) = 2.3E[(\mathbf{X} + a)^2] = 2.3\{E[(\mathbf{X} - \mu_{\mathbf{X}})^2] + (\mu_{\mathbf{X}} + a)^2\}$$

where $\mu_{\mathbf{X}}$, the *(population) mean* of $\mathbf{X}$, is defined by

$$\mu_{\mathbf{X}} = E(\mathbf{X}).$$

But the above equation states that $L(a)$ is at least equal to $2.3E[(\mathbf{X} - \mu_{\mathbf{X}})^2]$ and is greater than that except when $a = -\mu_{\mathbf{X}}$. Hence the value of a which minimizes $L(a)$ is $a = -\mu_{\mathbf{X}}$. Hence, in this example, the mean of the population is a fundamental parameter. The corresponding expected loss is 2.3 times $E[(\mathbf{X} - \mu_{\mathbf{X}})^2]$ which is called the *(population) variance* and is denoted by $\sigma_{\mathbf{X}}^2$.

Exercise 4.26. Mr. Sheppard has several other rifles. On what basis should he select one of these rifles?

It is only fair for the authors to say a few words about why Mr. Sheppard felt he could choose $l(\mathbf{Y}) = 2.3\mathbf{Y}^2$, and why this example plays a very important part in the theory of statistics.

Previously we pointed out that, if a small section of a smooth curve were magnified tremendously, it would look like a straight line. Related to this statement is the following. Suppose l is a function whose graph is a smooth curve. Suppose this curve never goes below the horizontal axis but touches it when the abscissa is zero. In other words, the function l assumes its minimum at zero. See Figure 4.8 for an example.

For most graphs with this property, encountered in practical experience, the following can be said. If the curve is magnified near the minimum, it looks very much like a straight line. To the extent that it is not straight, it looks like a " parabola " which is given by some function f of the form

$$f(x) = cx^2$$

where c is some positive number. To illustrate, we plot a parabola next to a curve in Figure 4.8.

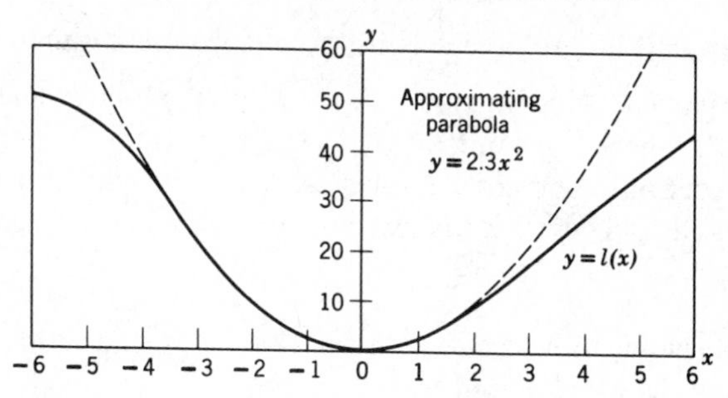

Figure 4.8. The graph of a smooth function l attaining its minimum at (0, 0) approximated by a parabola.

Now the loss of utility suffered by Mr. Sheppard when he misses by $\mathbf{Y}$ ordinarily would be expected to behave like a curve with the above-described properties. It should be smooth and reach its minimum at $\mathbf{Y} = 0$. Hence, for small misses, Mr. Sheppard is approximately correct in approximating his loss by $c\mathbf{Y}^2$ for some number c. Mr. Sheppard insists that he does not intend to make any big misses and so he is willing to consider his approximation a good one. Strictly speaking, the important issue is the following. Does the method which minimizes the expectation of his approximation give a good result in terms of the *true* loss? As long as the approximation is very good for the misses which are liable to occur, it is generally reasonable to assume that the answer is yes. It should be kept in mind that *minimizing the expectation of the approximation yields an approximation to the best action.*

What is the importance of this problem in statistics? Generally, statisticians have not been extremely interested in Mr. Sheppard's problem. However, it is similar to one which is very interesting to statisticians, namely, the problem of *estimation*. Frequently a statistician is interested in estimating a parameter θ (property of a probability distribution). He bases this estimate on a statistic (property of the observations) $\mathbf{T}$. This statistic $\mathbf{T}$ is a random variable and the measure of its "goodness" is in terms of how much $\mathbf{T}$ tends to differ from θ. For many examples, the expected loss of utility associated with using $\mathbf{T}$ as an estimate of θ is approximated by

$$cE[(\mathbf{T} - \theta)^2].$$

The analogy between this problem and Mr. Sheppard's is clear and more will be said about it in the chapter on estimation.

Example 4.4. Mr. Butregs owns a cow which he wishes to compare with a new one being offered to him for trade. Since his farm can support only one cow, he wishes to know which is better. If he keeps his old cow for a year, her total production will be

$$\mathbf{T} = \mathbf{X}_1 + \mathbf{X}_2 + \cdots + \mathbf{X}_{365}$$

where $\mathbf{X}_i$ is her milk production on the ith day, and his resulting utility will be represented by

$$\mathbf{G} = f(\mathbf{T}) = f(\mathbf{X}_1 + \mathbf{X}_2 + \cdots + \mathbf{X}_{365})$$

where f is an increasing function. That is to say, as $\mathbf{T}$ increases $f(\mathbf{T})$ increases. If he trades the old cow for the new one, the new one will have total production

$$\mathbf{T}^* = \mathbf{Y}_1 + \mathbf{Y}_2 + \cdots + \mathbf{Y}_{365}$$

where $\mathbf{Y}_i$ is her production on the ith day, and his resulting utility will be represented by

$$\mathbf{G}^* = f(\mathbf{T}^*) = f(\mathbf{Y}_1 + \mathbf{Y}_2 + \cdots + \mathbf{Y}_{365}).$$

If we assume that the $\mathbf{X}_i$ have a common distribution, that of $\mathbf{X}$, and that the $\mathbf{Y}_i$ have the distribution of $\mathbf{Y}$, then

$$\frac{\mathbf{X}_1 + \mathbf{X}_2 + \cdots + \mathbf{X}_{365}}{365}$$

will be close to $\mu_{\mathrm{X}} = E(\mathbf{X})$. Similarly,

$$\frac{\mathbf{Y}_1 + \mathbf{Y}_2 + \cdots + \mathbf{Y}_{365}}{365}$$

will be close to $\mu_{\mathrm{Y}} = E(\mathbf{Y})$. It follows that, if $\mu_{\mathrm{X}} > \mu_{\mathrm{Y}}$, $\mathbf{T}$ will almost surely be larger than $\mathbf{T}^*$ and, therefore, $\mathbf{G}$ will be larger than $\mathbf{G}^*$. In other words, the choice of cow should be made according to which cow has *greater expected daily milk production.*

This same principle was used to tell us that, if we have a choice of playing a game many times, we should do so only if the expected money gain is positive.

Example 4.5. Mr. Heath wishes to build a warehouse on El Camino Real. El Camino Real is a major highway in California along which many communities are built. These communities cluster very closely to the highway and, in combination, resemble a community consisting of a long line. Suppose that a prospective customer for the warehouse will be in position **X** along the highway. If the warehouse is at position a, the cost of delivering is proportional to the absolute value of $\mathbf{X} - a$. That is to say, we assume that, except for the fixed loading and unloading costs, the main cost of delivering is the time it takes to travel along the highway to the destination and back. Assuming that his utility is a decreasing function of the total amount of distance traveled to his customers, Mr. Heath applies the argument of Example 4.4. Accordingly he decides to select a so as to minimize

$$E(|\mathbf{X} - a|)$$

since $|\mathbf{X} - a|$ is the distance he must travel to his customer.[1] What value of a will he select? In the Appendix E_4 it is shown that the minimizing value of a is the (population) *median* $\nu_{\mathbf{X}}$ of the probability distribution of **X**. It is clear that for Mr. Heath a relevant measure of the variability of **X** is given by the *absolute deviation* $E(|\mathbf{X} - \nu_{\mathbf{X}}|)$.[2]

Exercise 4.27. Draw a graph of the function f given by $f(x) = |x|$ and compare it with that of g where $g(x) = x^2$. Note that both are minimized at 0 but that the first is not smooth while the second is.

Exercise 4.28*. Illustrate that the median minimizes $E(|\mathbf{X} - a|)$ as follows. Let **X take on the possible values -3, -1, 0, 3, 5 with probabilities 0.2, 0.4, 0.1, 0.1, and 0.2 respectively. Compute $E(|\mathbf{X} - a|)$ for $a = -2, -1, 0, 1$, and any other values you wish to try. What is the median of **X**? Do these computations corroborate the stated result?

Example 4.6. Mr. Heath was originally in the shoe manufacturing business. He made his first entry into business by buying an old machine which could manufacture shoes. But once it was adjusted, it would take over a week to readjust for a different size. Rather than put out shoes of various sizes at the cost of production

[1] We use $|a|$ to denote the positive value of a. Thus $|-2| = |2| = 2$. It is usually called the absolute value of a.

[2] The median of the probability distribution of a random variable **X** is a number $\nu_{\mathbf{X}}$ for which it is true that

$$P\{\mathbf{X} < \nu_{\mathbf{X}}\} \leq 0.5 \leq P\{\mathbf{X} \leq \nu_{\mathbf{X}}\}.$$

For some random variables, there may be an interval of numbers, any one of which will serve as median.

delays, Mr. Heath decided to choose one size and manufacture that size exclusively.

Mr. Heath's utility function was clearly an increasing function of the number of shoes he could sell in a given period (till his mortgage payment was due). He figured that, if he put out a size a, he would be able to sell to a customer whose foot size was between $a - 1/2$ and $a + 1/2$. He assumed that a large number, n, of customers would go to his retail outlet and be interested in the style he manufactured. Out of these, $\mathbf{m}$ would have size between $a - 1/2$ and $a + 1/2$. If we assume that the foot size of a customer is a random variable $\mathbf{X}$ then $\mathbf{m}/n$ will be close to $P\{a - 1/2 \leq \mathbf{X} \leq a + 1/2\}$. Hence he would maximize his utility by selecting the value of a which maximizes $P\{a - 1/2 \leq \mathbf{X} \leq a + 1/2\}$. Suppose $\mathbf{X}$ has a continuous distribution whose density is given by a smooth curve.

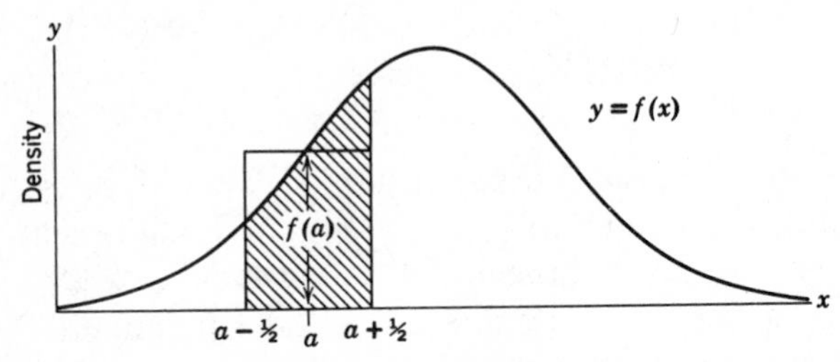

Figure 4.9. The distribution of foot size. Shaded area represents the probability that a random customer will be fit by a shoe of size a. This area is approximated by the area of the rectangle with base from $a - 1/2$ to $a + 1/2$ and height $f(a)$.

See Figure 4.9. Then $P\{a - 1/2 \leq \mathbf{X} \leq a + 1/2\}$ is the area between the graph and the horizontal axis for abscissas between $a - 1/2$ and $a + 1/2$. It is clear that one can approximately maximize this quantity by letting a be that abscissa for which the density is greatest. This abscissa is called the *mode* of the distribution of $\mathbf{X}$.

In these examples we have illustrated that there are problems where the following characteristics of a probability distribution are of fundamental importance: (1) *Mean*: $\mu_{\mathbf{X}} = E(\mathbf{X})$. (2) *Variance*: $\sigma^2_{\mathbf{X}} = E[(\mathbf{X} - \mu_{\mathbf{X}})^2]$. (3) *Median*. (4) *Mode*. For other problems other characteristics will be important. If a large sample is available, we can approximate these characteristics of the population by taking an analogous property of the sample. Thus for large samples

$\bar{\mathbf{X}}$ is close to $\mu_{\mathbf{X}}$, $\mathbf{d}_{\mathbf{X}}^2$ is close to $\sigma_{\mathbf{X}}^2$, the median of the sample is close to the median of the population and the mid-point of the interval of the histogram with the largest frequency will be close to the mode.

For the time being we will interpret our interest in Chapter 2 in these sample properties as an attempt to approximate the corresponding population properties. Thus, if Mr. Heath had available a large sample of shoe sizes, he would not necessarily compute the sample mean or variance. He may be interested mainly in the sample mode.

Exercise 4.29. In Figure 4.10 we plot utility u against p, the probability that a random customer will be fit by his shoe. We know (but Mr. Heath does not) that the shoe size $\mathbf{X}$ of the random customer is almost normally distributed with $\mu = 14$ and $\sigma = 1.5$. What is Mr. Heath's utility if the shoe size he manufactures is 13? 14? 15? Construct a graph giving $U(a)$ which is the utility if he decides to manufacture size a.

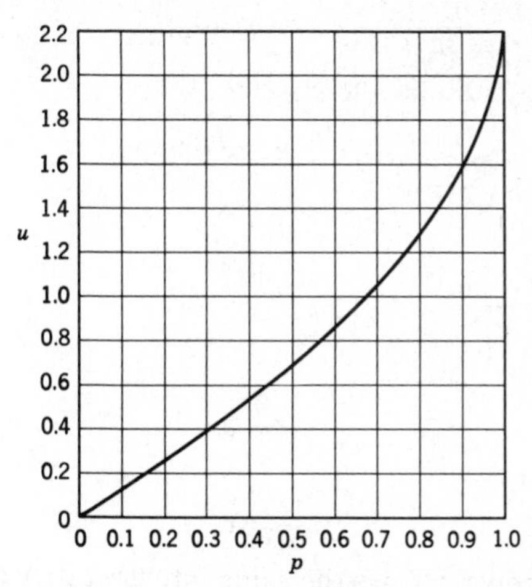

Figure 4.10. The utility u corresponding to the probability p that a random customer will be fit by a shoe manufactured by Mr. Heath.

Exercise 4.30. Locate the mode and median for the distribution given in Figure 3.10.

7. THE MEAN AND VARIANCE

In Chapter 2 we indicated some properties relevant to the theory and computation of the sample mean and sample variance. Analogous properties hold for the population mean and variance. We designate

1. The mean of **X** by $\mu_X = E(\mathbf{X})$.
2. The variance of **X** by $\sigma_X^2 = E[(\mathbf{X} - \mu_X)^2]$.
3. The standard deviation of **X** by σ_X.

Referring to Example 4.3, it is clear that μ_X is a measure of the "center" of the distribution of **X** and σ_X is a measure of the tendency of **X** to vary. The smaller σ_X, the more accurate is Mr. Sheppard's rifle or, more precisely, the greater is his utility. Interest in the *precision of an instrument* (other than a rifle) often leads to primary attention focusing on the standard deviation.

7.1 Some Properties of Mean, Variance, and Standard Deviation[1]

1. $E(\mathbf{X} - \mu_X) = 0$.
2. $\sigma_X^2 = E(\mathbf{X}^2) - \mu_X^2$.
3. If $\mathbf{Y} = a + b\mathbf{X}$ and b is positive, then

$$\mu_Y = a + b\mu_X$$

and

$$\sigma_Y = b\sigma_X.$$

4. If $\mathbf{Y} = \dfrac{\mathbf{X} - \mu_X}{\sigma_X}$

then

$$\mu_Y = 0$$

and

$$\sigma_Y = 1.$$

Note that Property 3 has the same interpretation as its analogue for sample properties. That is, if you multiply a random variable by a positive constant, you multiply both its mean and its standard deviation by the same constant. If you add a constant to a random variable, you add a constant to its mean but leave its standard deviation unaffected. The operation which gives $\mathbf{Y} = a + b\mathbf{X}$ is a

[1] See Appendix E_4.

combination of first multiplying by b and then adding a, and the above results are only reasonable.

The random variable **Y** of Property 4 is obtained from **X** by subtracting the mean and dividing by the standard deviation. This operation is sometimes called "standardizing." It essentially shifts the mean to zero and brings the standard deviation to one. Sometimes when it is desired to compare the shapes of two probability density functions, it pays to standardize the corresponding random variables first.

We illustrate how to compute expectations, means, and variances for discrete random variables. Suppose that **X** can equal 3, 5, 6, 9, and 10 with probabilities 0.2, 0.4, 0.2, 0.1, and 0.1 respectively.

TABLE 4.1

COMPUTATION OF EXPECTATIONS

	Probabilities	Possible Values of			
	p	$\mathbf{X}$	$\mathbf{X}^2$	$(\mathbf{X}-5.7)^2$	$2\mathbf{X}-1$
	0.2	3	9	7.29	5
	0.4	5	25	0.49	9
	0.2	6	36	0.09	11
	0.1	9	81	10.89	17
	0.1	10	100	18.49	19
Expectation		5.7	37.1	4.61	10.4

Here we obtained 5.7 by cumulating the products $0.2 \times 3 + 0.4 \times 5 + 0.2 \times 6 + 0.1 \times 9 + 0.1 \times 10$. The other expectations were obtained similarly. Since $\mu_{\mathbf{X}} = 5.7$, it follows that

$$\sigma_{\mathbf{X}}^2 = E[(\mathbf{X} - 5.7)^2] = 4.61.$$

This result could also be obtained by the formula

$$\sigma_{\mathbf{X}}^2 = E(\mathbf{X}^2) - \mu_{\mathbf{X}}^2 = 37.1 - (5.7)^2 = 4.61.$$

We computed $E(2\mathbf{X} - 1) = 10.4$ to verify the formula

$$E(2\mathbf{X} - 1) = 2E(\mathbf{X}) - 1.$$

Observe that an expectation is not a random quantity. It is a number which is completely determined by the probability distribution.

**Exercise 4.31.* Identify each of the following expressions as random, a constant not necessarily zero, or zero.

(a) $\sum_{i=1}^{n} (X_i - \bar{X})$ (d) $E(XY)$

(b) $E\left(\frac{X_1 + X_2}{2}\right)$ (e) $XE(Y)$

(c) σ_X^2 (f) $\sum_{i=1}^{n} (X_i - \mu_X)$

***Exercise 4.32.* The random variable X can equal −1, 2, 6, and 9 with probabilities 0.2, 0.3, 0.3, and 0.2 respectively. Compute the mean, variance, and standard deviation of X. Compute the mean and variance of X^2.

For the normal distribution, μ and σ do actually represent the mean and standard deviation. In fact, if X has a normal distribution with mean 0 and standard deviation 1, $E(X) = 0$, $E(X^2) = 1$, $E(X^3) = 0$, $E(X^4) = 3$, $E(X^5) = 0$, $E(X^6) = 15, \cdots$. Furthermore, $Y = a + bX$ has a normal distribution with mean a and standard deviation b if b is positive.

Exercise 4.33. If Y has a normal distribution with mean a and standard deviation b, compute $E(Y^3)$.

Hint: $Y^3 = (a + bX)^3 = a^3 + 3a^2bX + 3ab^2X^2 + b^3X^3$, where X is normally distributed with mean 0 and standard deviation 1.

Exercise 4.34. Let X be a random digit. Compute $E(X)$ and σ_X.

Exercise 4.35. (a) Let X be the outcome of the throw of an ideal die. Compute $E(X)$, σ_X^2, and σ_X. (b) Let Y be the sum of the faces showing upon the roll of two ideal dice. Compute $E(Y)$, σ_Y^2, and σ_Y.

Exercise 4.36. Compute $E(X)$ and σ_X^2 for X equal to the number of heads resulting from *one* toss of a bent coin with probability p of falling heads.

Exercise 4.37. Compute $E(X)$ and σ_X^2 for X equal to the number of heads resulting from n tosses of an ideal coin for $n = 1$, 2, and 3.

°*Exercise 4.38.* Compute $E(X)$ and σ_X for X equal to the outcome of the well-balanced dial.[1]

°*Exercise 4.39.* Compute $E(X)$ and σ_X for X the outcome of the St. Petersburg game against the bookie with only 2^{25} cents.

[1] This problem requires calculus. See the discussion after Consequence 6 of Appendix E_4.

8. SUMMARY

If many observations $\mathbf{X}_1, \mathbf{X}_2, \cdots, \mathbf{X}_n$ are taken (under similar circumstances) on a random variable $\mathbf{X}$, the average $\bar{\mathbf{X}}$ of these observations will tend to be very close to $E(\mathbf{X})$. If $\mathbf{X}$ has a discrete distribution and takes on the values $x_1, x_2, \cdots,$ with probabilities $p_1, p_2, \cdots,$ then

$$E(\mathbf{X}) = p_1x_1 + p_2x_2 + \cdots = \sum_i p_ix_i.$$

If a gambler with a large bankroll continually plays a game where each play yields a relatively small money gain of $\mathbf{X}$ dollars to him, this game will be profitable to him in the long run (his utility will increase) if $E(\mathbf{X}) > 0$. If $E(\mathbf{X}) < 0$, this game will be unprofitable to him in the long run. If time is worth money and $E(\mathbf{X}) = 0$, the game will also be unprofitable eventually.

If the gambler has only the option of playing the game once, and $\mathbf{X}$ is relatively large compared to the bankroll, then $E(\mathbf{X})$ is not the only relevant parameter of the distribution. The variability of $\mathbf{X}$ would tend to be important and, in general, the entire probability distribution is relevant. Essentially, what must be done in this situation is to consider utility. Applying certain mild assumptions, it can be shown that a person can attach to his *prospect* P a number $u(P)$ called the utility of the prospect.

The utility function has the properties:

1. $u(P_1) > u(P_2)$ if and only if P_1 is preferred to P_2.
2. $u(P) = E[u(\mathbf{P})]$ if P is the prospect of facing $\mathbf{P}$ where $\mathbf{P}$ may be random.
3. The utility function is bounded.

The second property essentially states that, even though the expectation of money may not be completely relevant, the expectation of *utility* is.

In certain fundamental applications of various degrees of importance, it was shown that utility is essentially determined by certain parameters (properties of the relevant probability distributions). The *descriptive parameters* discussed in the various examples were : the mean, the variance, the median, and the mode.

The discussion of this chapter is based on the assumption that the state of nature and, hence, all relevant probability distributions are known. The only ignorance treated is that due to randomness.

Exercise 4.40. What is the expected sum obtained in tossing five well-balanced dice?

**Exercise 4.41.* Let $\mathbf{X}_i = 1$ if the ith voter polled in a sample of 100 Phiggindians favors prohibition. Let $\mathbf{X}_i = 0$ otherwise. Let $\mathbf{Y}$ = the number of voters in the sample who favor prohibition. Then $\mathbf{Y} = \mathbf{X}_1 + \mathbf{X}_2 + \cdots + \mathbf{X}_{100}$. Suppose that 40% of the entire population of Phiggindian voters favor prohibition. Find: (a) $E(\mathbf{X}_i)$; (b) $E(\mathbf{Y})$; (c) $E(\overline{\mathbf{X}})$; (d) $\sigma^2_{\mathbf{X}_i}$. (*Hint*: see Exercise 4.36.)

°*Exercise 4.42.* Compute the mean and standard deviation of **X** in Exercise 3.41.[1]

°*Exercise 4.43.* A well-balanced coin is tossed until heads appears. Let **X** be the number of tosses. Compute the mean and standard deviation of **X**.

Exercise 4.44. Mr. Jones' utility for money is given by $u = 10x - x^2$ for $0 \leq x \leq 3$, where x is money in millions of dollars. He has $1 million. Does it pay for him to take " fair " bets? (Ignore income tax considerations.)

Exercise 4.45. Mr. Jones in the above exercise has an opportunity to invest his money so that he will end up with **X** million dollars, where **X** has density given by $f(0)=0.1$, $f(0.5)=0.2$, $f(1.0)=0.1$, $f(2.0) = 0.4$, and $f(3.0) = 0.2$. Should he invest? Is the game " fair "?

SUGGESTED READINGS

[1] Luce, R. D., and Howard Raiffa, *Games and Decisions*, John Wiley & Sons, New York, 1957.

[2] Von Neumann, John, and Oskar Morgenstern, *Theory of Games and Economic Behavior*, Princeton University Press, Princeton, first edition, 1944, second edition, 1947.

Reference [2] was the first book on the modern theory of games and has a detailed discussion of utility theory, but it is written from a mathematically advanced point of view. The more recent book of Reference [1] is elementary and comprehensive. Considerable space is devoted to a critical discussion of the ideas and concepts in utility, game, and decision theory.

[3] Davidson, D., Patrick Suppes, and Sidney Siegel, *Decision Making: An Experimental Approach*, Stanford University Press, Stanford, 1957.

This book is concerned with the experimental measurement of utility.

[1] This problem requires calculus. See discussion after Consequence 6, Appendix E_4.

CHAPTER 5

Uncertainty due to Ignorance of the State of Nature

1. INTRODUCTION

In Chapter 4 we discussed the case where the only uncertainty was due to randomness. The state of nature was assumed to be known. In this chapter we shall discuss the case where there are several possible states of nature, and it is not known which is the true state. The contents of this chapter could be very adequately presented without reference to any graphs. On the other hand, we shall supplement our discussion with a good many pictorial representations. These will be presented on the assumption that they add to the understanding and that they contribute a useful point of view.

2. TWO STATES OF NATURE—AN EXAMPLE

In this section we shall treat in great detail a problem involving two states of nature. The problem selected is deliberately simplified so that all possible strategies can be examined and compared. Briefly, we shall allow only one observation with three possible values and only three possible actions. As we shall see, this will allow only 27 possible strategies which will be enumerated.

Example 5.1. Mr. Nelson, a visitor to North Phiggins was informed that when it rains in North Phiggins, it *really* rains. In fact, it seems that on rainy days it starts pouring at 11 A.M. and does not stop until 11 P.M. Hence, before leaving his room in the morning, he was advised to refer to a rain indicator before deciding which outfit of clothes to wear. The rain indicator can indicate (1) fair, (2) dubious, and (3) foul. Mr. Nelson has three outfits—a fair weather outfit, one with a raincoat, and one with raincoat, boots, umbrella, and rain hat. The rain indicator states how often rainy days are preceded by indications of fair, dubious, and foul, and how often clear days are preceded by these indications. However, Mr. Nelson has no idea about how often it rains in North Phiggins.

In detail, the situation is the following. There are two possible states of nature. These are:

θ_1: Today is a sunny day.
θ_2: Today is a rainy day.

There are three available actions. These are:

a_1: Wear fair-weather outfit.
a_2: Wear outfit with raincoat.
a_3: Wear outfit with raincoat, boots, rain hat, and umbrella.

Mr. Nelson, having been informed of how heavily it rains and considering the burden of carrying heavy coats and boots and the embarrassment of being improperly clothed, has constructed a table of losses of utility (Table 5.1).

TABLE 5.1

MR. NELSON'S LOSS OF UTILITY $l(\theta, a)$

θ \ a	a_1	a_2	a_3
θ_1 (no rain)	0	1	3
θ_2 (rain)	5	3	2

It is customary to denote by zero the loss for the most favorable combination of θ (state of nature) and a (action) and then to compare all other (θ, a) with this best combination. In this way, we have the slight convenience that none of the losses are negative. Please note that the use of losses instead of actual utilities is quite arbitrary. It should have no influence on the problem of decision making. We use losses to be consistent with statistical traditions which seem to have been founded on a somewhat conservative and pessimistic point of view.

Being a reasonable person, Mr. Nelson plans to base his action on the evidence concerning θ given by the weather indicator. His experiment is the rather trivial one of observing the indicator before dressing. The random variable observed is $\mathbf{X}$, the weather indication, and it can assume one of the following three possible values:

x_1: Fair weather indicated.
x_2: Dubious.
x_3: Foul weather indicated.

The probability distribution of **X** depends on the state of nature. The probabilities are given in Table 5.2 which is attached to the weather indicator.

TABLE 5.2

PROBABILITY OF OBSERVING A WEATHER INDICATION x WHEN θ IS THE TRUE STATE OF NATURE; $f(x \mid \theta) = P\{\mathbf{X} = x \mid \theta\}$

	x_1	x_2	x_3
θ_1	0.60	0.25	0.15
θ_2	0.20	0.30	0.50

It may be remarked that a really ideal weather indicator would be one where the 0.60 and 0.50 were replaced by one, and all other entries by zero. Because this indicator is far from ideal, North Phiggindians occasionally find themselves improperly dressed.

Now Mr. Nelson must decide how he will react to the weather indicator. This plan of reaction is a strategy. Mr. Nelson lists all possible strategies. These are given in Table 5.3. Many of these

TABLE 5.3

A LIST OF ALL POSSIBLE STRATEGIES. EACH STRATEGY s INDICATES THE ACTION $\mathbf{A} = s(\mathbf{X})$ TAKEN IN RESPONSE TO AN OBSERVATION **X**

Strategy, s / Possible Observations, x	s_1	s_2	s_3	s_4	s_5	s_6	s_7	s_8	s_9	s_{10}	s_{11}	s_{12}	s_{13}	s_{14}	s_{15}
x_1	a_1	a_1	a_1	a_1	a_1	a_1	a_1	a_1	a_1	a_2	a_2	a_2	a_2	a_2	a_2
x_2	a_1	a_1	a_1	a_2	a_2	a_2	a_3	a_3	a_3	a_1	a_1	a_1	a_2	a_2	a_2
x_3	a_1	a_2	a_3	a_1	a_2	a_3	a_1	a_2	a_3	a_1	a_2	a_3	a_1	a_2	a_3

	s_{16}	s_{17}	s_{18}	s_{19}	s_{20}	s_{21}	s_{22}	s_{23}	s_{24}	s_{25}	s_{26}	s_{27}
x_1	a_2	a_2	a_2	a_3	a_3	a_3	a_3	a_3	a_3	a_3	a_3	a_3
x_2	a_3	a_3	a_3	a_1	a_1	a_1	a_2	a_2	a_2	a_3	a_3	a_3
x_3	a_1	a_2	a_3	a_1	a_2	a_3	a_1	a_2	a_3	a_1	a_2	a_3

strategies seem ridiculous while others seem reasonable. For example, s_{22} is the contrary man's strategy. It is the strategy where the action taken is the opposite of the one that would seem indicated. The strategies s_1, s_{14}, and s_{27} ignore the data. They correspond to the possible strategies of an absent-minded person who forgets to look at the indicator. On the other hand, s_6, s_9, s_{15}, and s_{18} are reasonable strategies of various degree of conservatism.

Now Mr. Nelson decides to evaluate all of these strategies. He computes the expected loss of utility $L(\theta, s)$ corresponding to each state of nature θ and each strategy s.

TABLE 5.4

EXPECTED LOSS OF UTILITY, $L(\theta, s)$

State of Nature, θ \ Strategy, s	s_1	s_2	s_3	s_4	s_5	s_6	s_7	s_8	s_9
θ_1	0.00	0.15	0.45	0.25	0.40	0.70	0.75	0.90	1.20
θ_2	5.00	4.00	3.50	4.40	3.40	2.90	4.10	3.10	2.60
	s_{10}	s_{11}	s_{12}	s_{13}	s_{14}	s_{15}	s_{16}	s_{17}	s_{18}
θ_1	0.60	0.75	1.05	0.85	1.00	1.30	1.35	1.50	1.80
θ_2	4.60	3.60	3.10	4.00	3.00	2.50	3.70	2.70	2.20
	s_{19}	s_{20}	s_{21}	s_{22}	s_{23}	s_{24}	s_{25}	s_{26}	s_{27}
θ_1	1.80	1.95	2.25	2.05	2.20	2.50	2.55	2.70	3.00
θ_2	4.40	3.40	2.90	3.80	2.80	2.30	3.50	2.50	2.00

The entries in Table 5.4 are easily computed in the manner used in Chapter 1 for computing expected losses. We illustrate by indicating the action probabilities and expected losses for the strategies s_5 and s_{22} in Table 5.5 and by computing the row corresponding to θ_1 and s_5. Then we take action a_1, if $\mathbf{X} = x_1$, which has probability 0.60. We take action a_2, if $\mathbf{X} = x_2$ or x_3, which has probability $0.25 + 0.15 = 0.40$. We never take action a_3. Thus actions a_1, a_2, and a_3 which give losses of 0, 1, and 3 are taken with probabilities 0.60, 0.40, and 0, yielding an expected loss of $(0.60)0 + (0.40)1 + (0)3 = 0.40$.

TABLE 5.5

LOSSES, ACTION PROBABILITIES AND EXPECTED LOSSES FOR STRATEGIES s_5 AND s_{22}

State of Nature	Losses			Action Probabilities			Expected Loss
	For $s_5 = (a_1, a_2, a_2)$						
	a_1	a_2	a_3	a_1	a_2	a_3	
θ_1	0	1	3	0.60	0.40	0	0.40
θ_2	5	3	2	0.20	0.80	0	3.40
	For $s_{22} = (a_3, a_2, a_1)$						
	a_1	a_2	a_3	a_1	a_2	a_3	
θ_1	0	1	3	0.15	0.25	0.60	2.05
θ_2	5	3	2	0.50	0.30	0.20	3.80

Exercise 5.1. Compute the action probabilities and losses for strategy s_{10}.

Exercise 5.2. Use the table of random digits to select an observation according to the probabilites corresponding to θ_1. Apply s_{10} to this observation and list the corresponding action and loss. Repeat ten times and average the ten losses.

**Exercise 5.3.* If Mr. Nelson replaced his weather meter by a crystal ball which told him the state of nature, θ, reasonable use of this information would provide him with *ideal action probabilities.* Present this table of ideal action probabilities. These are called ideal because with ordinary information no strategy could do this well.

Since there are two expected losses for each strategy, one corresponding to each θ, we can represent a strategy by a point on a graph the coordinates of which are these two expected losses. The abscissa will be the expected loss when θ_1 is the state of nature and the ordinate the expected loss when θ_2 is the state of nature. The strategies correspond to the labeled points in Figure 5.1.

A desirable strategy is one where both expected losses are small, i. e., where the point representing it is as far to the left and as low as possible. Thus it is clear from the figure that, of all the strategies, the only ones entitled to serious consideration are s_1, s_2, s_5, s_6, s_9, s_{15}, s_{18}, and s_{27}. The other strategies are *inadmissible* because

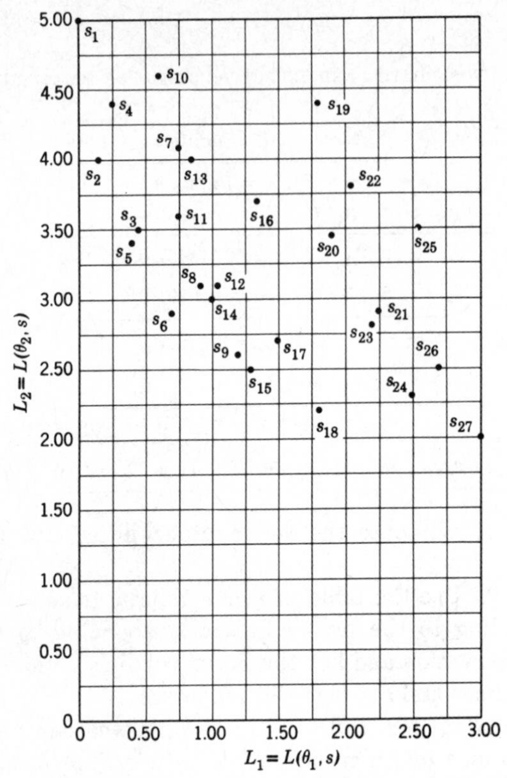

Figure 5.1. Expected loss points $(L_1, L_2) = (L(\theta_1, s), L(\theta_2, s))$ for 27 strategies in Mr. Nelson's rain problem.

each of them is *dominated* by one of these. For example, s_3, which, on reference to Table 5.3 seems to be a reasonable strategy, is dominated by s_5. That is to say, if θ_1 were the state of nature, the expected loss $L(\theta_1, s_3) = 0.45$ would be larger than $L(\theta_1, s_5) = 0.40$, and if θ_2 were the state of nature, the expected loss $L(\theta_2, s_3) = 3.5$ would be larger than $L(\theta_2, s_5) = 3.4$. Hence, no matter what the state of nature, s_5 will yield a smaller expected loss than s_3. Graphically, the point representing s_5 is both below and to the left of the one representing s_3.

Although we have reduced our problem to considering only the eight strategies indicated above, it is not clear which of these is to be preferred over the others. In fact, Mr. Nelson has decided that he likes s_6 and s_{15}, but he cannot decide between them. Finally in desperation he decides to toss a coin. If it falls heads he will apply s_6 and if it falls tails he will apply s_{15}. In his desperation Mr. Nelson has uncovered a new type of strategy. This strategy is called a *mixed* or *randomized* strategy. The expected losses corresponding to this strategy are easily evaluated. For example, if θ_1 were the state of nature, he would with probability 1/2 select s_6 with expected loss of utility of 0.70, and with probability 1/2, he would select s_{15} and have an expected loss of 1.30. Hence, his expected loss for this mixed strategy would be $(1/2)(0.70) + (1/2)(1.30) = 1$. On the other hand, if θ_2 were the state of nature, his expected loss would be $(1/2)(2.9) + (1/2)(2.5) = 2.7$. Graphically, this strategy is represented by a point midway between those represented by s_6 and s_{15}.

This mixed strategy may be interpreted from a slightly different point of view. Both strategies s_6 and s_{15} lead to a response of a_2 if x_2 is observed, and a_3 if x_3 is observed. They differ only if x_1 is observed. Hence, the mixed strategy may be described as follows. If x_1 is observed, toss a coin to decide between actions a_1 and a_2. If x_2 is observed, action a_2 is taken, and if x_3 is observed, a_3 is taken. This strategy has the somewhat disconcerting property that the actions taken may depend not only on the outcome of the experiment but *may also depend in part on the irrelevant toss of a coin.* Nevertheless, whatever criterion is to be used in judging this strategy, that criterion should involve not the way this strategy was selected but its consequences, i.e., the expected losses in both states of nature.

Mr. Nelson's strategy of taking s_6 and s_{15}, each with probability 1/2, is not the only possible mixed strategy. For example, he could have used a strategy which yields s_{15} with probability 1/4, and s_6 with probability 3/4. His expected losses would have been

$$(3/4)(0.70) + (1/4)(1.30) = 0.85$$

and

$$(3/4)(2.9) + (1/4)(2.5) = 2.8$$

which gives the point 1/4 of the way from s_6 toward s_{15} (see Figure 5.1). In general, if s and s' are any two strategies mixed or otherwise, and s' is selected with probability p and s with probability $1 - p$, then this mixture of s and s' is represented by a point which is p part of the way from s to s'. Thus every point on the line segment connecting the points representing s and s' represents some strategy. But this means that the set of points representing strategies is *convex*. (A convex set is a set which contains all line segments connecting points of the set.) In fact, the set of points representing strategies is the smallest convex set containing the original *unmixed* or *pure* strategies. See Figure 5.2.

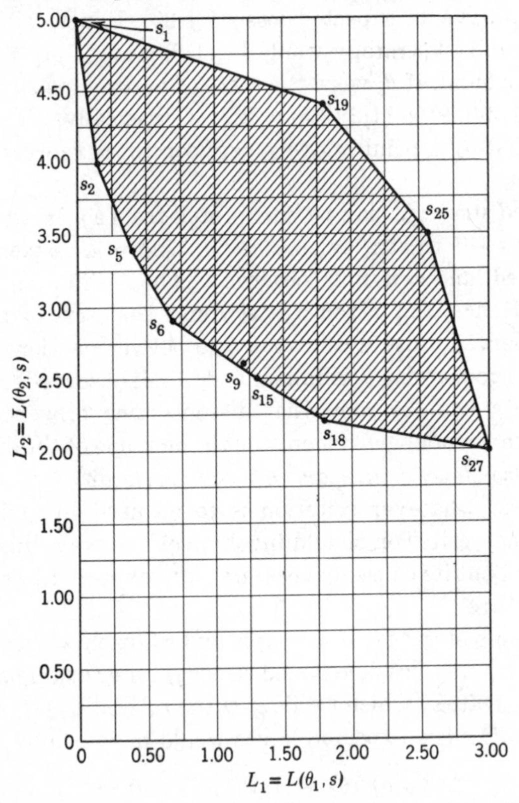

Figure 5.2. Expected loss points $(L_1, L_2) = (L(\theta_1, s), L(\theta_2, s))$ for all strategies mixed and pure in Mr. Nelson's rain problem.

One interesting phenomenon indicated by Figure 5.2 is that s_9 lies above the line segment connecting s_6 and s_{15} and, hence, is dominated (inadmissible). For example, taking s_6 with probability 1/4 and s_{15} with probability 3/4 yields the point with coordinates

$$(1/4)(0.70) + (3/4)(1.30) = 1.15$$

and

$$(1/4)(2.9) + (3/4)(2.5) = 2.6$$

which is directly to the left of the point representing s_9.

**Exercise 5.4.* Suppose that there is rain in North Phiggins about 2/3 of the days. Then the expected loss corresponding to a strategy is $(1/3)L(\theta_1, s) + (2/3)L(\theta_2, s)$. Find the best pure strategy. That is, the pure strategy which minimizes this expected loss. Now represent the lines $(1/3)L_1 + (2/3)L_2 = 1$, $(1/3)L_1 + (2/3)L_2 = 1.5$, and $(1/3)L_1 + (2/3)L_2 = 2$ on a graph. Suggest a graphical method of locating the best among all strategies (pure and randomized). Is this strategy pure or randomized?

Exercise 5.5. For each strategy s there is a maximum expected loss which is the greater one of $L(\theta_1, s)$ and $L(\theta_2, s)$. Denote this by $max\ (L(\theta_1, s), L(\theta_2, s))$. Find the minimax pure strategy, that is, the pure strategy which minimizes $max\ (L(\theta_1, s), L(\theta_2, s))$. Now represent the sets $\{(L_1, L_2) : max\ (L_1, L_2) = 1\}$, $\{(L_1, L_2) : max(L_1, L_2) = 1.5\}$, and $\{(L_1, L_2) : max\ (L_1, L_2) = 2\}$ on a graph. Suggest a graphical method of locating the minimax strategy among all strategies. Is this strategy pure or randomized?

3. TWO STATES OF NATURE: CONVEX SETS AND LINES

We shall use Mr. Nelson's problem and the discussion of Section 2 to motivate our general treatment of the class of problems with two possible states of nature. First we deal with *pure* strategies. *A strategy s is said to be pure or non-randomized if it assigns an action to each of the possible observations.* In other words, a pure strategy is a function on the set of possible observations to the set of possible actions. As was pointed out in Section 2, there are two points of view in looking at randomized strategies. For one of these we first look at our observation and then use a random device (depending on the observation) to decide which action to take. An example given was to take a_1 or a_2 each with probability 1/2 if x_1 is observed. If x_2 is observed, take a_2, and if x_3 is observed, take a_3. For convenience we shall concentrate on the other point of view. According to this, a *mixed* or *randomized strategy s is a choice of one of the set of pure strategies where the choice is made with a* *random device.* Thus the above example could be regarded as a choice of s_6 or s_{15}, each with probability 1/2.

Now one may ask about mixtures of randomized strategies. It can be shown that *random mixtures of randomized strategies may be regarded as randomized strategies.* Exercise 5.6 illustrates this point.

Exercise 5.6. Let s consist of selecting s_1 with probability 1/2 and s_2 with probability 1/2. Let s' consist of selecting s_1 with probability 1/4 and s_3 with probability 3/4. Let s'' consist of selecting s with probability 1/2 and s' with probability 1/2. If s'' is used, what are the probabilities of applying s_1, s_2, and s_3?

Each strategy should be evaluated in terms of its consequences. The relevant aspects of these consequences are measured by the expected losses. Thus a strategy s is to be evaluated in terms of the values of $L(\theta_1, s)$ and $L(\theta_2, s)$. It follows that *it suffices to represent a strategy s by a point in the plane whose coordinates are $L(\theta_1, s)$ and $L(\theta_2, s)$ respectively.* It should be noted that several strategies can be represented by the same point. For example, s_9 is represented by the same point as the strategy which selects s_6, s_{15}, and s_{23} with probabilities 6/27, 20/27, and 1/27 respectively. It is interesting to note that, while s_9 leads to action a_3 if x_2 is observed, each of s_6, s_{15} and s_{23} leads to a_2 if x_2 is observed. Hence this randomized strategy and s_9 are quite different in strategic content although they have equivalent expected losses and are represented by the same point. Let s_0 and s_1 be any two strategies and consider the mixture s which consists of selecting s_0 with probability $1 - w$ and s_1 with probability w. The stategy s is represented by $(L(\theta_1, s), L(\theta_2, s))$, where

$$\begin{aligned} L(\theta_1, s) &= (1 - w)\, L(\theta_1, s_0) + w\, L(\theta_1, s_1) \\ L(\theta_2, s) &= (1 - w)\, L(\theta_2, s_0) + w\, L(\theta_2, s_1). \end{aligned} \tag{5.1}$$

We shall now discuss one of two important representations of lines with a view to seeing that the point representing s is on the line segment connecting the points representing s_0 and s_1. Suppose (x_0, y_0) and (x_1, y_1) are two points in the plane. Let (x, y) be another point on the line through these points. Suppose that (x, y) is 1.7 times as far from (x_0, y_0) as is (x_1, y_1). (Figure 5.3.) We draw a horizontal line through $A = (x_0, y_0)$ and vertical lines through $B=$

(x_1, y_1), and $C = (x, y)$ which intersect the horizontal line at $B' = (x_1, y_0)$ and $C' = (x, y_0)$. The triangles ABB' and ACC' are similar and hence the lengths $\overline{AB'}$, $\overline{BB'}$, and $\overline{AB}$ are proportional to $\overline{AC'}$, $\overline{CC'}$, and $\overline{AC}$ respectively. That is to say

$$\frac{\overline{AC'}}{\overline{AB'}} = \frac{\overline{CC'}}{\overline{BB'}} = \frac{\overline{AC}}{\overline{AB}}.$$

But $\overline{AC'} = x - x_0$, $\overline{AB'} = x_1 - x_0$, $\overline{CC'} = y - y_0$, $\overline{BB'} = y_1 - y_0$, and $\overline{AC}/\overline{AB} = 1.7$. Thus

$$\frac{x - x_0}{x_1 - x_0} = \frac{y - y_0}{y_1 - y_0} = 1.7$$

and

$$x = x_0 + 1.7(x_1 - x_0) = (-0.7)x_0 + (1.7)x_1$$
$$y = y_0 + 1.7(y_1 - y_0) = (-0.7)y_0 + (1.7)y_1.$$

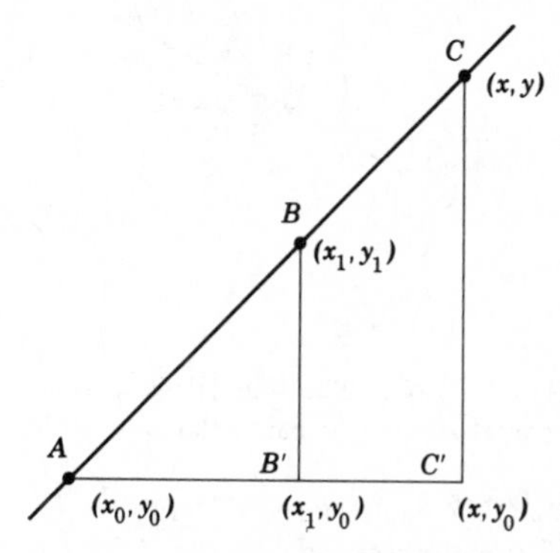

Figure 5.3. A diagram illustrating the derivation of Equation (5.2).

More generally, if (x, y) were w times as far from (x_0, y_0) as is (x_1, y_1), then

(5.2)
$$x = (1 - w)x_0 + wx_1$$
$$y = (1 - w)y_0 + wy_1.$$

The expressions for x and y are so similar that it is customary to abbreviate these two equations. Let $\bar{u}_0$, $\bar{u}_1$, and $\bar{u}$ represent the points (x_0, y_0), (x_1, y_1), and (x, y). We write

(5.3)
$$\bar{u} = (1 - w)\bar{u}_0 + w\bar{u}_1.$$

Strictly speaking, $\bar{u}$, $\bar{u}_0$, and $\bar{u}_1$ are not numbers, and this equation is not an ordinary equation. We shall interpret it to mean that each coordinate of $\bar{u}$ is obtained by adding $(1 - w)$ times the corresponding coordinate of $\bar{u}_0$ to w times the corresponding coordinate of $\bar{u}_1$. In other words, we interpret this equation to mean the same as Equations (5.2).

What happens to the point $\bar{u}$ as w varies? Referring to Figure 5.4, we note that, as w varies from 0 through positive values, $\bar{u}$ moves from $\bar{u}_0$ in the direction of $\bar{u}_1$. At $w = 1$, it becomes $\bar{u}_1$, and as w exceeds 1, $\bar{u}$ moves beyond $\bar{u}_1$. On the other hand, as w varies from 0 through negative values, $\bar{u}$ moves from $\bar{u}_0$ in the direction opposite $\bar{u}_1$. *The line through* $\bar{u}_0$ *and* $\bar{u}_1$ *may be represented by*

$$(5.4) \qquad \{\bar{u} : \bar{u} = (1 - w)\bar{u}_0 + w\bar{u}_1, \quad w \text{ is a number}\}.$$

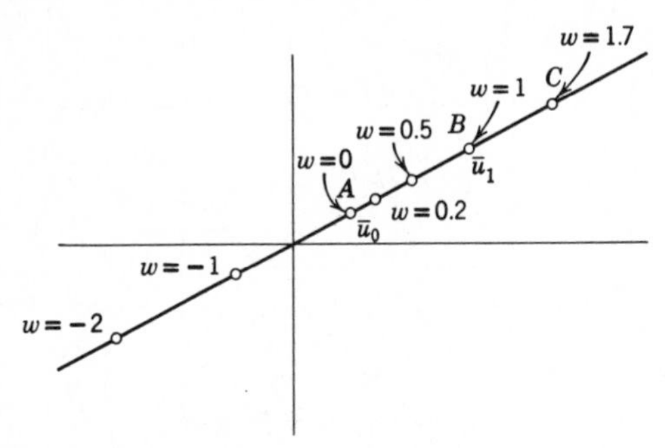

Figure 5.4. Points on a line. The line $\{\bar{u}: wu_0 + (1 - w)\bar{u}_1\}$ showing points corresponding to particular values of w.

The line segment from $\bar{u}_0$ *to* $\bar{u}_1$ *corresponds to the values of* w *between* 0 *and* 1 *and is represented by*

$$(5.5) \qquad \{\bar{u} : \bar{u} = (1 - w)\bar{u}_0 + w\bar{u}_1, \quad 0 \le w \le 1\}.$$

Notice that the numbers $1 - w$ and w which multiply $\bar{u}_0$ and $\bar{u}_1$ are both non-negative and add up to one for $\bar{u}$ on the line segment. If $0 \le w \le 1$, the expression $(1 - w)x_0 + wx_1$ is called a *weighted average* of x_0 and x_1 corresponding to the weights $(1 - w)$ and w. Similarly, if $0 \le w \le 1$, $(1 - w)\bar{u}_0 + w\bar{u}_1$ is called a weighted average of $\bar{u}_0$ and $\bar{u}_1$. Physically speaking, it has the following interpretation. If a weight of w pounds is at $\bar{u}_1$ and a weight of $(1 - w)$ pounds is at $\bar{u}_0$, the center of gravity of the pair of weights is at $(1 - w)\bar{u}_0 + w\bar{u}_1$. The ordinary average corresponds to the case $w = 1/2$, where both points have equal weights. The closer w gets to one, the closer to $\bar{u}_1$ is the weighted average or center of gravity.

A convex set is a set of points which has the property that, if $\bar{u}$ *and* $\bar{v}$ *are points of the set, the line segment connecting* $\bar{u}$ *and* $\bar{v}$ *is contained in the set.* In other words, S is convex if, when $\bar{u} \in S$ ($\bar{u}$ is an element of S) and $\bar{v} \in S$, then $(1 - w)\bar{u} + w\bar{v} \in S$ for all w between 0 and 1.

The convex set generated by a set A is the smallest convex set which contains A. Thus the convex set generated by three points $\bar{u}_1$, $\bar{u}_2$, $\bar{u}_3$ is the triangle with these points at the vertices. (See Figure 5.5.)

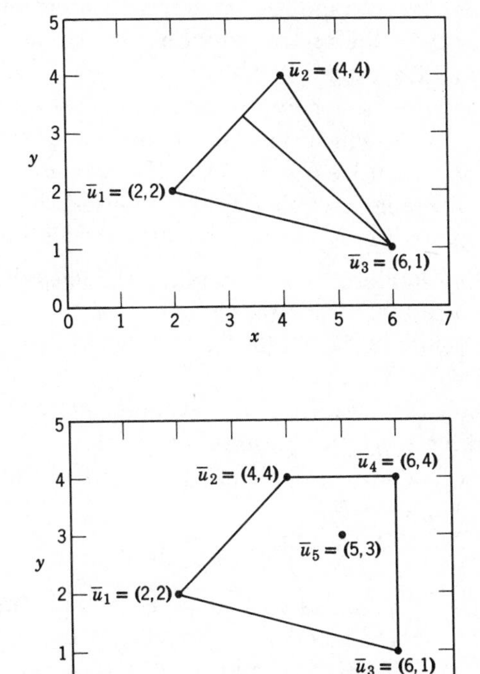

Figure 5.5. Convex sets generated by $\bar{u}_1$, $\bar{u}_2$ and $\bar{u}_3$, and by $\bar{u}_1$, $\bar{u}_2$, $\bar{u}_3$, $\bar{u}_4$ and $\bar{u}_5$.

The notion of weighted average can be extended to three or more quantities. Thus, $w_1x_1 + w_2x_2 + w_3x_3 + \cdots$ is a weighted average of the numbers $x_1, x_2, x_3, \cdots$ if the w_i are non-negative numbers adding up to 1. Similarly, $w_1\bar{u}_1 + w_2\bar{u}_2 + w_3\bar{u}_3 + \cdots$ is a weighted average of the points $\bar{u}_1, \bar{u}_2, \bar{u}_3, \cdots$ if the w_i are non-negative numbers adding up to 1. This expression is to be interpreted, as previously, as the point (x, y), where

$$x = w_1x_1 + w_2x_2 + w_3x_3 + \cdots$$
$$y = w_1y_1 + w_2y_2 + w_3y_3 + \cdots.$$

The triangle is the set of weighted averages of $\bar{u}_1$, $\bar{u}_2$, and $\bar{u}_3$. Similarly, it is true that the smallest convex set containing the points $\bar{u}_1, \bar{u}_2, \cdots, \bar{u}_n$ is the set of weighted averages of these points. (Again a weighted average has the physical interpretation of the center of gravity of the system where a weight of w_i pounds is put at the point $\bar{u}_i$. The center of gravity gets close to $\bar{u}_i$ if w_i is close to 1.) This set will be a polygon whose vertices are those points among $\bar{u}_1, \bar{u}_2, \cdots, \bar{u}_n$ which cannot be expressed as weighted averages of the others. (See Figure 5.5.) Not all convex sets are polygonal. For example, a circle (with its interior) is a convex set. It is the set generated by the points on the circumference.

Exercise 5.7. Represent in set notation the lines through:

(a) The points (2, 2) and (4, 4);
(b) The points (2, 2) and (6, 1);
(c) The points (6, 1) and (4, 4).

Exercise 5.8. Plot the weighted averages of (2, 2), (4, 4), and (6, 1) which give these three points the weights:

(a) 1/2, 1/2, and 0;
(b) 0.8, 0.1, and 0.1;
(c) 0.1, 0.8, and 0.1;
(d) 1/3, 1/3, and 1/3.

Exercise 5.9. Represent each point of Exercise 5.8 as a weighted average of (6, 1) and a point on the line connecting (2, 2) and (4, 4).

Exercise 5.10. Express the point (4.4, 2.4) as a weighted average of the three points (6, 1), (2, 2), and (4, 4).

Exercise 5.11. Let S be a circle (circumference and interior). Delete one point P. Is the remaining set convex if P is on the circumference? If P is in the interior?

Exercise 5.12. Let S be a square (including its interior). Delete all four vertices (corners). Is the remaining set convex?

Exercise 5.13. Any point of the convex set generated by A can be expressed as the weighted average of at most three points of

A. Sometimes less than three points are sufficient. Indicate how many points are the most required for the following *A*.

(a) *A* consists of the single point (1, 1);
(b) *A* consists of the two points (1, 1), (2, 2);
(c) *A* consists of the three points (1, 1), (2, 2), (3, 3);
(d) *A* consists of the three points (2, 2), (4, 4), (6, 1);
(e) *A* consists of the circumference of a circle;
(f) *A* consists of the boundary of a square;
(g) *A* consists of the vertices (corners) of a square.

Exercise 5.14. Which of the following sets are convex?

(a) $\{(x, y) : max\,(x, y) = 1\}$;
(b) $\{(x, y) : max\,(x, y) \geq 1\}$;
(c) $\{(x, y) : max\,(x, y) \leq 1\}$;
(d) $\{(x, y) : max\,(x, y) < 1\}$.

Exercise 5.15. The notion of a convex set extends easily to three-dimensional space. Thus a sphere with its interior is a convex set, but a doughnut is not. Describe the convex set generated by a doughnut. Describe the convex set generated by a coil spring.

Let us return to Equations (5.1). Since a probability is between 0 and 1 and all numbers between 0 and 1 can be probabilities, it follows that the point representing s is on the line segment connecting the points representing s_0 and s_1. Conversely, every point on this line segment represents some strategy which is a mixture of s_0 and s_1. Since s_0 and s_1 can be any two strategies, it follows that *S, the set of points representing all possible strategies* (pure and randomized), *is convex*. In fact, *S is the convex set generated by the points representing the pure strategies.*

We say that *the strategy s dominates* s^* *if* $L(\theta_1, s) \leq L(\theta_1, s^*)$ *and* $L(\theta_2, s) \leq L(\theta_2, s^*)$, *but the strategies are not represented by the same point*. In this case the point representing s is either: (a) to the left and below; or (b) directly below; or (c) directly to the left of the point representing s^*.

A strategy s is admissible if it is not dominated by any other strategy (pure or randomized). Here we may recall that, although s_9 would have been admissible among all pure strategies, it cannot be considered admissible now that randomized strategies are available. The class of admissible strategies corresponds to a portion of the boundary of the convex set S.

This portion of the boundary will be called the *admissible part of the boundary.*

In Mr. Nelson's problem, no matter what criterion he will apply in deciding on a strategy, it seems clear that he should eventually take one of the admissible strategies. Suppose that he confines his attention only to admissible strategies. This is a reasonable notion and is often applicable. However, there are occasionally somewhat pathological problems where there are no or very few admissible strategies. For example, suppose that there were available an infinite number of strategies s_1, s_2, $\cdots$ and

$$\begin{aligned}
L(\theta_1, s_1) &= L(\theta_2, s_1) = 1\\
L(\theta_1, s_2) &= L(\theta_2, s_2) = 1/2\\
L(\theta_1, s_3) &= L(\theta_2, s_3) = 1/3\\
&\vdots\\
L(\theta_1, s_i) &= L(\theta_2, s_i) = 1/i\\
&\vdots
\end{aligned}$$

These strategies would be represented by the points in Figure 5.6. Then there is no admissible strategy. For no matter which strategy is selected, there is a better one. The convex set representing all strategies including the mixed ones is the line segment from (0, 0) to (1, 1) which does not contain the end point (0, 0). The difficulty arises from the fact that although one can get strategies the expected losses of which are arbitrarily close to (0, 0), one cannot achieve (0, 0). This difficulty is troublesome theoretically but not very serious otherwise since we can get arbitrarily close.

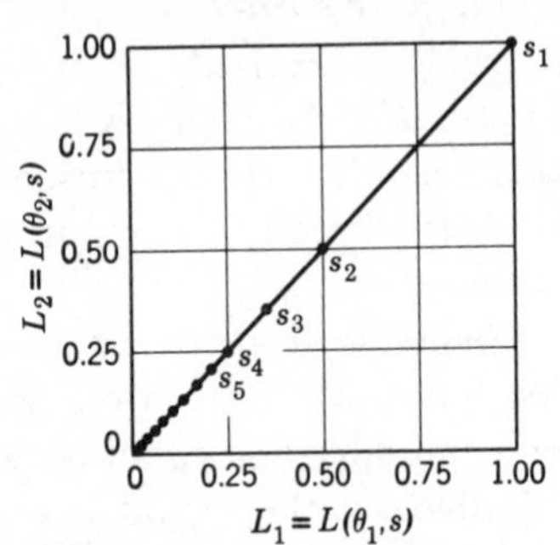

Figure 5.6. Losses for a special example with infinitely many strategies.

In problems where there are only a finite number of pure strategies, the convex set of interest is the smallest convex set containing the points corresponding to the pure strategies. This set will contain its boundary points, and the above-mentioned difficulty will not arise. In fact, we shall assume hereafter that, for our problems, the set representing the expected losses of our strategies contains its boundary points.

In summary, we discussed four important concepts in this section. First, a straight line through $\bar{u}_0$ and $\bar{u}_1$ may be represented in set notation as $\{\bar{u} : \bar{u} = (1 - w)\bar{u}_0 + w\bar{u}_1, \quad w \text{ is a number}\}$. Second, a convex set has the property that the line segment connecting two points of the set is in the set. Third, the class of all strategies may be represented by the convex set S generated by the points representing the pure strategies. Finally, the class of admissible strategies is represented by part of the boundary of S.

Exercise 5.16. Is it always true that a mixture of two admissible strategies is admissible ? Illustrate with a diagram.

Exercise 5.17. In the rain problem, the strategy s_{17} is inadmissible. Are there mixtures of s_6 and s_{18} which dominate s_{17} ? If so, specify one.

Exercise 5.18. When Mr. Smith arrives home, Mrs. Smith tells him that a dozen cookies disappeared from the cookie jar. Mr. Smith feels that, if his son John has been naughty and has eaten the cookies without permission he should be punished. His loss of utility table is given as follows.

	a_1 (Punish)	a_2 (Do not Punish)
θ_1 (Naughty)	1	2
θ_2 (Innocent)	4	0

Since other children had access to the cookie jar, Mr. Smith decides to base his action on the outcome of the experiment which consists of observing whether his son eats heartily at supper (z_1), eats moderately at supper (z_2), or barely eats (z_3). His estimate of the probability distribution of the data is given as follows.

	z_1	z_2	z_3
θ_1	0.1	0.4	0.5
θ_2	0.2	0.6	0.2

List and evaluate all pure strategies. Represent them graphically. Indicate the class of all admissible strategies. (Students are advised to keep a copy of the expected losses since this problem is referred to in the exercises of later sections.)

4. TWO STATES OF NATURE: BAYES STRATEGIES AND SUPPORTING LINES

Suppose that Mr. Nelson is informed that in North Phiggins it rains on the average two days out of three. In other words, the state θ_1 has a probability 1/3 while θ_2 has a probability 2/3. When such probabilities can be attached to the states of nature prior to experimentation, they are called *a priori* probabilities. Suppose strategy s_6 were applied. About 1/3 of the time the expected loss L would be 0.7 and 2/3 of the time it would be 2.9. This would lead to an expected L of

$$\mathscr{L}(s_6) = \frac{1}{3}(0.7) + \frac{2}{3}(2.9) = \frac{6.5}{3}.$$

One may evaluate $\mathscr{L}(s)$ for all strategies s and take a strategy which minimizes this weighted average of expected losses. Such a strategy is called a Bayes strategy corresponding to the *a priori* probabilities 1/3 and 2/3. (Bayes was an English mathematician and clergyman of the 18th century.) In this example the Bayes strategy is s_{18}. In general *a Bayes strategy corresponding to the a priori probabilities* $1 - w$ *and* w *is a strategy* s *which minimizes*

$$\mathscr{L}(s) = (1 - w)\, L(\theta_1, s) + w\, L(\theta_2, s). \tag{5.6}$$

Every strategy is represented by a point of the convex set S. For every such point (L_1, L_2), there is a corresponding value of $(1 - w)L_1 + wL_2$. To find the Bayes strategies, we must locate the points of S which minimize $(1 - w)L_1 + wL_2$. We propose to show that the Bayes strategies can be located by shifting an appropriate line parallel to itself until it touches S. To do so, we shall study a second important representation of a line.

Refer to Figure 5.3. We seek an expression relating the coordinates x and y of a point C on the line which passes through the given points A and B. (The coordinates of those given points are (x_0, y_0) and (x_1, y_1) respectively.) In the figure we note that the triangles ABB' and ACC' are similar and therefore the lengths $\overline{BB'}$

and $\overline{AB'}$ are proportional to the lengths $\overline{CC'}$ and $\overline{AC'}$. That is to say

$$\frac{\overline{BB'}}{\overline{AB'}} = \frac{\overline{CC'}}{\overline{AC'}}$$

or

$$\frac{y_1 - y_0}{x_1 - x_0} = \frac{y - y_0}{x - x_0}.$$

This equation relating the coordinates x and y can be put in more convenient form by multiplying both sides by $(x_1 - x_0)(x - x_0)$. This gives

$$(x - x_0)(y_1 - y_0) = (y - y_0)(x_1 - x_0)$$

and thus also

$$x(y_1 - y_0) + y(x_0 - x_1) = x_0y_1 - y_0x_1.$$

If we now let a denote $(y_1 - y_0)$, b denote $(x_0 - x_1)$, and c denote $(x_0y_1 - y_0x_1)$, we may write the relation above as

$$ax + by = c. \tag{5.7}$$

This is an equation which relates the coordinates x and y of any point (such as C) on the line through $A = (x_0, y_0)$ and $B=(x_1, y_1)$. (It should be noted that, because (x_0, y_0) and (x_1, y_1) are different points, at least one of the numbers a and b must be different from zero.) Conversely, it is true that, if x and y are such that $ax+by=c$, then (x, y) is on the line. The above equation is our second important representation of the line. More precisely, *the points on the line may be represented by*

$$\{(x, y) : ax + by = c\}.$$

Furthermore, it is not difficult to prove that, *as long as a and b are any two numbers which are not both zero, the set of (x, y) for which $ax + by = c$ is a line.*

Consider the function f defined for all points (x, y) in the plane by

$$f(x, y) = 2x + 3y.$$

We have shown that the set of points for which $f(x, y)$ has the value 5 is a line (Figure 5.7). Consider the value of f as a point moves up from the line. Then x stays fixed, but y increases, and

$f(x, y)$ increases. Thus $2x + 3y > 5$ above the line and similarly, $2x + 3y < 5$ below the line. In general, *if a and b are not both zero, $ax + by = c$ represents a line, and $ax + by > c$ on one side of the line and $ax + by < c$ on the other side.* Moreover, since the points on the line $2x + 3y = 6$ lie above the line $2x + 3y = 5$, these lines never intersect and must be parallel. In general, *increasing c has the effect of moving the line $ax + by = c$ parallel to itself.*

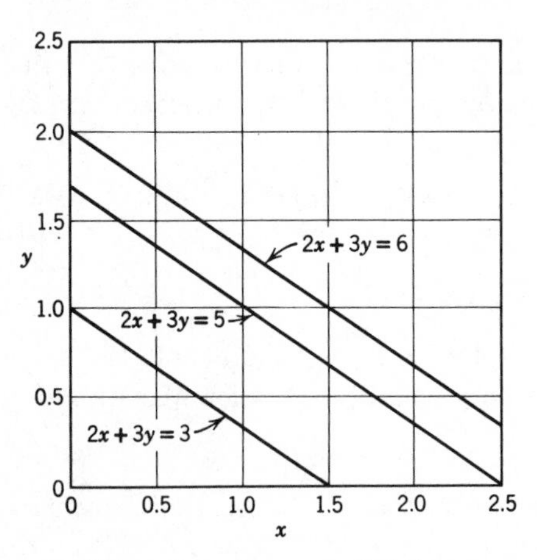

Figure 5.7. Some parallel lines.

The above statement indicates that the *coefficients* a, b, and c have certain relations to the line represented by $ax + by = c$. Another such relation is observed when we note that, if (x, y) is such that $2x + 3y = 5$, then $4x + 6y = 10$. In other words, if a, b, and c are doubled, the line is unaltered. More generally if a, b, and c are all multiplied by the same nonzero number, the line is unaltered.[1]

Two very special cases arise when a is zero and when b is zero. In the first case, the line is represented by the equation $by = c$ and is clearly a horizontal line c/b units above the horizontal axis. In the second case, the line is represented by $ax = c$ and is clearly a vertical line.

Suppose now that the line is not vertical, that is to say $b \neq 0$. Then for points on the line

[1] If a, b, and c were multiplied by zero, we would have the equation $0 = 0$ which is true for all (x, y) and not only for the line.

$$by = -\ ax + c$$

$$y = \frac{-a}{b}x + \frac{c}{b}.$$

Let

$$m = \frac{-a}{b}$$

and

$$e = \frac{c}{b}.$$

Then

$$y = mx + e \tag{5.8}$$

is a slightly modified version of the above representation which applies for nonvertical lines. We may regard the Equation (5.8) as one which represents a function f where $f(x) = mx + e$.

What more can be said about the numbers a, b, c, m and e, which are called the coefficients for the line ? Suppose a and b have different signs. That is, suppose a is negative and b is positive, or vice versa. Then $m = -a/b$ is positive. But if m is positive, an increase in x will lead to an increase in y. Similarly, if a and b have the same sign, $m = -\ a/b$ is negative and an increase in x leads to a decrease in y ; see Figure 5.8. Notice that as m increases from one positive value to another, the line rises more steeply. For this reason, m is sometimes called the slope of the line. Finally, if

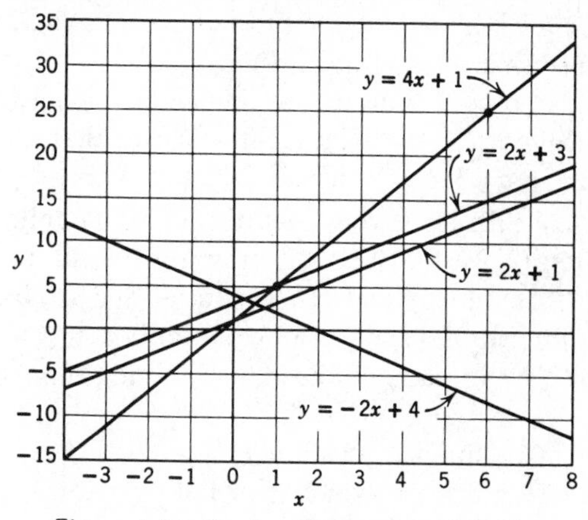

Figure 5.8. Graphs of several straight lines.

$x = 0$, $y = e$. Hence e represents the height of the point where the line crosses the vertical axis.

Suppose now that m is kept fixed and e is changed. Then the line is changed to a new line which is parallel to the old line. This is, of course, related to the fact that, if a and b are fixed but c is changed, the line shifts to a parallel one. Hence by changing c or e, we can shift the line parallel to itself as far as we wish.

Exercise 5.19. Draw the lines given by the following equations:

(a) $y = 3x - 2$
(b) $y = 3x + 1$
(c) $y = -2x + 1$
(d) $y = -2x + 3$
(e) $y = 2$
(f) $x = 5$

Exercise 5.20 Draw the lines given by the following equations:

(a) $2x + y = 4$
(b) $2x + y = 6$
(c) $2x + y = 8$
(d) $4x + 2y = 8$
(e) $2x - y = 4$
(f) $2x - 2y = 6$
(g) $-x - 2y = 4$
(h) $-x - 2y = 6$

**Exercise 5.21.* The sum of a and b in the line $2x + 3y = 5$ is 5.

(a) Change a, b, and c to a', b', and c' so as to leave the line unaltered but so that $a' + b' = 1$.

(b) Do the same for $-2x - 5y = 3$ and graph the line.

(c) Do the same for $3x - y = 1$ and graph the line.

(d) Why cannot the same be done for $x - y = 1$?

Exercise 5.22. Represent each of the above lines in the form $y = mx + e$.

Exercise 5.23. For each of the lines of Exercise 5.21 answer the following questions:

(a) Let (x, y) be a point on the line and (x, y^*) a point directly above it. Is $ax + by$ greater or less than $ax + by^*$?

(b) Is $a'x + b'y$ greater than or less than $a'x + b'y^*$?

Exercise 5.24. Draw seven lines through the following seven pairs of points: (2, 0) and (0, 2); (2, 0) and (0, 5); (4, 0) and (0, 10); (4, 0) and (0, 15); (0, 15) and (15, 0); (5, 0) and (5, 5); (0, 3) and (3, 3). Each line corresponds to an equation $wx + (1 - w)y = c$.

(a) Identify two pairs of these lines having common values of w.

(b) Identify a line for which $w = 0$.

(c) Identify a line for which $w = 1$.

(d) Identify a line for which $c = 1$.

(e) Find the equation of the line through (4, 0) and (0, 10).

(f) Which of the above lines is parallel to the line $15x+4y=3$?

A line is said to be a supporting line of a set S at the boundary point $\bar{u}$ if $\bar{u}$ is a boundary point of S and (1) *the line passes through $\bar{u}$ and* (2) *S is completely on one side of the line* (S may touch the line).

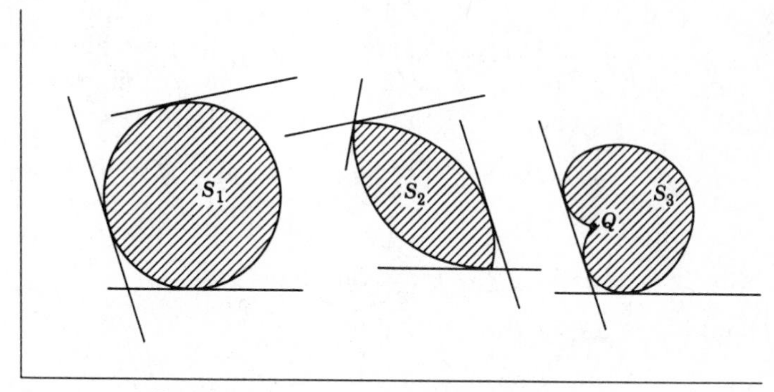

Figure 5.9. Some sets and supporting lines.

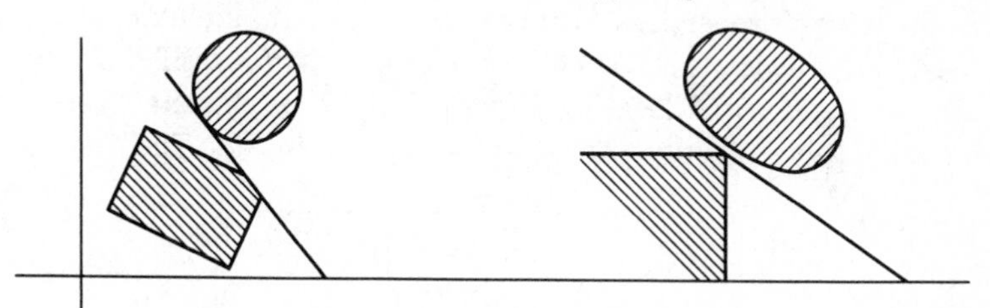

Figure 5.10. Separating lines for convex sets.

A set S is said to be bounded if it can be completely enclosed in a sufficiently large circle. Thus a triangle is bounded. On the other hand, the set of points on a parabola is not. A line segment is bounded but a line is not.

If S is a bounded set and L is a line, it is possible gradually to shift the line parallel to itself until it becomes a supporting line at some boundary point or points of S. See Figure 5.9. *A striking property of convex sets is that given* any *boundary point $\bar{u}$ of a convex set S, there is a supporting line of S at $\bar{u}$.* In Figure 5.9 this property is illustrated for two convex sets and, for a third nonconvex set, it is clear that it fails (at the point Q, for instance). Note that for S_1 each boundary point has only one possible supporting line.

On S_2, there are several points where the boundary is not smooth, and where several supporting lines may be drawn. Another striking property of convex sets, illustrated in Figure 5.10 is the following. *If two convex sets have no points in common a line can be drawn separating them* in the sense that all points of one set lie on one side of the line and all points of the other set lie on the other side of the line.

The definition of a supporting line can be interpreted algebraically. Suppose that the line $L = \{(x, y) : ax + by = c\}$ supports S at $\bar{u}_0$. Since S is on one side of the line, we have either

$$ax + by \geq c \qquad \text{for all } (x, y) \in S \tag{5.9}$$

or

$$ax + by \leq c \qquad \text{for all } (x, y) \in S, \tag{5.9a}$$

and since $\bar{u}_0$ is on the line, we have

$$ax_0 + by_0 = c \qquad \text{where } (x_0, y_0) = \bar{u}_0. \tag{5.9b}$$

The above properties of convex sets and their algebraic interpretation have special relevance for the problems of classifying admissible strategies and locating Bayes solutions graphically. Let s_0 be an admissible strategy represented by a point $\bar{u}_0$. Then no point

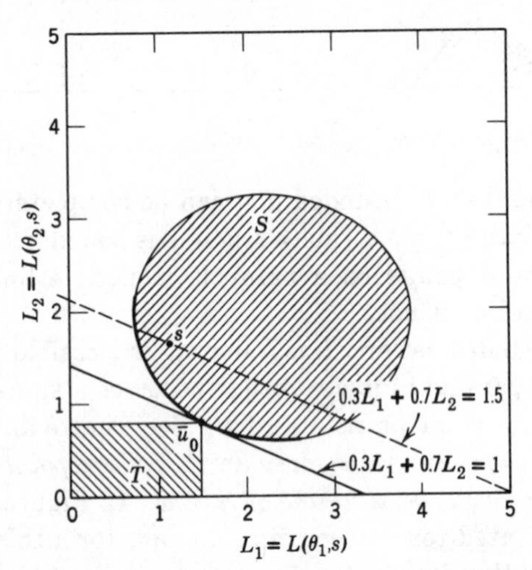

Figure 5.11. Convex set S representing all strategies for some problem. s_0 is an admissible strategy represented by $\bar{u}_0$.

of S lies in the convex set T of points which lie both below and to the left of $\bar{u}_0$. (In Figure 5.11, T is the interior of the shaded region below and to the left of $\bar{u}_0$. Although $\bar{u}_0$ is on the boundary of T it is not a point of T.) Clearly any line separating S and T must go through $\bar{u}_0$ and, hence, is a supporting line of S and T at $\bar{u}_0$. Since it supports T at $\bar{u}_0$, it must either (1) have a negative slope, (2) be vertical, or (3) be horizontal.[1] In none of these cases do a and b have different signs. Thus their sum $a + b$ is not zero and has the same sign as a and b. We can divide the coefficients by $a + b$ and represent the line by $\{(L_1, L_2) : a'L_1 + b'L_2 = c'\}$ where $a' = a/(a + b)$ and $b' = b/(a + b)$ are both non-negative numbers adding up to one. Then we can write $a' = 1 - w$ and $b' = w$ where $1 - w$ and w may be regarded as weights between 0 and 1. Thus, for any admissible strategy s_0 represented by $\bar{u}_0$, there is a pair of weights $1 - w$ and w so that $(1 - w)L_1 + wL_2 = c'$ supports S at $\bar{u}_0$. Furthermore, since T is below or to the left of the line, S is above or to the right of the line, hence, $(1 - w)L_1 + wL_2 \geq c'$ for all points of S, and $(1 - w)L_1 + wL_2 = c'$ for $\bar{u}_0$. But this means that s_0 is the Bayes strategy corresponding to the *a priori* probabilities $1 - w$ and w. In summary, *every admissible strategy is a Bayes strategy for some a priori probabilities* $1 - w$ *and* w. This fundamental classification of admissible strategies derives mainly from the fact that two convex sets with no points in common can be separated by a line.

A partial converse is also true. *If* $1 - w$ *and* w *are positive a priori probabilities, the corresponding Bayes strategies are admissible.* This is intuitively obvious because a strategy cannot be best on the average and worse than another for all states of nature. The last sentence is a slight exaggeration since a dominated strategy need not be *worse* than another for *all* states of nature. It could be worse for only one state of nature and equal for all others. We can gain some insight concerning the above converse by interpreting the Bayes strategy geometrically. For a strategy s_0 represented by $\bar{u}_0$ to be Bayes, $\bar{u}_0$ must be a point of S minimizing $(1 - w)L_1 + wL_2$. To find such Bayes strategies, we draw a line $(1 - w)L_1 + wL_2 = c$ below S. We increase c, thereby moving this line with negative slope upward parallel to itself until it touches S (see

[1] The truth of this can be seen by considering the slopes of those lines which can "pivot" on the corner of T; no such line can have positive slope.

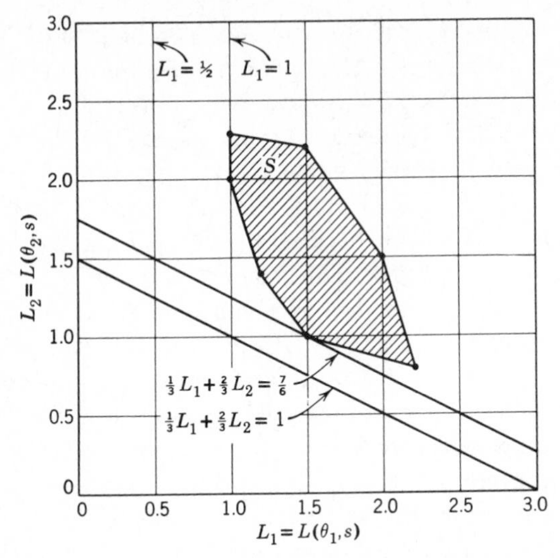

Figure 5.12. Graphic representation of a Bayes strategy.

Figure 5.12). *The points where the line* $(1 - w)L_1 + wL_2 = c$ *supports S (from below) are the Bayes strategies corresponding to the a priori probabilities* $1 - w$ *and* w. Since all points of S are on or above the line, and all points which dominate a point of support are below the line, no point of S can dominate a point of support. This means that no strategy can dominate one of these Bayes strategies, i.e., each Bayes strategy corresponding to positive *a priori* probabilities is admissible.

What happens when one of the *a priori* probabilities is zero? Suppose $w = 0$. Then the Bayes strategies minimize

$$(1 - w)L(\theta_1, s) + wL(\theta_2, s) = L(\theta_1, s)$$

and are represented graphically by the points at which a vertical line will support S (from the left). If there are several such points of support, they lie on a vertical line and only the lowest represents an admissible strategy. Statistically this means that, if we *knew* θ_2 were impossible and θ_1 were the state of nature, we would be willing to accept any strategy which minimizes $L(\theta_1, s)$ no matter how large $L(\theta_2, s)$. Thus we would not mind such inadmissible strategies in this *degenerate* case. Similarly if $w = 1$, Bayes strategies minimize $L(\theta_2, s)$ and are represented by points on the horizontal line of support (from below).

In Mr. Nelson's problem, it would be very reasonable for someone who was well acquainted with the frequency of rain in North Phiggins to apply the corresponding Bayes strategy. But in many problems, the state of of nature cannot be regarded as random, and it does not make sense to talk of the probability or frequency with which a certain state θ occurs. For example, suppose that θ is the velocity of light. Presumably θ is some fixed number which is estimated by various experiments the outcomes of which have probability distributions depending on θ. It would be unreasonable to treat the speed of light as random. Even so it makes sense to consider Bayes strategies. The reason for this is that, first, the class of all Bayes strategies (corresponding to all *a priori* probabilities) contains the class of admissible strategies and, therefore, one may reasonably restrict one's attention to the class of Bayes strategies. Second, the Bayes strategies will be shown to be relatively easy to obtain. Thus, by studying Bayes strategies, the statistician can often reduce his problem to choosing from among a relatively small class of easily obtainable "reasonable" strategies.

The fact that all admissible strategies are Bayes, and almost vice versa, was made clear from purely graphical considerations. These yield some additional results also. For example, at least one of the points where a supporting line touches a convex set is not a weighted average of other points. This means that at least one of the Bayes strategies corresponding to the *a priori* probabilities $(1 - w, w)$ is a *pure* strategy. In fact, if there are several non-equivalent Bayes strategies corresponding to $(1 - w, w)$, they consist of certain pure strategies and their mixtures. In the rain example, s_{15} and s_{18} and all their mixtures are the Bayes strategies corresponding to the *a priori* probabilities 6/16 and 10/16. This result implies that so long as we would be satisfied with any Bayes strategy corresponding to the *a priori* probabilities $1 - w$ and w, we need not bother with the randomized strategies. Since there are many problems with only a finite number of pure strategies but with infinitely many randomized strategies, this result is sometimes computationally helpful in that we may focus our computations on the pure strategies. In any case, it is pleasing to some philosophically minded statisticians who object to using irrelevant data, as they must if they are to use randomized strategies.

Graphically we may note that, as w increases, the supporting line $(1 - w)L_1 + wL_2 = c$ tends to become horizontal. The corresponding boundary point moves down and to the right. Hence, as w approaches 1, the corresponding Bayes strategy tends to minimize $L(\theta_2, s)$ and tends to disregard $L(\theta_1, s)$. This result is quite reasonable for, if one were almost certain that θ_2 were the state of nature, one would desire to minimize $L(\theta_2, s)$.

Exercise 5.25. Draw the lines

(a) $0.9x + 0.1y = 2$ (d) $0.3x + 0.7y = 2$
(b) $0.7x + 0.3y = 2$ (e) $0.1x + 0.9y = 2$
(c) $0.5x + 0.5y = 2$ (f) $0.5x + 0.5y = 4$

Exercise 5.26. Represent the line through the points (1.8, 2.2) and (1.3, 2.5) in the related forms $y = mx + e$ and $(1 - w)x + wy = c$. Hint : $2.2 = 1.8m + e$ and $2.5 = 1.3m + e$.

Exercise 5.27. Mr. Nelson is reliably informed that it rains 40% of the time in North Phiggins. Locate the appropriate Bayes strategy graphically. The weather meter reads "dubious." What action should Mr. Nelson take?

Exercise 5.28. Find the Bayes strategy for Mr. Smith's problem of Exercise 5.18 where the *a priori* probabilities of θ_1 and θ_2 are 0.5 and 0.5.

Exercise 5.29. Suppose that Mr. Nelson was informed that in North Phiggins local ordinances forbade the wearing of a raincoat except if worn with boots, umbrella, and rain hat. The effect of this is to reduce his choice of actions to only two — a_1 and a_3. This also reduces him to only eight pure strategies. List these strategies and the corresponding expected losses. Represent them graphically. Find (both graphically and by direct computation) the Bayes strategies corresponding to *a priori* probabilities $w = 0.1, 0.4, 0.8$.

In the following three exercises, S is convex set representing all possible strategies in a statistical problem.

Exercise 5.30. Can a line with a positive slope support S at an admissible strategy? Illustrate with a drawing.

Exercise 5.31. Can S lie below a line with negative slope which supports S at an admissible strategy? Illustrate with a drawing.

Exercise 5.32. Let $1 - w$ and w be *a priori* probabilities. Prove that the abscissa of the point where the corresponding supporting line intersects the line $L_1 = L_2$ is the minimized value of $(1 - w)L_1 + wL_2$.

Exercise 5.33. Indicate two convex sets which have one point in common and cannot be separated by a straight line.

Exercise 5.34. Indicate a convex set which cannot be supported by a vertical line. *Hint*: This set must be unbounded.

5. TWO STATES OF NATURE: MINIMAX STRATEGIES

Mr. Nelson was leaving his room wearing his raincoat when in came an old friend of his, Mr. Lancaster, who, also being aware of the possibility of rainy days, cautioned Mr. Nelson to guard against the vagaries of the weather. At first he proposed that Mr. Nelson

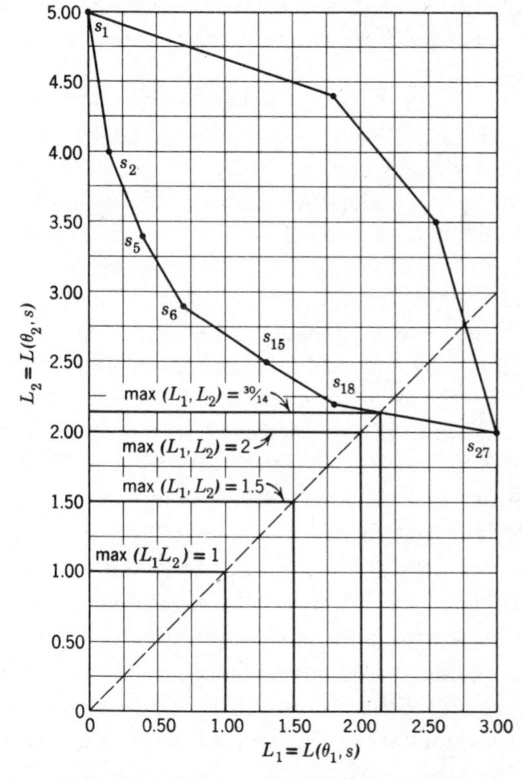

Figure 5.13. Graphic representation of the minimax expected loss strategy for Mr. Nelson's rain problem.

assume that the weather was bound to be bad and just wear his complete rain outfit. Then his loss would be only two. But Mr. Nelson, falling into the spirit of pessimism, replied that, surely if he wore his complete outfit, the sun would shine and his embarrassment would contribute to a loss of three. If, however, he selected strategy s_{18}, his expected loss would be no more than 2.2, no matter which of the θ's represented the state of nature. Mr. Lancaster who had never heard of statistics and strategies was quite enthused by this argument. Finally he extended it as follows. He claimed that, whatever strategy was applied, the θ which turned up would undoubtedly be the one which would would maximize the loss. Then clearly one should apply that strategy for which the maximum loss was as small as possible. In this way Mr. Lancaster formulated the principle of applying the criterion of *minimizing the maximum expected loss,* which we shall abbreviate as *minimax expected loss.*

Referring to Figure 5.13, they decided to tackle the problem graphically. The set of points (L_1, L_2) for which the larger of the two coordinates is one is designated by $\{(L_1, L_2) : max\,(L_1, L_2) = 1\}$ and is the set of points on two half-lines. One is the horizontal half-line for which $L_2 = 1$ and $L_1 < 1$. The other is the vertical half-line for which $L_1 = 1$ and $L_2 < 1$. The combined figure resembles a rather blunt wedge with an apex at (1, 1). The set of points for which $max\,(L_1, L_2) = c$ is similar except that its vertex is at (c, c). As c increases, this wedge moves upward and to the right. *The points where the wedge first meets S is the point corresponding to the minimax expected loss strategy.* In our example this point has coordinates (30/14, 30/14) and corresponds to mixing s_{18} and s_{27} with probabilities 10/14 and 4/14 respectively.

It is no accident that for the minimax strategy the two expected losses $L(\theta_1, s)$ and $L(\theta_2, s)$ were equal. As long as the apex of the wedge is the point which touches the set S, this will be the case. When this is known to be the case, it often helps in the computational task of finding the minimax strategy. If the apex is not the point or not the only point where the wedge first touches S, then either the horizontal part or the vertical part of the wedge touches S, and this part is then part of a supporting line to S at the minimax point; see Figure 5.14 for examples. This happens if S

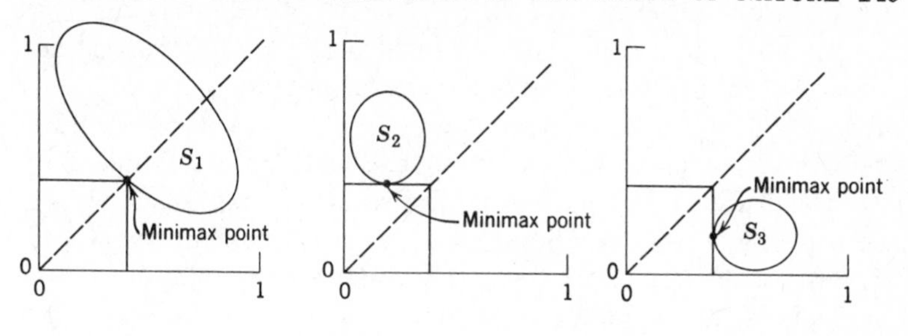

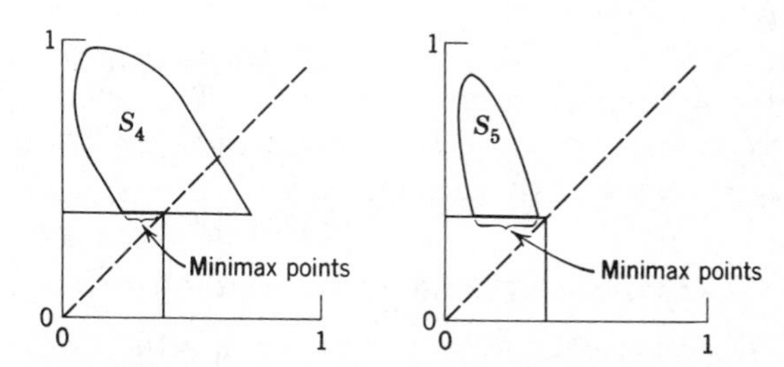

Figure 5.14. Minimax points for various convex sets.

lies completely below or completely above the line $L_1 = L_2$. When this occurs, the minimax strategies coincide with some of the Bayes strategies for the *a priori* probabilities (1, 0) or (0, 1). Here, again, we may have inadmissible strategies which minimize the maximum expected loss.

Often the *minimax expected loss* will be a mixed strategy. This is the case in Mr. Nelson's problem. Both he and Mr. Lancaster were pleased with this criterion. Mr. Lancaster had succeeded in minimizing the effect of nature's apparent dislike for him and Mr. Nelson had the opportunity to justify the use of a randomized strategy. The authors are not especially impressed by either point of view. One attitude which has been expressed to justify the use of this criterion is that, in many examples, the *minimax expected loss* strategy does guarantee a rather small expected loss no matter what the state of nature. In that case, there is not much to lose in applying this criterion.

**Exercise 5.35.* If Mr. Nelson applies the minimax expected loss strategy and observes a reading of "dubious," what action does he take? Apply the table of random digits if necessary.

Exercise 5.36. In a decision making problem, there are four pure strategies represented by (0.5, 3.0), (1.0, 1.5), (2.0, 1.0), and (3.0, 0.8). Graph the set S corresponding to all strategies. Find the Bayes strategy corresponding to the *a priori* probabilities 1/3 and 2/3. Locate the minimax expected loss strategy graphically.

Exercise 5.37. Locate the minimax expected loss strategy graphically for Mr. Smith's problem of Exercise 5.18.

Exercise 5.38. Locate the minimax expected loss strategy graphically for Mr. Nelson's problem of Exercise 5.29.

Exercise 5.39. Compute the minimax expected loss and the corresponding strategy for Mr. Nelson's problem of Exercise 5.29. *Hint*: For the minimax strategy $L_1 = L_2$. The solution can be obtained by expressing L_1 and L_2 as mixtures of the losses for the two strategies whose mixture is minimax.

6. TWO STATES OF NATURE: REGRET

Just as Nelson and Lancaster were leaving the room wearing their complete rain outfits, in came Mr. Crump. He was much impressed with their explanation of minimax expected loss but, after some thought, he argued for a modification. His theory was that, if it rains, the loss in utility is bound to be at least two. If it rains and the loss were three, because in their ignorance they wore only raincoats, then their *regret due to not having guessed* θ *correctly* would be only one. He claimed that they should consider the *regrets*, and possibly minimax the *expected regret* which we label *risk*. While Mr. Lancaster could not agree with him, Mr. Nelson was quite interested and decided to evaluate the *minimax risk* strategy. The minimum losses attainable under θ_1 and θ_2 respectively are 0 and 2. Hence the regrets, $r(\theta, a)$, are obtained by subtracting 0 and 2 from $l(\theta_1, a)$ and $l(\theta_2, a)$ respectively, as in Table 5.6. That the *risks* or average regrets may be similarly obtained from the average losses can be seen as follows. A strategy s leads to the random action $\mathbf{A}$ with regret $r(\theta, \mathbf{A}) = l(\theta, \mathbf{A}) - 2$. Taking expectations of both sides, we have $R(\theta, s) = L(\theta, s) - 2$, where R is the *risk*. Thus, to obtain the risks, we subtract the minimum losses 0 and 2 from the expected losses $L(\theta_1, s)$ and $L(\theta_2, s)$

TABLE 5.6

LOSS AND REGRET FOR MR. NELSON*

Loss = $l(\theta, a)$

	a_1	a_2	a_3	Min. Loss
θ_1	0	1	3	0
θ_2	5	3	2	2

Regret = $r(\theta, a)$

	a_1	a_2	a_3
θ_1	0	1	3
θ_2	3	1	0

* $r(\theta, a) = l(\theta, a)$ − minimum loss for that value of θ.

TABLE 5.7

EXPECTED LOSS L AND EXPECTED REGRET OR RISK R FOR MR. NELSON'S PURE STRATEGIES

$L(\theta, s)$

Strategy, s / State of Nature, θ	s_1	s_2	s_3	s_4	s_5	s_6	s_7	s_8	s_9
θ_1	0.00	0.15	0.45	0.25	0.40	0.70	0.75	0.90	1.20
θ_2	5.00	4.00	3.50	4.40	3.40	2.90	4.10	3.10	2.60

	s_{10}	s_{11}	s_{12}	s_{13}	s_{14}	s_{15}	s_{16}	s_{17}	s_{18}
θ_1	0.60	0.75	1.05	0.85	1.00	1.30	1.35	1.50	1.80
θ_2	4.60	3.60	3.10	4.00	3.00	2.50	3.70	2.70	2.20

	s_{19}	s_{20}	s_{21}	s_{22}	s_{23}	s_{24}	s_{25}	s_{26}	s_{27}
θ_1	1.80	1.95	2.25	2.05	2.20	2.50	2.55	2.70	3.00
θ_2	4.40	3.40	2.90	3.80	2.80	2.30	3.50	2.50	2.00

$R(\theta, s)$

	s_1	s_2	s_3	s_4	s_5	s_6	s_7	s_8	s_9	s_{10}	s_{11}	s_{12}	s_{13}
θ_1	0.00	0.15	0.45	0.25	0.40	0.70	0.75	0.90	1.20	0.60	0.75	1.05	0.85
θ_2	3.00	2.00	1.50	2.40	1.40	0.90	2.10	1.10	0.60	2.60	1.60	1.10	2.00

	s_{14}	s_{15}	s_{16}	s_{17}	s_{18}	s_{19}	s_{20}	s_{21}	s_{22}	s_{23}	s_{24}	s_{25}	s_{26}	s_{27}
θ_1	1.00	1.30	1.35	1.50	1.80	1.80	1.95	2.25	2.05	2.20	2.50	2.55	2.70	3.00
θ_2	1.00	0.50	1.70	0.70	0.20	2.40	1.40	0.90	1.80	0.80	0.30	1.50	0.50	0.00

respectively; see Table 5.7. The risks may be represented graphically by moving S vertically until the horizontal axis is a supporting line, and horizontally until the vertical axis is a supporting line;

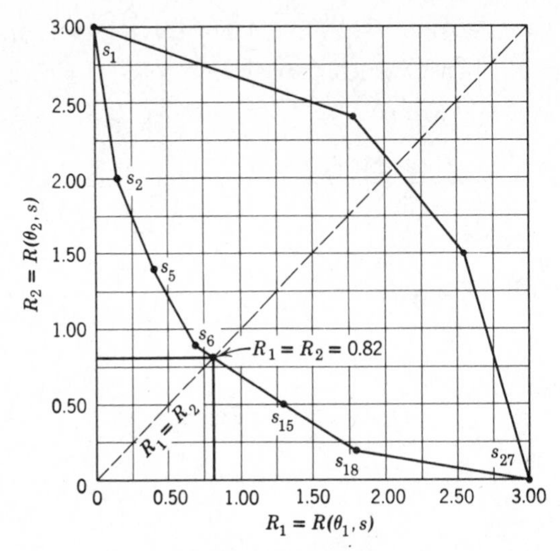

Figure 5.15. Graphical representation of the minimax risk strategy in Mr. Nelson's rain problem.

see Figure 5.15. (In our our example, S need not be moved horizontally.) Then the wedge method is applied to obtain the *minimax risk* strategy. This strategy has risks (0.82, 0.82), expected losses (0.82, 2.82), and mixes strategies s_6 and s_{15} with probabilities 0.8 and 0.2 respectively. Mr. Nelson is rather convinced when Mr. Crump illustrates (see Exercise 5.40) how slight differences in a catastrophic state will almost completely control the minimax expected loss strategy.

Exercise 5.40. Compute the minimax expected loss strategy and the minimax risk strategy in the following example with two states of nature and two pure strategies. (You should represent the problem graphically.)

Expected Losses

	s_1	s_2
θ_1	0	99
θ_2	100	99

Exercise 5.41. Just as Messrs. Nelson, Lancaster, and Crump were about to reach a decision, they were joined by Mr. Montgomery who had a different point of view. His policy, it turned out, was

to assume that things would turn out well and, accordingly, he selected the strategy which *minimized* the *minimum expected loss*. Compute the *minimin expected loss* and *minimin risk* strategies in Mr. Nelson's problem.

Exercise 5.42. When Mr. Montgomery's proposal was disregarded, he proposed the following problem with two states of nature and four pure strategies.

Expected Losses

	s_1	s_2	s_3	s_4
θ_1	200	168	120	0
θ_2	0	48	120	300

Clearly s_3 is the minimax risk strategy. However, suppose s_4 consists of always taking a certain action which is found to be illegal. Suppose also that s_1, s_2, and s_3 never use this action. Then what is the minimax regret strategy for the problem remaining after s_4 is discarded? Does it make sense to have a "criterion of optimality" which selects s_3 when s_4 is allowed but which rejects s_3 when the "undesirable" and irrelevant strategy s_4 is no longer allowed?

Exercise 5.43. Locate the minimax risk strategy graphically for Mr. Smith's problem of Exercise 5.18.

Exercise 5.44. Locate the minimax risk strategy graphically for Mr. Nelson's problem of Exercise 5.29.

Although many statisticians do not approve of the minimax risk criterion, most of them are willing to admit that for an intelligent choice of strategies it suffices to consider only the risks. (See Appendix E_6 for a discussion of this issue.) They may merely disagree on how the risks should be used.

The Bayes strategy corresponding to the *a priori* probabilities $1 - w$ and w minimizes the weighted average of expected losses $\mathscr{L}(s) = (1 - w)\, L(\theta_1, s) + w\, L(\theta_2, s)$. Suppose that the risks are obtained by subtracting the minimum losses $m(\theta_1)$ and $m(\theta_2)$ from $l(\theta_1, a)$ and $l(\theta_2, a)$. Then $L(\theta_1, s) = R(\theta_1, s) + m(\theta_1)$, $L(\theta_2, s) = R(\theta_2, s) + m(\theta_2)$, and

$$\mathscr{L}(s) = [(1 - w)\, R(\theta_1, s) + w\, R(\theta_2, s)] + [(1 - w)\, m(\theta_1) + w\, m(\theta_2)]\,.$$

The second part of the right-hand side does not involve s. Hence, the Bayes strategy minimizes the *weighted average of risks,*

$$\mathscr{R}(s) = (1 - w)\, R(\theta_1, s) + w\, R(\theta_2, s).$$

Thus minimizing the weighted average of the risks yields the same strategies as minimizing the weighted average of expected

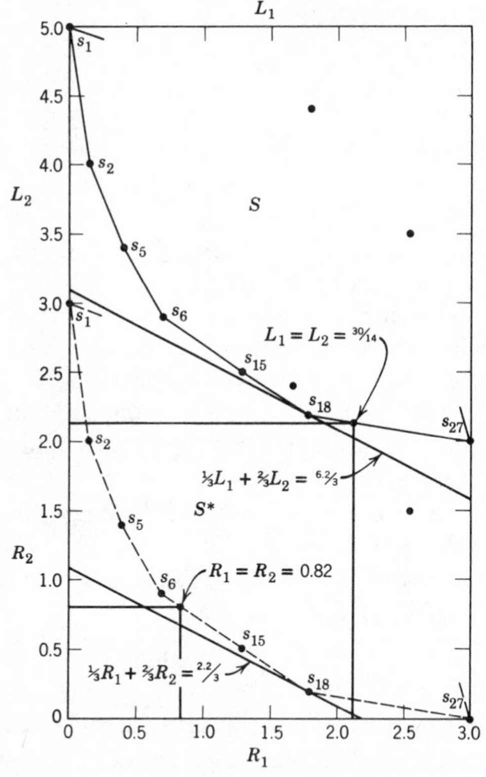

Figure 5.16. The set S of expected loss points $(L_1, L_2) = [L(\theta_1, s), L(\theta_2, s)]$ and the set S^* of risk points $(R_1, R_2) = [R(\theta_1, s), R(\theta_2, s)]$ corresponding to all strategies in Mr. Nelson's rain problem.

losses. Geometrically this means that, if the set of expected losses S is moved into S^* by shifting it till it touches both axes, the points where S and S^* touch a supporting line with direction determined by $1 - w$ and w are points corresponding to the same strategies. (See Figure 5.16.) It is customary for statisticians to compute automatically regrets instead of losses and not even to bother with expected losses.

7. LINES, PLANES, AND CONVEX SETS IN HIGHER DIMENSIONS

The theory of lines and convex sets in the plane proved helpful in discussing and understanding the problem of decision making in the case of two states of nature. In the contractor problem, Example 1.1, there were three states of nature. Then corresponding to any strategy s, there were three relevant expected losses $L(\theta_1, s)$, $L(\theta_2, s)$, and $L(\theta_3, s)$ which can be collected as one triple of numbers

$$\bar{L}(s) = (L(\theta_1, s), L(\theta_2, s), L(\theta_3, s)). \tag{5.10}$$

Just as a pair of numbers (x, y) can be represented by a point in the plane, we can represent a triple (x, y, z) by a point in three-dimensional space. The coordinates represent distances in each of three mutually perpendicular directions. The lines through the *origin* in these directions may be called the x, y, and z axes. Because of the correspondence between triples and points in space, theorems in geometry can be translated into certain algebraic relations. Very often these theorems are proved by first establishing the algebraic relations. One could proceed purely algebraically but, generally, the geometer is motivated by the intuitions which are developed by looking or thinking in terms of pictures.

For a problem which involves four or more states of nature, one must consider quartets or larger groups of numbers. Here, again, many algebraic relations can be considered to have geometric meaning even though it is difficult to visualize four-dimensional space. As far as statistics is concerned, the algebraic relations are of fundamental importance, but these relations are often thought of because of ordinary visualizations in the plane and space. We shall indicate properties of lines, planes, and convex sets in higher dimensions. One may visualize their counterparts in two or three dimensions to help one's intuition, but from our point of view these properties are merely algebraic relations. For the sake of convenience we shall confine ourselves to illustrations in three dimensions, but all our statements will have meaning and be true in k dimensions for $k \geq 3$.

In the plane we discussed the fact that a line through two points (x_0, y_0) and (x_1, y_1) could be represented by the set of points (x, y) for which

$$x = (1 - w)x_0 + wx_1$$
$$y = (1 - w)y_0 + wy_1.$$

If we represented the points (x_0, y_0), (x_1, y_1), and (x, y) by $\bar{u}_0$, $\bar{u}_1$, and $\bar{u}$, we could write

$$\bar{u} = (1 - w)\bar{u}_0 + w\bar{u}_1.$$

In space (three-dimensional) we have points represented by $\bar{u} = (x, y, z)$; see Figure 5.17.

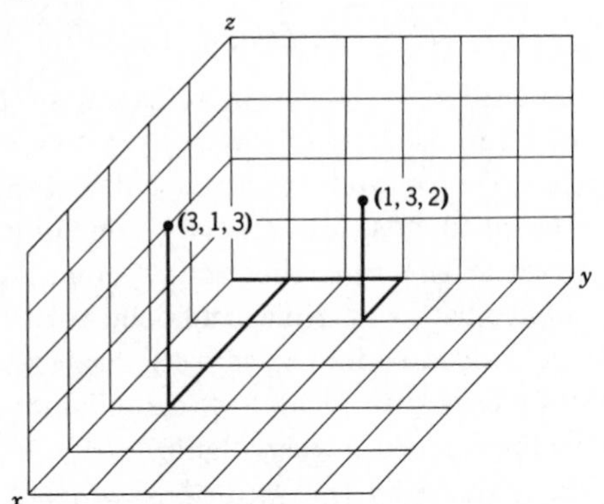

Figure 5.17. Points in space.

The argument which applied to give the representation of a line in the plane (Figure 5.3) can easily be extended to give us the following. The *line* through the two points $\bar{u}_0 = (x_0, y_0, z_0)$ and $\bar{u}_1 = (x_1, y_1, z_1)$ is represented by

$$\{\bar{u} : \bar{u} = (1 - w)\bar{u}_0 + w\bar{u}_1, \ w \text{ is a number}\} \tag{5.11}$$

where $(1 - w)\bar{u}_0 + w\bar{u}_1$ is to be interpreted as before as the point (x, y, z) each coordinate of which is obtained by adding $(1 - w)$ times the corresponding coordinate of $\bar{u}_0$ to w times the corresponding coordinate of $\bar{u}_1$. In other words,

$$x = (1 - w)x_0 + wx_1$$
$$y = (1 - w)y_0 + wy_1$$
$$z = (1 - w)z_0 + wz_1.$$

Similarly, the line segment connecting $\bar{u}_0$ and $\bar{u}_1$ is the set

$$\{\bar{u} : \bar{u} = (1 - w)\bar{u}_0 + w\bar{u}_1,\ 0 \leq w \leq 1\}. \tag{5.12}$$

Once more $\bar{u}$ is w times as far from $\bar{u}_0$ as is $\bar{u}_1$. If $w_1, w_2, \cdots, w_m$ are m non-negative numbers adding up to one, the point

$$\bar{u} = w_1\bar{u}_1 + w_2\bar{u}_2 + \cdots + w_m\bar{u}_m \tag{5.13}$$

is a weighted average of the points $\bar{u}_1, \bar{u}_2, \cdots, \bar{u}_m$. A *convex set* is defined, as before, as a set S which contains the line segment connecting $\bar{u}$ and $\bar{v}$ if it contains the points $\bar{u}$ and $\bar{v}$. The smallest convex set containing a set A is called the *convex set generated by* A. The convex set generated by m points $\bar{u}_1, \bar{u}_2, \cdots, \bar{u}_m$ is the set of weighted averages of these points. In two dimensions, the convex set generated by A may be thought of as the region enclosed in a rubber band set down to embrace all of A. Similarly the convex set generated by a three-dimensional set A may be visualized as the region contained within an elastic wrapper around A. It is a fact that, for every point $\bar{u}$ in the convex set generated by A, there can be found four points of A of which $\bar{u}$ is a weighted average. (In k dimensions, four would be replaced by $k + 1$. The reader may illustrate this fact for himself for $k = 2$.) The convex set generated by a finite number of points is a polyhedron, the vertices of which are some (or all) of the original points.

In three-dimensional space, a plane is represented by a *linear equation*. That is to say, corresponding to any plane P, there are four numbers a, b, c, d such that a, b, and c are not all zero and

$$P = \{(x, y, z) : ax + by + cz = d\}. \tag{5.14}$$

In k dimensions a $k - 1$ *dimensional "hyperplane"* would be given by

$$P = \{(x_1, x_2, \cdots, x_k) : b_1x_1 + b_2x_2 + \cdots + b_kx_k = d\} \tag{5.14a}$$

where $b_1, b_2, \cdots, b_k$ are not all zero. In two dimensions, the $k - 1$ dimensional hyperplane is a line. This is why the line has two rather distinct representations in the plane. One corresponds to the line in k dimensions and the other to the $k - 1$ dimensional hyperplane in k dimensions. Multiplying the coefficients a, b, c, and d by a given nonzero number does not affect the plane. Changing d shifts the plane parallel to itself. If a, b, and c are positive, an increase in d shifts the plane in the direction in which x, y, and z increase. If $a = 0$, the plane is parallel to the x-axis. If a and b are zero, the plane parallels the x- and y-axes. In Figure 5.17 it would be a horizontal plane. Suppose the coefficients a, b, and c are non-negative. Since they do not all vanish, we can divide by $a + b + c$, which is positive. Thereby we obtain an equation for the plane in which the coefficients of x, y, z are non-negative and add up to one. Then we could express the plane by

$$w_1x + w_2y + w_3z = d'$$

where $w_1 = a/(a + b + c)$, $w_2 = b/(a + b + c)$, $w_3 = c/(a + b + c)$, and $d' = d/(a + b + c)$. The left-hand side of this expression is a weighted average of the coordinates x, y, and z.

A plane P is a supporting plane of S at the boundary point $\bar{u}_0$ of S, if $\bar{u}_0$ is a point of P and if S lies completely on one side of P. Algebraically the supporting plane properties are

$$(5.15) \qquad ax + by + cz \geq d \qquad \text{for all } (x, y, z) \in S$$

or

$$(5.15a) \qquad ax + by + cz \leq d \qquad \text{for all } (x, y, z) \in S$$

and

$$(5.16) \qquad ax_0 + by_0 + cz_0 = d \qquad \text{where } \bar{u}_0 = (x_0, y_0, z_0).$$

If S is a bounded set and P is a plane, then P can be moved parallel to itself until it becomes a supporting plane of S. If S is a convex set and $\bar{u}_0$ is a boundary point of S, then there is a supporting plane P of S at $\bar{u}_0$. Two convex sets having no points in common may be separated by a plane.

A point $\bar{u}$ is said to be dominated by a point $\bar{v}$ if each coordinate of $\bar{u}$ is at least as large as the corresponding coordinate of $\bar{v}$ and $\bar{u} \neq \bar{v}$. The admissible part of the boundary of a convex set S consists of those boundary points which are not dominated by any points of S. If S is a convex set and $\bar{u}_0$ is on the admissible part of the boundary of S, then S lies above a supporting plane $ax+by+cz=d$ through $\bar{u}_0$ whose coefficients a, b, c are non-negative. Conversely, if $ax + by + cz = d$ is a supporting plane of the convex set S (from below) with a, b, c positive, then the boundary points of S which touch the plane are on the admissible part of the boundary. If any of the coefficients a, b, c are zero, i.e., if the plane is parallel to one of the three axes, then at least one of the boundary points of S which touch the plane is on the admissible part of the boundary. However, there may be other boundary points of S which touch the plane and are not on the admissible part of the boundary.

If these remarks are interpreted algebraically after dividing a, b, c, and d by $a + b + c$, we obtain the following:

If $\bar{u}_0$ is on the admissible part of the boundary of the convex set, then there are non-negative weights w_1, w_2, w_3 adding up to one such that the minimum value of the weighted average $w_1x + w_2y + w_3z$ for $(x, y, z) \in S$ is achieved at $\bar{u}_0$. If w_1, w_2, w_3 are positive weights adding up to one, any point $\bar{u}_0$ of S at which the weighted average $w_1x + w_2y + w_3z$ is minimized is on the admissible boundary of S. If any of the weights are zero, then at least one of the points $\bar{u}_0$ which minimize the weighted average is on the admissible part of the boundary.

We repeat that these results and notions are extensible in a rather obvious fashion to higher than three dimensions.

8. THREE OR MORE UNKNOWN STATES OF NATURE

In this section we shall assume that there are k possible states of nature. The statements that we make will follow from the k-dimensional extension of the relations of the type established in Section 7. The results discussed here will constitute brief extensions of similar results in the case where there are two states of nature. Because of the very few pure strategies available in the two-states-of-nature rain example discussed in Sections 2 through 6, and because of the ease of graphic representations in two dimensions, we were able to use the graphic representations to actually locate various types of "optimal" strategies. In higher dimensions or with more pure strategies available, it becomes impractical to use graphic representations for this purpose. Nevertheless they can be used to understand certain important properties of various types of strategies which will be useful in actually computing these strategies.

Corresponding to each strategy s, there are k expected losses given by $\bar{L}(s) = (L(\theta_1, s), L(\theta_2, s), \cdots, L(\theta_k, s))$. If s and s^* are strategies which yield expected loss points $\bar{L}(s)$ and $\bar{L}(s^*)$, then mixing s and s^* with probabilities $1 - w$ and w will yield the expected loss point $(1-w)\bar{L}(s) + w\bar{L}(s^*)$. But this means that all points on the line segment connecting $\bar{L}(s)$ and $\bar{L}(s^*)$ are attainable with the use of mixed strategies. In fact, *the set of expected loss points obtained with the use of mixed (randomized) strategies is the convex set generated by the points representing the pure strategies.*

A strategy s is at least as good as s^* if

$$L(\theta_1, s) \leq L(\theta_1, s^*), L(\theta_2, s) \leq L(\theta_2, s^*), \cdots, L(\theta_k, s) \leq L(\theta_k, s^*).$$

A strategy s is equivalent to s^* if

$$L(\theta_1, s) = L(\theta_1, s^*), L(\theta_2, s) = L(\theta_2, s^*), \cdots, L(\theta_k, s) = L(\theta_k, s^*).$$

A strategy s dominates a strategy s^* if s is at least as good as s^* and s is not equivalent to s^*. In this case, each coordinate of the point representing s is less than or equal to the corresponding coordinate of s^*, and at least one coordinate of s is less than the corresponding coordinate of s^*.

A strategy s is admissible if there is no strategy which dominates it. As before, we shall assume that the convex set of expected losses attained by using all strategies, pure and randomized, contains all of its boundary points. Then it is clear that it suffices to confine one's attention to the class of admissible strategies.

In the contractor problem, Example 1.1, suppose that the contractor discovers that in this community 30% of the new home-owners have peak loads of 15 amp, 45% have peak loads of 20 amp and 25% have peak loads of 30 amp. In other words, he can assume *a priori* probabilities of 0.30, 0.45, and 0.25 for θ_1, θ_2, and θ_3. Then he would desire to minimize his weighted average of expected losses $\mathscr{L}(s) = 0.30L(\theta_1, s) + 0.45L(\theta_2, s) + 0.25L(\theta_3, s)$. But $\mathscr{L}(s)$ is a weighted average of the coordinates of $\bar{L}(s)$ with weights 0.30, 0.45, and 0.25. Suppose that w_1, w_2, and w_3 are weights (≥ 0) adding up to one. *A strategy which minimizes the weighted average $\mathscr{L}(s) = w_1 L(\theta_1, s) + w_2 L(\theta_2, s) + w_3 L(\theta_3, s)$ is called a Bayes strategy for the a priori probabilities w_1, w_2, and w_3.*

As a consequence of the relations of the preceding section, it follows that *any admissible strategy is a Bayes strategy for some a priori probabilities.* Furthermore, *if all the weights or a priori probabilities are positive, the corresponding Bayes strategies are admissible. If some of the weights are zero, at least one of the Bayes strategies will be admissible.* Hence it makes sense to confine one's attention to the class of Bayes strategies even if it does not make sense to assume that the states of nature are random variables.

One advantage of applying this result is that it is often relatively easy to compute Bayes strategies. Reasons for this will be discussed in Chapter 6.

As in the two-dimensional case, geometric considerations make it obvious that for any set of weights the corresponding Bayes strategies consist of certain pure strategies and their mixtures. If it makes sense to apply certain *a priori* probabilities to the possible states of nature, then there is a pure strategy which is Bayes, and one need not use mixed strategies.

The conservative statistician like Mr. Lancaster may wish to apply that strategy s for which the largest of $L(\theta_1, s)$, $L(\theta_2, s)$, $\cdots$, $L(\theta_k, s)$ is as small as possible. This criterion is called *minimax expected loss.* Geometrically the set of points (x, y, z) for which $max\,(x, y, z) = c$ consists of parts of three planes parallel to the axes;[1] see Figure 5.18. These are:

$$P_1 = \{(x, y, z) : x = c,\ y \leq c,\ z \leq c\}$$
$$P_2 = \{(x, y, z) : x \leq c,\ y = c,\ z \leq c\}$$

and

$$P_3 = \{(x, y, z) : x \leq c,\ y \leq c,\ z = c\}.$$

P_1, P_2, and P_3 have the one apex point (c, c, c) in common. P_1 and P_2 have the half-line

$$L_{12} = \{(x, y, z) : x = c,\ y = c,\ z \leq c\}$$

in common. Similar remarks apply for L_{13} and L_{23}. The minimax expected loss strategy may be obtained by moving the set

$$\{(x, y, z) : max\,(x, y, z) = c\}$$

by increasing c until it touches the boundary of S. If it touches at the apex, the minimax strategy s will have the property that

$$L(\theta_1, s) = L(\theta_2, s) = L(\theta_3, s).$$

Otherwise the boundary will lie on some but not all half-planes P_1, P_2, P_3. Suppose the boundary lies on P_1 and P_2 (and hence L_{12}) but not on P_3. Then the minimax strategy is a Bayes strategy for some

[1] The largest of the three numbers (x, y, z) is represented by $max\,(x, y, z)$.

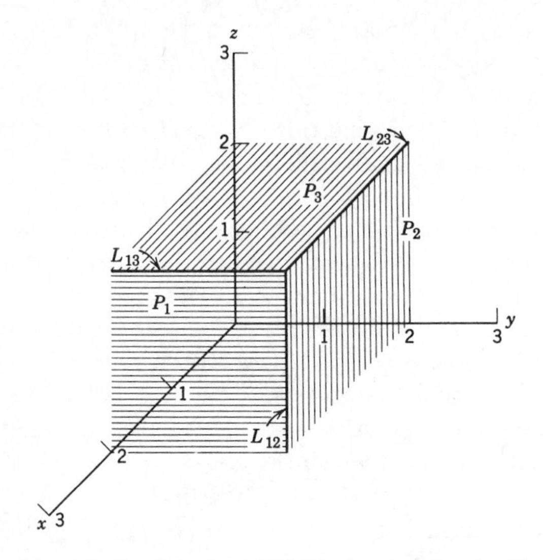

Figure 5.18. Diagram of $\{(x, y, z)\colon \max(x, y, z) = 2\}$.

a priori probabilities with $w_3 = 0$. Furthermore, $L(\theta_1, s) = L(\theta_2, s)$. If the boundary lies on P_1 but not on P_2 or P_3, the minimax expected loss strategy is a Bayes strategy for the weights (1, 0, 0). In other words, the minimax strategy then minimizes $L(\theta_1, s)$.

In general, for k states of nature, the situation is as follows. The minimax strategy is Bayes for some *a priori* probabilities $w_1, w_2, \cdots, w_k$. The expected losses $L(\theta_i, s)$ corresponding to the non-zero weights w_i among $w_1, w_2, \cdots, w_k$ are all equal to each other.

The *minimax risk* strategy is obtained by applying the minimax criterion to the *risks* or *expected regrets*. The risks are obtained by subtracting the smallest possible loss for θ_i from $L(\theta_i, s)$. Geometrically this corresponds to shifting S in each direction.

Although many statisticians approve of confining their attention to risks and usually do so, some of them are reluctant to accept the minimax risk criterion. (See Exercise 5.42.) For these statisticians who object to minimax risk, there is consolation in the fact that a Bayes strategy can be obtained not only by minimizing the *weighted average of expected losses*

$$(5.17)\quad \mathscr{L}(s) = w_1 L(\theta_1, s) + w_2 L(\theta_2, s) + \cdots + w_k L(\theta_k, s)$$

but also by minimizing the *weighted average of risks*

$$\mathscr{R}(s) = w_1 R(\theta_1, s) + w_2 R(\theta_2, s) + \cdots + w_k R(\theta_k, s) . \tag{5.18}$$

9. SUMMARY

The possibility of using randomized strategies makes it possible to apply geometric relations involving convex sets to the theory of decision making in the case of k possible states of nature. When it is possible to apply *a priori* probabilities to the states of nature, one can obtain *Bayes strategies.* A Bayes strategy is geometrically represented as a boundary point on the convex set S of possible expected loss points $(L(\theta_1, s), L(\theta_2, s), \cdots, L(\theta_k, s))$ such that a plane corresponding to the *a priori* probabilities is a supporting plane of S at this point. Even in the case where it does not make sense to consider the state of nature as random, it does pay to consider Bayes strategies since the class of Bayes strategies contains the admissible strategies. Furthermore, as we shall see in Chapter 6, Bayes strategies are relatively easy to compute. For the benefit of statisticians who prefer not to have decisions depend on the result of irrelevant experiments and, therefore, abhor randomized strategies, it is interesting to note that for any set of *a priori* probabilities there is at least one pure Bayes strategy.

The criteria *minimax expected loss* and *minimax risk* were discussed and are each subject to certain fundamental objections. One of the basic statistical problems which is still unanswered is that of determining what constitutes a reasonable criterion for selecting an optimal strategy.

In spite of this basic lack, the situation is not really very black. In practice there is ordinarily available a large quantity of data. Then, as is often the case, if *a priori* probabilities are picked rather arbitrarily and the Bayes strategy is compared with the minimax expected loss criterion or minimax risk criterion, all of these tend to be very similar strategies and correspond to neighboring risk points with very small risks. In fact, there is little reason to object to minimax risk if, in a particular example, this criterion gives the assurance of a very small risk.

One suggestion for selecting a strategy is basically the following. When a problem is approached, there is usually available a great deal of miscellaneous and somewhat relevant information which may lead one to think of certain states as more likely than others. With this information one may approximately measure one's "degree of belief" by some *a priori* probabilities of the θ's. From this point on, apply the corresponding Bayes strategy.

In this chapter we have spent a relatively great deal of time in examining the relations among strategies without examining the meaning of these strategies in the particular problem. It is sometimes possible to list all strategies and examine their risks without attempting to think of what there is about a strategy that makes it more reasonable than others. On the other hand, a general rule of thumb for a reasonable strategy is that it uses the data to guess at the state of nature and then takes an action which would be good if the guess were correct. Because a wrong guess may sometimes be catastrophic, it is important to hedge by trying to avoid actions which are very bad if the guess is wrong. Hence in the rain problem, the action a_2 is not especially good whether it rains or shines. If the data strongly suggest rain, one should use a_3. If they strongly suggest shine, one should use a_1. However, if they do not strongly suggest one or the other, it sometimes pays to hedge by taking the mediocre action a_2 which protects against the wrong guess. This principle is applied in Exercise 1.2 where the contractor example was expanded by introducing an action which is not optimal for any state of nature.

In many problems it will be quite difficult to compute strategies which are "optimal" in some sense. Frequently we shall have to be satisfied with suggesting strategies which seem reasonable and then computing the risks associated with these strategies. Often the loss associated with a given action a and the state of nature θ are not precisely known. Or the problem may be relevant for many of the statistician's customers, and they may have different loss tables. In such cases, we may merely list the *action probabilities* for each strategy s considered (see Table 5.5), and leave it to the customer to decide which strategy's action probabilities he likes best. When listing the action probabilities, it is not uncommon to compare them with *ideal action probabilities* (see Exercise 5.3).

Exercise 5.45. For a two-state, three-action problem, the losses and probability distribution of the data are given below.

Losses

	a_1	a_2	a_3
θ_1	4	3	1
θ_2	2	4	6

Probability Distribution

	z_1	z_2
θ_1	0.7	0.3
θ_2	0.5	0.5

List the ideal action probabilities, compute and graph the risks for all pure strategies, compute a Bayes strategy for *a priori* probabilities ($\frac{1}{2}$, $\frac{1}{2}$), and graphically represent the minimax risk strategy. Is there more than one Bayes strategy for the above *a priori* probabilities ? Evaluate the minimax risk and the weighted average of risks for the Bayes strategy.

Exercise 5.46. For a three-state, two-action problem, the losses and probability distribution of the data are given below.

Losses

	a_1	a_2
θ_1	4	1
θ_2	3	5
θ_3	2	6

Probability Distribution

	z_1	z_2
θ_1	0.7	0.3
θ_2	0.5	0.5
θ_3	0.4	0.6

List the ideal action probabilities, compute the risks for all pure strategies, and compute a Bayes strategy for *a priori* probabilities (0.3, 0.3, 0.4). Is there more than one Bayes strategy for these *a priori* probabilities ?

SUGGESTED READINGS

[1] Blackwell, David, and M. A. Girshick, *Theory of Games and Statistical Decisions,* John Wiley and Sons, New York, 1954, Dover Publications, Inc., New York, 1979. An excellent textbook for graduate students in mathematical statistics.

[2] Luce, R. D., and Howard Raiffa, *Games and Decisions,* John Wiley and Sons, New York, 1957.

[3] Savage, L. J., *The Foundations of Statistics,* John Wiley and Sons, New York, 1954, Dover Publications, Inc., New York, 1972.

A rather advanced book which explores the rationale for selecting ''psychological'' *a priori* probabililties.

[4] Wald, Abraham, *Statistical Decision Functions,* John Wiley and Sons, New York, 1950, Dover Publications, Inc., New York, 1973.

This is the first book on the subject by the originator of decision theory. It is very advanced.

A brief discussion of game theory will be found in Appendix F_1. This discussion is related to the text of Chapter 5.

CHAPTER 6

The Computation of Bayes Strategies

1. A POSTERIORI PROBABILITY AND THE NO-DATA PROBLEM

As Mr. Nelson and his friends were squabbling about strategy, the hotel owner, Mr. Solomon, came to the door and politely suggested that they make less noise. They apologized and explained their difficulties to him, whereupon he suggested that they make use of the fact that in North Phiggins it rains only 40% of the days. After a little computing, they all agreed that s_6 was the appropriate Bayes strategy. However, Mr. Lancaster, who was put out by the general rejection of his criterion, raised a new issue. In his hometown, rainy days are more likely to be followed by rainy days than are sunny days.

Mr. Solomon admitted that this was true in North Phiggins too. In fact, he had observed that 70% of rainy days were followed by rainy days, whereas only 20% of sunny days were followed by rainy days. "Then," claimed Mr. Lancaster, "if yesterday was a rainy day, we should apply the Bayes strategy corresponding to the *a priori* probabilities 0.3 for sunny and 0.7 for rainy. If yesterday was a sunny day, we should use probabilities 0.8 and 0.2. In neither case should we use 0.6 and 0.4." The others had to admit that this seemed reasonable and, noting that it had been rainy the day before, they all used the Bayes strategy for *a priori* probabilities 0.3 and 0.7. This led them all to apply s_{18} even though their loss tables differed somewhat. Because the rain indicator read "dubious" they all left the hotel with their complete rain outfits, much to Lancaster's pleasure.

Being in a somewhat reflective mood, Mr. Solomon was intrigued by the following notion. Two pieces of data (observations) had been used. One was yesterday's weather, which might have been rainy or sunny. The other was the rain indicator reading, which could have been "fair," "dubious," or "foul." Altogether, there were six possible pairs of data. These were: (rainy yesterday, "fair" reading), (rainy yesterday, "dubious" reading), (rainy yesterday, "foul" reading), (sunny yesterday, "fair" reading), (sunny yesterday, "dubious" reading), and (sunny yesterday, "foul" reading). For each of these six possible pairs, there were three possible actions. Altogether, there were $3 \times 3 \times 3 \times 3 \times 3 \times 3 = 3^6 = 729$ possible pure strategies. But by using yesterday's weather to compute a modified *a priori* or rather an *a posteriori probability*, his guests had apparently been satisfied that they had used that information to the fullest extent possible and could now treat the problem as one having the new probabilities for θ_1 and θ_2 and with the rain indicator reading as the only datum (and thus with only 27 pure strategies). Two questions intrigued Mr. Solomon. First, was the procedure used by the guests correct? That is, must the Bayes strategy for the two observation problem with its 729 possible strategies be the same as the Bayes strategy obtained by "digesting" one of the observations into the *a priori* probability and then solving the reduced 27-strategy problem? Second, could the other observation (indicator reading) also be "digested" into the *a priori probability*, further reducing the problem to a more trivial one with no data? As we shall see later, the answer to each question is "yes." Let us discuss how to find Bayes strategies for *no-data* problems, i.e., for problems where no data are available but the *a priori* probabilities are known.

Suppose then that Mr. Nelson had no weather meter but knew that the probabilities of sunny day and rainy day were 0.6 and 0.4 respectively. Since he has no data, his available pure strategies are either: (1) to take action a_1; (2) to take action a_2; or (3) to take action a_3. If he takes action a_1, his regret (see Table 5.6) is 0 under θ_1 and 3 under θ_2, and his risk is $(0.6)0 + (0.4)3 = 1.2$. Similarly, for action a_2, his risk is $(0.6)1 + (0.4)1 = 1.0$, and for action a_3 it will be $(0.6)3 + (0.4)0 = 1.8$. Thus the solution of his no-data problem is to take action a_2. In Table 6.1 we indicate a tabular presentation of this solution based on the following definitions. The risk $B(\overline{w}, a)$ corresponding to the *a priori* probabilities given by $\overline{w} = (w_1, w_2, \cdots, w_k)$ and action a is defined by

$$(6.1) \quad B(\overline{w}, a) = w_1 r(\theta_1, a) + w_2 r(\theta_2, a) + \cdots + w_k r(\theta_k, a).$$

TABLE 6.1

TABULAR PRESENTATION OF THE NO-DATA PROBLEM

	Regrets			*A Priori* Probabilities
	a_1	a_2	$\cdots$	$\bar{w}$
θ_1	$r(\theta_1, a_1)$	$r(\theta_1, a_2)$	$\cdots$	w_1
θ_2	$r(\theta_2, a_1)$	$r(\theta_2, a_2)$	$\cdots$	w_2
$\vdots$	$\vdots$	$\vdots$	$\vdots$	$\vdots$
θ_k	$r(\theta_k, a_1)$	$r(\theta_k, a_2)$	$\cdots$	w_k
$B(\bar{w}, a)$	$B(\bar{w}, a_1)$	$B(\bar{w}, a_2)$	$\cdots$	

Bayes risk = $\mathscr{B}(\bar{w})$
Bayes action = action which minimizes $B(\bar{w}, a)$

	Rain Example: Regrets			
	a_1	a_2	a_3	$\bar{w}$
θ_1	0	1	3	0.6
θ_2	3	1	0	0.4
$B(\bar{w}, a)$	1.2	1.0	1.8	

$\mathscr{B}(\bar{w}) = 1.0$
Bayes action = a_2

The *Bayes action* for the no-data problem with *a priori* probabilities given by $\bar{w}$ is the action a which gives the smallest value of $B(\bar{w}, a)$. The risk corresponding to this Bayes action is $\mathscr{B}(\bar{w})$ and is called the *Bayes risk* and is

$$\mathscr{B}(\bar{w}) = min\,(B(\bar{w}, a_1), B(\bar{w}, a_2), \cdots). \tag{6.2}$$

**Exercise 6.1.* In Example 1.1, suppose that the contractor has not gathered any of the data but that he is told that each state of nature has *a priori* probability 1/3. Which action should he take? What is his Bayes risk?

Exercise 6.2. Find the Bayes action and the Bayes risk for the no-data version of the problem you set up in Exercise 1.3 when the states of nature have equal *a priori* probability.

Exercise 6.3. Do the same for the example of Exercise 1.5.

Exercise 6.4. Do the same for the rain problem, Example 5.1.

Exercise 6.5. Do the same for the problem of Exercise 5.18.

Exercise 6.6. Do the same for the problem of Exercise 5.45.

**Exercise 6.7.* If Mr. Nelson had a crystal ball which told him what the state of nature was, he could take the proper action. How much should Mr. Nelson be willing to pay for the use of a crystal ball if he had no rain meter but knew that the *a priori* probability of rain was 0.4?

Some further insight may be gained by the following graphical representation of the solution of the no-data problem which we present for the case where there are two possible states of nature.[1] If the *a priori* probabilities are given by $\overline{w} = (w_1, w_2) = (1 - w, w)$, we may regard

$$B(\overline{w}, a) = (1 - w)\, r(\theta_1, a) + w\, r(\theta_2, a)$$

as the value of a function defined for $0 \le w \le 1$. This function is represented by a straight line. Thus $B(\overline{w}, a_1) = (1-w)0 + (w)3 = 3w$ is a straight line which goes through the points (0, 0) and (1, 3); see Figure 6.1. Similarly $B(\overline{w}, a_2) = (1 - w)1 + (w)1 = 1$ represents a horizontal straight line and $B(\overline{w}, a_3) = (1 - w)3 + (w)0 = 3 - 3w$ is a straight line.

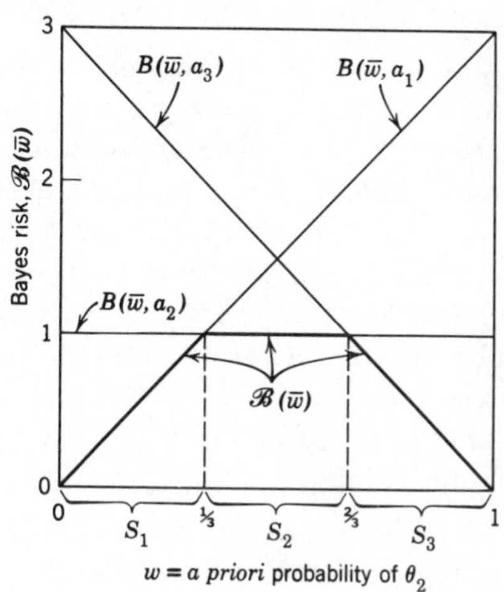

Figure 6.1. Bayes risk in the no-data version of the rain problem, Example 5.1.

The Bayes action is the a for which $B(\overline{w}, a)$ is a minimum, and this minimum value is the Bayes risk $\mathscr{B}(\overline{w})$ which is represented by the heavy lines for $0 \le w \le 1$. Thus we see that, in the rain problem, Example 5.1,

a_1 is the Bayes action if $0 \le w \le 1/3$;
a_2 is the Bayes action if $1/3 \le w \le 2/3$;
a_3 is the Bayes action if $2/3 \le w \le 1$.

Let S_1 represent the set of w's for which a_1 would be the Bayes action; S_2 the set of w's for which a_2 would be the Bayes action;

[1] These graphical ideas are extensible, with some difficulty in visualization, to the case of more than two states of nature.

and S_3 the set of w's for which a_3 would be the Bayes action. In the rain problem, $S_1 = \{w: 0 \leq w \leq 1/3\}$; $S_2 = \{w: 1/3 \leq w \leq 2/3\}$; and $S_3 = \{w: 2/3 \leq w \leq 1\}$. If $w = 1/3$, either a_1, a_2, or a randomized mixture of these can be taken, and if $w = 2/3$, either a_2, a_3, or a randomized mixture of these is appropriate.

Exercise 6.8. Draw figures showing $B(\overline{w}, a_1)$, $B(\overline{w}, a_2)$, $B(\overline{w}, a_3)$, and $\mathscr{B}(\overline{w})$ for the problem of Exercise 5.45.

Exercise 6.9. Draw figures showing $B(\overline{w}, a_1)$, $B(\overline{w}, a_2)$, and $\mathscr{B}(\overline{w})$ for the problem of Exercise 5.18.

2. CONDITIONAL PROBABILITY

In Section 1 it was proposed that the problem of finding a Bayes strategy can be reduced to that of solving a no-data problem by using the data to convert the *a priori* probabilities to *a posteriori* probabilities. These so called *a posteriori* probabilities will be seen to be special cases of *conditional probabilities* which we shall study in this section. Conditional probability arises in dealing with the following type of problem. Suppose that an experiment is performed and A and B are two sets of possible outcomes. If someone who saw the outcome reported that A had occurred, how would this information affect the probability of B? To be more specific, let us consider the following example.

Example 6.1. Mr. Sharp recently took a job as a cook in a restaurant. Owing to considerable research in the past, he noticed that customers could be classified in two ways. They were either affluent (as indicated by being well dressed) or not affluent. Also, they were either big eaters (as indicated by having double orders) or not big eaters. He also observed that in the long run the proportion of customers in these various classifications could be described approximately by Table 6.2.

TABLE 6.2

PROBABILITY TABLE FOR CUSTOMERS AMONG VARIOUS CLASSES

	B Big Eater	$\tilde{B}$ Not Big Eater	
Affluent A	0.2	0.1	0.3
Not affluent $\tilde{A}$	0.2	0.5	0.7
	0.4	0.6	1.0

In other words, 20% of the customers were affluent and big eaters, 50% were neither, 40% were big eaters, etc.

For some time Mr. Sharp used to add to his income by betting the other cook that the next customer would not be a big eater. After a few weeks the other cook finally realized that he was losing money and insisted on odds of 3 to 2. Mr. Sharp agreed that these odds of 3 to 2 against big eaters were reasonable but he insisted that he should be allowed to select which side he favored. He then bribed the waiter to wink at him before announcing the order if the customer was affluent. Mr. Sharp figured that, since in the long run 20% of the customers were affluent and big eaters and only 10% were affluent and not big eaters, if he confined his attention only to affluent customers, then 2/3 of them were big eaters. Then he would do well to bet on big eaters. On the other hand, only 2/7 of nonaffluent customers were big eaters. When the waiter winked, Mr. Sharp bet that the customer was a big eater. Otherwise he bet the other way.

Now let us analyze this example more formally. At first Mr. Sharp considers the entry of a customer into the restaurant as an experiment. The outcome of the experiment is the customer. The set A is the set of all possible customers who are affluent. B is the set of possible customers who are big eaters. We define $\{A \text{ and } B\}$ as the set of all elements which are in both A and B. In our example $\{A \text{ and } B\}$ is the set of all possible customers who are both affluent and big eaters.

Similarly $\{A \text{ and } \tilde{B}\}$ is the set of all possible customers who are affluent and not big eaters. The probability, $P\{A \text{ and } B\}$, of the customer being affluent and a big eater was estimated by Mr. Sharp to be 0.2. In general the entries of Table 6.2 are approximations of those in Table 6.3.

TABLE 6.3

PROBABILITY TABLE FOR TWO-WAY CLASSIFICATIONS

	B	$\tilde{B}$	
A	$P\{A \text{ and } B\}$	$P\{A \text{ and } \tilde{B}\}$	$P\{A\}$
$\tilde{A}$	$P\{\tilde{A} \text{ and } B\}$	$P\{\tilde{A} \text{ and } \tilde{B}\}$	$P\{\tilde{A}\}$
	$P\{B\}$	$P\{\tilde{B}\}$	

Let us call an experiment *relevant* if its outcome is an element of A. It seems reasonable to define $P\{B|A\}$, the *conditional probability of B given A,* as the long-run proportion of *relevant* experiments where the outcome is in B. Thus, in Mr. Sharp's example, out of 1,000,000 experiments about 300,000 (1,000,000$P\{A\}$) yield affluent customers and are therefore relevant. Of these relevant 300,000 experiments about 2/3 or 200,000 (1,000,000$P\{A \text{ and } B\}$) yield big eaters. Hence the proportion of relevant experiments where we obtain a big eater is

$$\frac{2}{3} = \frac{1{,}000{,}000P\{A \text{ and } B\}}{1{,}000{,}000P\{A\}} = \frac{P\{A \text{ and } B\}}{P\{A\}}.$$

Thus a reasonable definition of the *conditional probability of B given A* is

$$P\{B|A\} = \frac{P\{A \text{ and } B\}}{P\{A\}} \qquad \text{if } P\{A\} \neq 0, \tag{6.3}$$

or, equivalently, we may write

$$P\{A \text{ and } B\} = P\{A\}\,P\{B|A\}. \tag{6.4}$$

Thus, in Mr. Sharp's problem,

$$P\{B|A\} = \frac{P\{A \text{ and } B\}}{P\{A\}} = \frac{0.2}{0.3} = 0.667$$

and

$$P\{B|\tilde{A}\} = \frac{P\{\tilde{A} \text{ and } B\}}{P\{\tilde{A}\}} = \frac{0.2}{0.7} = \frac{2}{7} = 0.286$$

while $P\{B\} = 0.4$.

Now let us consider the following example.

Example 6.2. What is the probability that 2 socks will be black if they are taken in the dark from a drawer containing 5 black and 7 green socks? This problem can be solved, if we assume every pair (mixed or otherwise) equally likely to be selected, by counting the total number of pairs and the number of pairs which are both black. Since there are 66 possible pairs of socks and only 10 possible black pairs, the desired probability is $10/66 = 5/33$. Although there are various methods of counting the numbers of such pairs, another approach is the following. Let A be the set of pairs for which the first sock is black. We abbreviate $A =$ {first sock is black}. Let $B =$ {second sock is black}. It is easy to see that $P\{A\} = 5/12$ since there are 5 black out of 12 socks equally likely to be picked first. Similarly $P\{B|A\} = 4/11$ since, after a black sock is picked, there are 4 remaining black socks out of 11 equally likely remaining socks. Thus, $P\{A \text{ and } B\} = 5/12 \times 4/11 = 5/33$.

Example 6.3. There are six men labeled 1 to 6 lined up at random. What is the probability that number 1 is at an end and number 2 is next to him? $P\{1 \text{ is at an end}\} = 2/6$, $P\{2 \text{ next to } 1 | 1 \text{ is at an end}\} = 1/5$, $P\{1 \text{ is at an end and } 2 \text{ is next to } 1\} = 2/6 \times 1/5 = 1/15$.

In Appendix E_7 we extend Equation (6.4) to obtain the rather intuitively obvious Equations (6.4a), (6.4b):

$$P\{A \text{ and } B \text{ and } C\} = P\{A\}\, P\{B|A\}\, P\{C|A \text{ and } B\}, \tag{6.4a}$$

$$P\{A \text{ and } B \text{ and } C \text{ and } \cdots\} = P\{A\}\, P\{B|A\}\, P\{C|A \text{ and } B\} \cdots. \tag{6.4b}$$

Example 6.4. What is the probability that the birthdays of three people selected at random will fall on different days of the year (forgetting February 29)? The probability that the second will fall on a date different from the first is 364/365. The conditional probability that a third will *then* fall on still another date is 363/365. The desired probability is [(364)(363)]/[(365)(365)]. Similarly, the probability that 25 people selected at random all have different birthdays is

$$\frac{(364)(363)\cdots(341)}{(365)(365)\cdots(365)} = 0.43$$

which may seem surprisingly small since it follows that it is more likely than not for at least 2 people out of 25 people to have the same birthday.

Exercise 6.10. In Example 6.3 find the probability that man 1 is not at an end and man 2 is next to him. Find the probability that 1 and 2 are together (irrespective of where 1 is).

Exercise 6.11. What is the probability that 3 cards drawn in succession from a pack of 52 playing cards will be a heart, a spade, and a heart respectively?

Exercise 6.12. What is the probability that 2 cards drawn from a pack of cards will form a pair (have the same number)?

Exercise 6.13. In Mr. Sharp's betting on big eaters (Example

6.1), suppose that he had to bet one way or the other on each customer. If he bets on big eater, he wins 60 cents or loses 40 cents. If he bets against big eater, he wins 40 cents or loses 60 cents. What is his expected winning on the next customer to come into the restaurant, assuming that Mr. Sharp will connive with the waiter?

Exercise 6.14. In Table 6.3, eight parameters are listed. However, some of these are related to others. How many parameters are required to enable one to know all the others? (For example if $P\{A\}$ is used, one would not need $P\{\tilde{A}\} = 1 - P\{A\}$, etc.)

Exercise 6.15. Construct the two-way classification probability table for the sets A and B of Example 6.2. Compute $P\{A|B\}$.

Exercise 6.16. Using the information in Section 1 about probability of rain in North Phiggins, construct the two-way classification probability table for $A =$ {sunny yesterday} and $B =$ {sunny today}.

Exercise 6.17. If a poker hand has 3 spades and 2 hearts, and 2 hearts are discarded, what is the probability that 2 spades will be drawn?

Exercise 6.18. Given that the roll of 2 dice yields a sum of 6, what is the probability that the result of the first die is a 1? a 3?

Exercise 6.19. Given that the roll of 2 dice yields a sum of 6, what is the probability that either one of the dice is a 1? a 3?

3. A POSTERIORI PROBABILITY

When we give states of nature *a priori* probabilities, we act as though the state of nature were random. Then just as the observation "affluent" led to a revised (conditional) probability, of big eater, the result of an experiment will similarly lead to revised probabilities of the states of nature. These new probabilities, conditional probabilities of the state of nature, given the result of the experiment, will be called *a posteriori* probabilities.

Example 6.5. Let us assume that the contractor of Example 1.1 is informed that among his potential customers 20% have peak load 15 amp, 50% have peak load 20 amp, and 30% have peak load 30 amp. Thus we may assume that θ_1, θ_2, and θ_3 have *a priori* probabilities 0.2, 0.5, and 0.3 respectively. When interviewed, his customer responds that he uses at most 15 amp (z_3). In view of this observation, what are the *a posteriori* probabilities of θ_1, θ_2, and θ_3?

In the evaluation of the *a posteriori* probabilities, the state of nature must be treated as random and is, therefore, designated by $\boldsymbol{\theta}$. Thus the contractor regards his experiment as one of selecting a random customer in state $\boldsymbol{\theta}$ and a random reply $\mathbf{Z}$ from this customer. The possible outcomes may be labeled (θ_1, z_1), (θ_1, z_2), (θ_1, z_3), (θ_1, z_4), (θ_2, z_1), $\cdots$, (θ_3, z_4). Note that the contractor observes $\mathbf{Z}$ but not $\boldsymbol{\theta}$. We are given the *a priori* probabilities

$$w_1 = P\{\boldsymbol{\theta} = \theta_1\} = 0.2,$$
$$w_2 = P\{\boldsymbol{\theta} = \theta_2\} = 0.5,$$

and

$$w_3 = P\{\boldsymbol{\theta} = \theta_3\} = 0.3.$$

What we called the probability distribution of the data, $f(z|\theta)$, may be regarded as the conditional distribution of $\mathbf{Z}$ given the state of nature θ. Thus,

$$f(z_3|\theta_1) = P\{\mathbf{Z} = z_3|\boldsymbol{\theta} = \theta_1\} = 0,$$
$$f(z_3|\theta_2) = P\{\mathbf{Z} = z_3|\boldsymbol{\theta} = \theta_2\} = 1/2,$$

and

$$f(z_3|\theta_3) = P\{\mathbf{Z} = z_3|\boldsymbol{\theta} = \theta_3\} = 1/3.$$

We are interested in the *a posteriori* probabilities, designated by $\mathbf{w}_i$, which are the conditional probabilities of the states of nature given the data $\mathbf{Z} = z_3$. These are

$$\mathbf{w}_1 = P\{\boldsymbol{\theta} = \theta_1|\mathbf{Z} = z_3\},$$
$$\mathbf{w}_2 = P\{\boldsymbol{\theta} = \theta_2|\mathbf{Z} = z_3\},$$

and

$$\mathbf{w}_3 = P\{\boldsymbol{\theta} = \theta_3|\mathbf{Z} = z_3\}.$$

Using the definition of conditional probability, Equation (6.3), we have

$$\mathbf{w}_i = P\{\boldsymbol{\theta} = \theta_i|\mathbf{Z} = z_3\} = \frac{P\{\boldsymbol{\theta} = \theta_i \text{ and } \mathbf{Z} = z_3\}}{P\{\mathbf{Z} = z_3\}}.$$

We evaluate the numerator

$$P\{\boldsymbol{\theta} = \theta_1 \text{ and } \mathbf{Z} = z_3\} = P\{\boldsymbol{\theta} = \theta_1\}\, P\{\mathbf{Z} = z_3|\boldsymbol{\theta} = \theta_1\}$$
$$= w_1 f(z_3|\theta_1) = 0,$$
$$P\{\boldsymbol{\theta} = \theta_2 \text{ and } \mathbf{Z} = z_3\} = P\{\boldsymbol{\theta} = \theta_2\}\, P\{\mathbf{Z} = z_3|\boldsymbol{\theta} = \theta_2\}$$
$$= w_2 f(z_3|\theta_2) = 0.25,$$

and

$$P\{\theta = \theta_3 \text{ and } \mathbf{Z} = z_3\} = P\{\theta = \theta_3\}\, P\{\mathbf{Z} = z_3 | \theta = \theta_3\}$$
$$= w_3 f(z_3 | \theta_3) = 0.10.$$

To evaluate the denominator $P\{\mathbf{Z} = z_3\}$, which we designate by $f(z_3)$, we note that the possible outcomes leading to $\mathbf{Z} = z_3$ are (θ_1, z_3), (θ_2, z_3), and (θ_3, z_3). Then,

$$P\{\mathbf{Z} = z_3\} = P\{\theta = \theta_1 \text{ and } \mathbf{Z} = z_3\} + P\{\theta = \theta_2 \text{ and } \mathbf{Z} = z_3\}$$
$$+ P\{\theta = \theta_3 \text{ and } \mathbf{Z} = z_3\},$$

and

$$f(z_3) = w_1 f(z_3 | \theta_1) + w_2 f(z_3 | \theta_2) + w_3 f(z_3 | \theta_3) = 0.35$$

is a weighted average of the probabilities of observing z_3 under the various states of nature. Thus we have the *a posteriori* probabilities

$$\mathrm{w}_1 = 0,$$
$$(6.5) \qquad \mathrm{w}_2 = \frac{0.25}{0.35} = \frac{5}{7},$$
$$\mathrm{w}_3 = \frac{0.10}{0.35} = \frac{2}{7}.$$

We have used the data $\mathbf{Z} = z_3$ to convert the *a priori* probabilities (0.2, 0.5, 0.3) to the *a posteriori* probabilities (0, 5/7, 2/7).

Obviously this computation can be extended to apply for any observation **Z**. In general we have the following expression for the *a posteriori* probabilities where there are three states of nature.

$$\mathrm{w}_1 = \frac{w_1 f(\mathbf{Z} | \theta_1)}{f(\mathbf{Z})}$$
$$(6.6) \qquad \mathrm{w}_2 = \frac{w_2 f(\mathbf{Z} | \theta_2)}{f(\mathbf{Z})}$$
$$\mathrm{w}_3 = \frac{w_3 f(\mathbf{Z} | \theta_3)}{f(\mathbf{Z})}$$

where

$$(6.7) \qquad f(\mathbf{Z}) = w_1 f(\mathbf{Z} | \theta_1) + w_2 f(\mathbf{Z} | \theta_2) + w_3 f(\mathbf{Z} | \theta_3).$$

It is important to note that the *a posteriori* probabilities which represent a conditional probability distribution on the states of nature given the data, involve only (1) the *a priori* probabilities w_i and (2) the density of only the observed data for the various states of nature, i.e., $f(\mathbf{Z}|\theta_1)$, $f(\mathbf{Z}|\theta_2)$, and $f(\mathbf{Z}|\theta_3)$.

To collect the ideas which have appeared in this section, we present the evaluation of *a posteriori* probabilities for the contractor example in tabular form. See Tables 6.4 and 6.5. This evaluation is carried out for each of the possible observations.

TABLE 6.4

TABULAR FORM FOR COMPUTATION OF *A POSTERIORI* PROBABILITIES

Possible Observation	z
$P\{\mathbf{Z}=z \text{ and } \boldsymbol{\theta}=\theta_1\}$	$w_1 f(z\|\theta_1)$
$P\{\mathbf{Z}=z \text{ and } \boldsymbol{\theta}=\theta_2\}$	$w_2 f(z\|\theta_2)$
$P\{\mathbf{Z}=z \text{ and } \boldsymbol{\theta}=\theta_3\}$	$w_3 f(z\|\theta_3)$
$P\{\mathbf{Z}=z\}=f(z)$	$w_1 f(z\|\theta_1)+w_2 f(z\|\theta_2)+w_3 f(z\|\theta_3)$
$P\{\boldsymbol{\theta}=\theta_1\|\mathbf{Z}=z\}=\mathbf{w}_1$	$\frac{w_1 f(z\|\theta_1)}{f(z)}$
$P\{\boldsymbol{\theta}=\theta_2\|\mathbf{Z}=z\}=\mathbf{w}_2$	$\frac{w_2 f(z\|\theta_2)}{f(z)}$
$P\{\boldsymbol{\theta}=\theta_3\|\mathbf{Z}=z\}=\mathbf{w}_3$	$\frac{w_3 f(z\|\theta_3)}{f(z)}$

TABLE 6.5

COMPUTATION OF *A POSTERIORI* PROBABILITIES FOR EXAMPLE 6.5

$f(z|\theta)$

	z_1 z_2 z_3 z_4	w		z_1	z_2	z_3	z_4
θ_1	$\frac{1}{2}$ $\frac{1}{2}$ 0 0	0.2	$w_1 f(z\|\theta_1)$	$\frac{1}{2}(0.2)=0.10$	$\frac{1}{2}(0.2)=0.10$	$0(0.2)=0.00$	$0(0.2)=0.00$
θ_2	0 $\frac{1}{2}$ $\frac{1}{2}$ 0	0.5	$w_2 f(z\|\theta_2)$	$0(0.5)=0.00$	$\frac{1}{2}(0.5)=0.25$	$\frac{1}{2}(0.5)=0.25$	$0(0.5)=0.00$
θ_3	0 0 $\frac{1}{3}$ $\frac{2}{3}$	0.3	$w_3 f(z\|\theta_3)$	$0(0.3)=0.00$	$0(0.3)=0.00$	$\frac{1}{3}(0.3)=0.10$	$\frac{2}{3}(0.3)=0.20$
			$f(z)$	0.10	0.35	0.35	0.20
			$\mathbf{w}_1$	$\frac{0.10}{0.10}=1$	$\frac{0.10}{0.35}=\frac{2}{7}$	$\frac{0.00}{0.35}=0$	$\frac{0.00}{0.20}=0$
			$\mathbf{w}_2$	$\frac{0.00}{0.10}=0$	$\frac{0.25}{0.35}=\frac{5}{7}$	$\frac{0.25}{0.35}=\frac{5}{7}$	$\frac{0.00}{0.20}=0$
			$\mathbf{w}_3$	$\frac{0.00}{0.10}=0$	$\frac{0.00}{0.35}=0$	$\frac{0.10}{0.35}=\frac{2}{7}$	$\frac{0.20}{0.20}=1$

The extension of these equations to more than three states of nature is obvious and need not be discussed.[1]

**Exercise 6.20.* Compute the *a posteriori* probabilities in Example 1.1, when the states of nature are equally likely. Do this for each of the four possible observations. (These results are used in Exercise 6.29.)

Exercise 6.21. Do the same as in Exercise 6.20 for your problem in Exercise 1.3. (These results are used in Exercise 6.30.)

Exercise 6.22. Do the same for the example of Exercise 1.5. (These results are used in Exercise 6.31.)

Exercise 6.23. Do the same for Example 5.1. (These results are used in Exercise 6.32.)

Exercise 6.24. Find the *a posteriori* probabilities for Example 5.1 if the *a priori* probability of rain is 0.4 and "dubious" is observed. What if the *a priori* probability of rain were replaced by 0.8? by 1.0?

Exercise 6.25. There are three bags labeled 1, 2, and 3. These bags contain respectively 3 white and 3 black balls, 4 white and 2 black balls, and 1 white and 2 black balls. Our experiment consists of selecting a bag at random (each bag is equally likely to be selected) and drawing a ball at random from this bag. Find the probability of selecting bag 2, and drawing a black ball. Find the probability of drawing a black ball. Given that a black ball has been drawn, what is the conditional probability that bag 2 had been selected? (It may be helpful to label the possible outcomes $(1, B)$, $(1, W)$, $(2, B)$, $(2, W)$, $(3, B)$, $(3, W)$, and to note that the set corresponding to drawing bag 2 is $\{(2, B), (2, W)\}$.)

Exercise 6.26. A drawer contains 7 red, 9 black, and 12 green socks. What is the probability that 2 socks drawn at random (without replacement) will match?

Exercise 6.27. Given that the two socks of Exercise 6.26 match, find the probability that they are red? black? green?

Exercise 6.28. Six men each toss a coin once. If all but one have the same result, that one is called the odd man. What is the probability that a specified man will be odd? What is the probability that there will be an odd man?

[1] It has been common to label this computation as Bayes theorem which is often described as a method of computing the probability of a *cause* (state of nature) given the *effect* (data).

4. COMPUTATION OF BAYES STRATEGIES

We are now in a position to apply the schemes suggested by Mr. Solomon to compute Bayes strategies. First we shall use the data to convert our *a priori* probabilities to *a posteriori* probabilities, and then we shall solve the no-data problem corresponding to these *a posteriori* probabilities.

The most important property of this scheme is that *it does yield the Bayes strategy* (see Appendix E_8). We illustrate the computation with Table 6.6.

TABLE 6.6

TABULAR FORM FOR THE COMPUTATION OF THE BAYES STRATEGY APPLIED TO EXAMPLE 5.1 WITH *A Priori* PROBABILITY 0.4 FOR RAIN AND ONE OBSERVATION (WEATHER METER READING)

$f(z \mid \theta)$

	z_1	z_2	z_3
θ_1	0.60	0.25	0.15
θ_2	0.20	0.30	0.50

$r(\theta, a)$

	a_1	a_2	a_3
θ_1	0	1	3
θ_2	3	1	0

$\overline{w}$		z_1 (fair)	z_2 (dubious)	z_3 (foul)
0.6	$w_1 f(z \mid \theta_1)$	0.36	0.15	0.09
0.4	$w_2 f(z \mid \theta_2)$	0.08	0.12	0.20
	$f(z)$	0.44	0.27	0.29
	$\mathbf{w}_1$	$\frac{0.36}{0.44}$	$\frac{0.15}{0.27}$	$\frac{0.09}{0.29}$
	$\mathbf{w}_2$	$\frac{0.08}{0.44}$	$\frac{0.12}{0.27}$	$\frac{0.20}{0.29}$
		a_1 a_2 a_3	a_1 a_2 a_3	a_1 a_2 a_3
	$B(\overline{\mathbf{w}}, a)$	$\frac{0.24}{0.44}$ $\frac{0.44}{0.44}$ $\frac{1.08}{0.44}$	$\frac{0.36}{0.27}$ $\frac{0.27}{0.27}$ $\frac{0.45}{0.27}$	$\frac{0.60}{0.29}$ $\frac{0.29}{0.29}$ $\frac{0.27}{0.29}$
	Minimizing action A	a_1	a_2	a_3
	$\mathscr{B}(\overline{\mathbf{w}})$	$\frac{0.24}{0.44}$	$\frac{0.27}{0.27}$	$\frac{0.27}{0.29}$
	Weighted average of risks corresponding to the Bayes strategy	$(0.44)\frac{0.24}{0.44}+(0.27)\frac{0.27}{0.27}+(0.29)\frac{0.27}{0.29}=0.78$		

Note: In the bottom half, the no-data problem is solved for several sets of *a priori* probabilities. To avoid rewriting the $r(\theta, a)$ table three times, the layout differs slightly from that of Table 6.1.

4.1. Remarks

1. *Crossing your bridge as you come to it.* If there were many possible observations, there would be many possible strategies. The basic advantage to this method is that it is not necessary to consider all possible strategies. One need only consider how to react to the particular data observed. For example, if z_2 (dubious) had been observed, it would have sufficed to carry out the computation in the second column alone to determine that the Bayes strategy calls for a_2. In this sense, this method permits you to "cross your bridge as you come to it" rather than phrase your detailed strategy in advance, thereby "crossing all possible bridges you might conceivably come to."

If it is desired to know what the Bayes strategy is before carrying out the experiment, then each column must be computed. In our example, the Bayes strategy is that which selects a_1, a_2, and a_3 corresponding to observations z_1, z_2, and z_3 respectively. This is s_6, which we previously obtained by laboriously evaluating *all* pure strategies first. *Using this table, there is no need to evaluate the other strategies.* The last row is the weighted average of risks corresponding to the Bayes strategy. This could also be computed by evaluating $\mathscr{R}(s_6) = w_1 R(\theta_1, s_6) + w_2 R(\theta_2, s_6)$. In the table, we used an alternative computation which is a by-product of the main part of the table. We take a weighted average of the $\mathscr{B}(\overline{\mathbf{w}})$ corresponding to the various observations where the weights are $f(z_1)$, $f(z_2)$, $\cdots$.

2. *Crossing bridges one at a time.* Mr. Solomon is interested in the following question. On one hand, he can compute the *a posteriori* probabilities $1 - \mathbf{w}$ and $\mathbf{w}$ based on his combined data (yesterday's weather and today's rain indicator reading). On the other hand, he could use the alternative method where:

(a) On the basis of yesterday's weather and the *a priori* probabilities $(1-w, w)$, he obtains *a posteriori* probabilities $(1-\mathbf{w}^*, \mathbf{w}^*)$; and

(b) on the basis of today's rain indicator reading and the *a posteriori* probabilities $1 - \mathbf{w}^*$, $\mathbf{w}^*$, he obtains *a posteriori* probabilities $1 - \mathbf{w}^{**}$, $\mathbf{w}^{**}$.

Do these final probabilities, $(1-\mathbf{w}^{**}, \mathbf{w}^{**})$, *coincide with* $(1-\mathbf{w}, \mathbf{w})$ *which he would have obtained if he lumped his data together*? The answer is shown to be "*yes*" in Appendix E_3 and, consequently, the *Bayes strategy or rather the a posteriori probability can be computed by "digesting" each piece of information one at a time.* In this sense, Mr. Solomon may be said to be "crossing bridges one at a time."

Graphically, this means that we start out with the abscissa w in Figure 6.1. This represents the *a priori* probabilities $(1 - w, w)$. After the first datum appears, w is replaced by an *a posteriori* probability $\mathbf{w}^*$, i.e., w jumps to location $\mathbf{w}^*$ which depends on the particular datum observed. When the second datum appears, $\mathbf{w}^*$ replaced by $\mathbf{w}^{**}$. If more data are obtained, the *a posteriori* probability goes through several successive values and finally ends up at some value $\mathbf{w}$. If this final value is on S_1, the set corresponding to action a_1 ($\{w: 0 \leq w \leq 1/3\}$ in this particular example), a_1 is an appropriate action. Similarly, if this final value falls on S_2 or S_3, the appropriate actions are a_2 and a_3 respectively. The Bayes risk is $\mathscr{B}(\overline{\mathbf{w}})$ where $\overline{\mathbf{w}} = (1 - \mathbf{w}, \mathbf{w})$ is the final *a posteriori* probability. As more and more data are compiled, the *a posteriori* probability tends to go toward the point representing the state of nature.[1] Thus, if θ_1 is the state of nature and considerable data are compiled, $\mathbf{w}$ will tend to be close to 0, and the Bayes risk will be small (see Figure 6.1). If θ_2 is the state of nature, $\mathbf{w}$ will tend to be close to 1, and the Bayes risk will also be small. It is possible, but unlikely, that after many observations the final *a posteriori* probability will be far from 0 even though θ_1 is the state of nature.

3. *Choice of experiments.* If we had available the choice of carrying out one of two possible experiments, which one should we carry out? Ideally, we would like to select an experiment which would send the *a posteriori* probability $\mathbf{w}$ to 0 if θ_1 is the state of nature, and to 1 if θ_2 is. Excluding the use of crystal balls, we may find it difficult to find such an experiment. Roughly speaking, the closer an experiment comes to accomplishing that feat, the better it is. Continuing in the Bayes point of view, we could evaluate an experiment as follows. Suppose we are at w before we perform the experiment. The location of $\mathbf{w}$ after the experiment is a random variable whose distribution depends on w and the experiment. The same can be said for $\mathscr{B} = \mathscr{B}(\overline{\mathbf{w}})$. A

[1] A proof of this statement is beyond the scope of this text.

good experiment to perform is one which makes $E(\mathscr{B})$ as small as possible.

4. *Does it pay to experiment?* Suppose that there is a choice of whether or not to perform an experiment at a certain cost. For example, take the rain problem with *a priori* probability of rain equal to 0.4. Then according to Table 6.1, the Bayes risk is 1.0. If a weather meter were available, the Bayes risk would be 0.78. Then it pays to use the weather meter if the cost of so doing is less than 0.22. In general, one should experiment if the cost of so doing is less than the consequent decrease in the Bayes risk.

This notion can be extended to the case where there is available a sequence of experiments. After the ith experiment is performed leading to the *a posteriori* probability $\overline{\mathbf{w}}^*$, one compares two quantities. The first is $\mathscr{B}(\overline{\mathbf{w}}^*)$, the Bayes risk of selecting an action with no more experimentation; the second is $\mathscr{C}(\overline{\mathbf{w}}^*)$, the cost of taking one more observation and proceeding thereafter in an optimal fashion. If $\mathscr{B}(\overline{\mathbf{w}}^*) \leq \mathscr{C}(\overline{\mathbf{w}}^*)$, *stop experimentation and take the appropriate action.* Otherwise continue experimentation. Generally, this type of comparison is mathematically unfeasible, but there are simple important examples where this idea permits us to classify optimal rules for deciding when to sample.

The problem of whether to continue experimentation indicates that, in the general decision making problem, a strategy should be a rule which decides after each observation:

(a) whether or not to continue experimentation;

(b) which experiment to take next if experimentation is continued; and

(c) which action to take if experimentation is stopped.

**Exercise 6.29.* Apply the results of Exercise 6.20 to compute the Bayes strategy.

Exercise 6.30. Apply the results of Exercise 6.21 to compute the Bayes strategy.

Exercise 6.31. Apply the results of Exercise 6.22 to compute the Bayes strategy.

Exercise 6.32. Apply the results of Exercise 6.23 to compute the Bayes strategy.

Exercise 6.33. Communities such as East Phiggins usually intensely like or dislike "rock and roll" music. On the basis of experience with other such communities, it is assumed that the proportion of the population liking the music is either 60% (θ_1) or 20% (θ_2). A booking agency which would like to determine whether or not to book a rock and roll act in East Phiggins (actions a_1 and a_2) has regrets given by

$r(\theta, a)$	a_1	a_2
θ_1	0	4
θ_2	10	0

The probability distribution corresponding to the observation of determining whether a random East Phiggindian likes rock and roll is of course given by

$f(z \mid \theta)$	z_1 (like)	z_2 (dislike)
θ_1	0.60	0.40
θ_2	0.20	0.80

Use a table of random numbers to simulate 10 random observations under the assumption, θ_1 is the state of nature. Start with *a priori* probabilities (1/2, 1/2). After the first observation, these are modified. After each observation, the previous probabilities are modified to new ones. Compute the first two successive values of these *a posteriori* probabilities. To what action would the Bayes strategy based on the two observations lead? Is this the correct action when we recall that $\theta = \theta_1$? Forgetting that we actually know $\theta = \theta_1$, compute the Bayes risk after the two observations, and determine how much utility was gained from the two observations. (Record the ten observations for use in Exercise 6.37.)

Exercise 6.34. Determine the Bayes strategies for Example 5.1 for *a priori* probability of rain, $w = 0.1, 0.3, 0.5, 0.7, 0.9$, and locate these strategies on a copy of Figure 5.2.

5. INDEPENDENCE

In Section 2 we discussed conditional probability. There we were able to use the information that the outcome was an element of A to re-evaluate the probability of B. Suppose now that this information does not affect the probability of B. That is, suppose that

$$P\{B|A\} = P\{B\}. \tag{6.8}$$

Then it seems reasonable to call the set B independent of the set A. But since

$$P\{A \text{ and } B\} = P\{A\}\, P\{B|A\}$$

this equation can be written

$$P\{A \text{ and } B\} = P\{A\}\, P\{B\} \tag{6.8a}$$

which is a more symmetric form. *We define A and B to be independent if Equation* (6.8a) *applies.*

It is customary (and sometimes hazardous) in statistical practice to hypothesize that two sets are independent if one can think of no reason why the fact that the outcome was an element in one set should reflect a tendency for the outcome to be in or not to be in the other set. For example, "heads" on the toss of a penny should not have any effect on the outcome of the toss of a nickel.

Thus, if $P\{A\} = 2/3$ and $P\{B\} = 3/8$, where $A = \{\text{nickel falls heads}\}$ and $B = \{\text{penny falls heads}\}$, then

$$P\{A \text{ and } B\} = P\{(H, H)\} = (2/3) \times (3/8) = 6/24.$$

See Table 6.7.

TABLE 6.7

PROBABILITIES FOR THE TOSS OF TWO BENT COINS

	Penny Falls Heads B	Penny Falls Tails $\tilde{B}$	
Nickel falls			
Heads A	6/24	10/24	2/3
Tails $\tilde{A}$	3/24	5/24	1/3
	3/8	5/8	1.0

The other probabilities of Table 6.7 are now readily computable:

$$\begin{aligned} P\{A \text{ and } \tilde{B}\} = P\{(H, T)\} &= P\{A\} - P\{A \text{ and } B\} \\ &= 2/3 - 6/24 = 10/24 \\ P\{\tilde{A} \text{ and } B\} = P\{(T, H)\} &= P\{B\} - P\{A \text{ and } B\} \\ &= 3/8 - 6/24 = 3/24 \end{aligned}$$

$$P\{\tilde{A}\} = 1 - P\{A\} = 1 - 2/3 = 1/3$$
$$P\{\tilde{B}\} = 1 - P\{B\} = 1 - 3/8 = 5/8$$
$$P\{\tilde{A} \text{ and } \tilde{B}\} = P\{\tilde{A}\} - P\{\tilde{A} \text{ and } B\} = 1/3 - 3/24 = 5/24.$$

Notice that

$$P\{A \text{ and } \tilde{B}\} = P\{A\}\, P\{\tilde{B}\}$$
$$P\{\tilde{A} \text{ and } B\} = P\{\tilde{A}\}\, P\{B\}$$

and

$$P\{\tilde{A} \text{ and } \tilde{B}\} = P\{\tilde{A}\}\, P\{\tilde{B}\}.$$

In other words, since A and B are independent, so are A and $\tilde{B}$, $\tilde{A}$ and B, and $\tilde{A}$ and $\tilde{B}$.

***Exercise 6.35.* Suppose a coin has probability 0.6 of falling heads. If it is tossed four times, what is the probability of obtaining H, H, T, H in that order?

Exercise 6.36. If a die is rolled twice, what is the probability that the first roll is a 4 and the second a 2 or 3?

Exercise 6.37. Generalize Exercise 6.35 to compute $f(\mathbf{Z} \mid \theta)$ for Exercise 6.33 where $\mathbf{Z}$ is the data consisting of the 10 observations which were recorded. Compute the *a posteriori* probability based on $\mathbf{Z}$. What action is called for? What is the Bayes risk corresponding to the *a posteriori* probability?

Exercise 6.38. Four men play the game "odd man pays for dinner" (see Exercise 6.28). If there is no odd man on the first set of tosses, the procedure is repeated until there is an odd man. Compute the probability distribution of **N**, the number of times the procedure must be repeated.

°*Exercise 6.39.* Compute $E(\mathbf{N})$ and $\sigma^2_{\mathbf{N}}$ in Exercise 6.38. Generalize to k men.

Exercise 6.40. The probability of a bomb hitting a target is 0.2. Assuming that all bombs are independently aimed, compute the probability that:

(a) If 3 bombs are dropped, all three hit the target;
(b) if 2 bombs are dropped, neither hits the target;
(c) if 5 bombs are dropped, all fail to hit the target;
(d) if 5 bombs are dropped, at least one hits the target.

Exercise 6.41. Messrs. A, B, and C are duck hunters whose probabilities of hitting a flying duck are 2/3, 3/4, and 1/4 respectively. A duck flies over and they fire simultaneously. What is the probability that the duck falls down hit?

°*Exercise 6.42.* The probability that a coin falls heads is p. What is the probability of exactly r heads out of n tosses? (*Hint*: Consider the number of ways r heads and $n - r$ tails can be arranged in order.)

°*Exercise 6.43.* In Exercise 6.42, suppose $p \to 0$, $n \to \infty$ and $np \to \lambda$. Compute the limit of the probability of exactly r heads.

Independence is especially important when it is applied to random variables. Two random variables $\mathbf{X}$ and $\mathbf{Y}$ are said to be independent if every set involving only restrictions on $\mathbf{X}$ is independent of every set involving only restrictions on $\mathbf{Y}$. Thus the number of heads in the coin-tossing experiment should be independent of the number rolled with a pair of dice. More often than not, statistical theory treats examples where successive observations in experiments are *independent* random variables. If $\mathbf{X}$ and $\mathbf{Y}$ are independent random variables, then

$$P\{\mathbf{X} = x_i \text{ and } \mathbf{Y} = y_j\} = P\{\mathbf{X} = x_i\}\, P\{\mathbf{Y} = y_j\}.$$

The function which gives probabilities of sets involving restrictions on $\mathbf{X}$ and $\mathbf{Y}$ together is called the *joint probability* distribution of $\mathbf{X}$ and $\mathbf{Y}$. Just as in the one-variable case, the joint probability distribution can be summarized by the cumulative distribution function defined by

$$F(a, b) = P\{\mathbf{X} \le a \text{ and } \mathbf{Y} \le b\}.$$

In the discrete case, there is a discrete density f such that

$$f(x, y) = P\{\mathbf{X} = x \text{ and } \mathbf{Y} = y\}. \tag{6.9}$$

In the continuous case, there is a density, f, such that volume under the surface $z = f(x, y)$ corresponds to probability. There are mixed cases where $\mathbf{X}$ is discrete and $\mathbf{Y}$ is continuous. These may be treated in a similar fashion, but we shall not discuss those cases here. If $\mathbf{X}$ has density g and $\mathbf{Y}$ has density h, then, in both the discrete and continuous cases, *independence of* $\mathbf{X}$ *and* $\mathbf{Y}$ *is equivalent to*

$$f(x, y) = g(x)\, h(y). \tag{6.10}$$

The notion of independent random variables is extensible to more than two random variables in an obvious way.

In the past, we have spoken of independent observations $X_1, X_2, \cdots$ on a random variable X. By this expression we meant that $X_1, X_2, \cdots$, etc., were independent random variables, each with the same probability distribution function as X. Whenever we speak of independent repetitions of an experiment, we mean that the outcomes are independent.

The following properties of the sample mean

$$\bar{X} = \frac{X_1 + X_2 + \cdots + X_n}{n}$$

of n independent observations on X play a fundamental role in evaluating the action probabilities of various procedures used in statistics.

(6.11) $$E(\bar{X}) = \mu_{\bar{X}} = \mu_X$$

(6.12) $$\sigma^2_{\bar{X}} = \sigma^2_X/n$$

(6.12a) $$\sigma_{\bar{X}} = \sigma_X/\sqrt{n}$$

(6.13) The probability distribution of $\bar{X}$ is approximately *normal* with mean μ_X and variance σ^2_X/n.

Equations 6.11 and 6.12 are established in Appendix E_3. Equation (6.11) states, for example, that for many repetitions the long-run average of $(X_1 + X_2 + X_3)/3$ is the same as the long-run average of X. Equation (6.12) states that, as the sample size n increases, the tendency of $\bar{X}$ to vary decreases like $1/\sqrt{n}$. Thus we measure the order of magnitude by which $\bar{X}$ tends to deviate from μ_X. On the other hand Sentence (6.13), which we shall call the *approximate normality* or *central limit* theorem, presents much more information. It furnishes an approximation to the probability distribution of $\bar{X}$. The amazing conclusion that, no matter what the probability distribution of X, the sample mean $\bar{X}$ tends to have an approximately *normal distribution* is one of the most elegant, powerful, and important results in the theory of probability.[1] A proof of this result is somewhat beyond the scope of this course.

[1] Strictly speaking, this result applies only if X has *finite* variance σ^2_X. All bounded random variables have finite variance.

Equations 6.11 and 6.12 can be regarded as applications of the basic equations

(6.14) $$\mu_{X+Y} = \mu_X + \mu_Y$$

(6.15) $$\sigma^2_{X+Y} = \sigma^2_X + \sigma^2_Y$$

when **X** and **Y** are independent. (See Appendix E_{10}.)

Exercise 6.44. Let **X** be a random digit. What are μ_X and σ^2_X? If $\overline{X}$ is the average of 10 random digits $X_1, X_2, \cdots, X_{10}$, what are $\mu_{\overline{X}}$ and $\sigma^2_{\overline{X}}$? If **Y** is the average of 50 independent $\overline{X}$'s (each $\overline{X}$ based on 10 digits), what are μ_Y and σ^2_Y? If **Z** is the average of 500 random digits, what are μ_Z and σ^2_Z? Check these results empirically. To do so, use a table of random numbers to compute 50 independent $\overline{X}$'s (each $\overline{X}$ based on 10 digits). Draw a histogram. Does this histogram resemble a normal density? Compute the sample mean and variance. These are $Y = \overline{(\overline{X})}$ and $s^2_{\overline{X}}$ respectively. Are they close to $\mu_{\overline{X}}$ and $\sigma^2_{\overline{X}}$? Does the deviation of **Y** from $\mu_{\overline{X}}$ seem reasonable in view of the value of σ^2_Y?

In many problems, one is interested in the probability of an event. The observations consist of "successes" or "failures". For example, a success might be identified with "heads" in a coin-tossing problem or with "cure of a patient" in a drug-testing problem. In such problems it is an advantage to replace the words "success," "heads," or "cure" with a numerical-valued random variable because so much of probability theory is devoted to such random variables. The usual convention is to define the random variable **X** such that

$$X = 1 \qquad \text{denotes a success}$$
$$X = 0 \qquad \text{denotes a failure.}$$

Suppose that $P\{X = 1\} = p =$ probability of success. Then

$$E(X) = (p)1 + (1 - p)0 = p$$
$$E(X^2) = (p)1^2 + (1 - p)0^2 = p$$

and

$$\sigma^2_X = E(X^2) - [E(X)]^2 = p - p^2 = p(1 - p).$$

The random variable **X** is called a Bernoulli or dichotomous random variable because there are only two possible values that it can take.

If independent successive trials give outcomes $\mathbf{X}_1, \mathbf{X}_2, \cdots, \mathbf{X}_n$, then the total number of successes is given by

$$\mathbf{X}_1 + \mathbf{X}_2 + \cdots + \mathbf{X}_n$$

and the observed proportion of successes is

$$\hat{\mathbf{p}} = (\mathbf{X}_1 + \mathbf{X}_2 + \cdots + \mathbf{X}_n)/n$$

Observe that $\hat{\mathbf{p}}$ is merely a special symbol for $\overline{\mathbf{X}}$, the average of n independent observations on the dichotomous random variable $\mathbf{X}$.

**Exercise 6.45.* A coin has probability 0.6 of falling heads. If $\hat{\mathbf{p}}$ is the proportion of heads in 100 tosses, use Equations (6.11) through (6.13) to approximate $P\{\hat{\mathbf{p}} \leq 0.5\}$.

Exercise 6.46. Mr. Sharp receives \$5 ($\mathbf{X} = 5$) if a random digit is 0 or 5. He loses \$1 ($\mathbf{X} = -1$) otherwise.

(a) Show that $E(\mathbf{X}) = 0.2$ and $\sigma_{\mathbf{X}} = 2.4$.

(b) Compute the probability that after 144 successive plays of this game he will have lost money.

(*Hint*: Apply the approximate normality theorem and observe that his total winnings are positive when his average winnings are positive.)

Exercise 6.47. In Mr. Sharp's problem of Example 6.1, the waiter's connivance would have done him no good if the sets "big eater" and "affluent" had been independent. A *measure* of the "dependence" would be given by Mr. Sharp's expected winnings in Exercise 6.13. Evaluate this measure in the general case, in terms of $P\{A\}$, $P\{B\}$, $P\{A \text{ and } B\}$, etc. Here we assume that, if he calls big eater, he wins $P\{\tilde{B}\}$ or loses $P\{B\}$, and if he calls against big eater he wins $P\{B\}$ or loses $P\{\tilde{B}\}$.

Exercise 6.48. If $\mathbf{X}$ and $\mathbf{Y}$ are independent normally distributed random variables, then $a\mathbf{X} + b\mathbf{Y} + c$ is normally distributed. If further $\mathbf{X}$ and $\mathbf{Y}$ have means $\mu_{\mathbf{X}}$ and $\mu_{\mathbf{Y}}$ and variances $\sigma^2_{\mathbf{X}}$ and $\sigma^2_{\mathbf{Y}}$, what can you say about $\mathbf{X} + \mathbf{Y}$, $\mathbf{X} - \mathbf{Y}$, and $2\mathbf{X} - 3\mathbf{Y}$?

Exercise 6.49. The pulling strength of a randomly selected West Phiggindian has mean 500 lb and standard deviation 100 lb. Compute the probability that a team of 49 men will outpull a team of 50 men if the teams are selected at random.

Exercise 6.50. The Jiffy accounting firm speeds up its work by rounding off all items to the nearest dime. If it is called upon to add up 10,000 items, what is the probability distribution of its error due to rounding? (Assume that the last figure in each nonrounded item is a random digit and that each number ending in 5 is rounded upward; e.g., 65 is rounded to 70.)

6. SUMMARY

The conditional probability of B given A is defined by

$$P\{B|A\} = P\{A \text{ and } B\}/P\{A\}$$

or

$$P\{A \text{ and } B\} = P\{A\}\, P\{B|A\}.$$

The sets A and B are said to be independent if

$$P\{A \text{ and } B\} = P\{A\}\, P\{B\}.$$

The random variables **X** and **Y** are said to be independent if each set involving restrictions on **X** alone is independent of each set involving restrictions on **Y** alone. If **X** and **Y** have probability density functions, then **X** and **Y** are independent if and only if

$$f(x, y) = g(x)\, h(y)$$

where f, g, and h are the densities of **X** and **Y** together, of **X**, and of **Y** respectively. The approximate normality theorem states that, if $\mathbf{X}_1, \mathbf{X}_2, \cdots, \mathbf{X}_n$ are independent observations on a random variable **X** with mean μ and variance σ^2, then *the distribution of*

$$\overline{\mathbf{X}} = \frac{\mathbf{X}_1 + \mathbf{X}_2 + \cdots + \mathbf{X}_n}{n}$$

is approximately normal with mean μ and variance σ^2/n. An especially important example (see Exercise 6.45) arises when we consider the proportion $\hat{\mathbf{p}}$ of successes in n independent trials of an event with probability p. Then $\hat{\mathbf{p}}$ has mean p and variance $p(1 - p)/n$. If n is large, $\hat{\mathbf{p}}$ is approximately normally distributed.

To compute Bayes strategies corresponding to given *a priori* probabilities, one may proceed as follows. Suppose there are k possible states $\theta_1, \theta_2, \cdots, \theta_k$ with *a priori* probability $\bar{w} = (w_1, w_2, \cdots, w_k)$. Suppose that the distribution of **Z** when θ_i is the state of nature is given by $f(z|\theta_i)$. Then the over-all probability distribution of **Z** taking $\bar{w}$ into account is given by

$$f(z) = w_1 f(z|\theta_1) + w_2 f(z|\theta_2) + \cdots + w_k f(z|\theta_k).$$

The a posteriori probabilities are given by $\bar{\mathbf{w}} = (\mathbf{w}_1, \mathbf{w}_2, \cdots, \mathbf{w}_k)$ *where*

$$\mathbf{w}_i = P\{\boldsymbol{\theta} = \theta_i | \mathbf{Z}\} = \frac{w_i f(\mathbf{Z}|\theta_i)}{f(\mathbf{Z})}.$$

If another observation is taken, a new *a posteriori* probability is obtained by replacing $\bar{\mathbf{w}}$ by $\bar{\mathbf{w}}^*$. In this way each observation leads to a new set of *a posteriori* probabilities. *If the a posteriori probabilities after the last observation are given by* $\bar{\mathbf{w}} = (\mathbf{w}_1, \mathbf{w}_2, \cdots, \mathbf{w}_k)$, *the action to be taken is the one for which the weighted average*

$$B(\bar{\mathbf{w}}, a) = \mathbf{w}_1 r(\theta_1, a) + \mathbf{w}_2 r(\theta_2, a) + \cdots + \mathbf{w}_k r(\theta_k, a)$$

is a minimum. This minimum value is the Bayes risk $\mathscr{B}(\bar{\mathbf{w}})$.

Effective experiments are those which tend to make the *a posteriori* probability tend to get close to $(1, 0, 0, \cdots, 0)$ if θ_1 is the state of nature and to $(0, 1, 0, \cdots, 0)$ if θ_2 is the state of nature, etc. By comparing the Bayes expected risk after the last experiment is carried out with the combined cost and the Bayes expected risk after more experiments are carried out, one can — at least in theory — determine whether it pays to continue experimentation.

It is worth noting that the *a priori* and *a posteriori* probabilities represent probability distributions on the set of possible states of nature. Furthermore, the above expression for $B(\bar{\mathbf{w}}, a)$ is essentially the expectation of $r(\boldsymbol{\theta}, a)$ where $\boldsymbol{\theta}$ is treated as a random variable with distribution determined by $\bar{\mathbf{w}}$. Keeping these notions in mind, the discussion in this chapter can be extended to problems where there are infinitely many possible values of θ.

In our general description of decision problems, a strategy was a detailed plan of how to react to all possible information which becomes available. In practical problems where each observation may take on one of many possible values and there are many observations taken, the detailed listing of one strategy may be extremely complex. Furthermore, the number of possible strategies may be immense. The beauty of Mr. Solomon's plan which yields Bayes solutions by using *a posteriori* probabilities to "digest" the data is that one need not consider in advance all possible observations. In other words, Mr. Solomon's plan is analogous to crossing only those bridges that you come to, compared to the alternative approach where a strategy must consider all possible bridges that you might conceivably come to. The computational advantages derived from Mr. Solomon's plan are great.

7. REVIEW AT THE END OF CHAPTER 6

At this point we have finished the basic ideas underlying statistical theory. Before we proceed to see how they are applied in more standard statistical problems, we shall briefly review the content of the first six chapters.

CHAPTER 1. This chapter consists essentially of the statement of the problem of decision making in the face of uncertainty. This statement is made via the contractor example.

CHAPTER 2. This chapter is a digression in the presentation of the basic statistical ideas. However, in introducing some standard methods of treating data, it prepares for the question of what are important properties of probability distributions.

CHAPTER 3. Here the basic notion of probability as a long-run proportion is introduced. The idea of probability was implicitly used in the contractor example.

CHAPTER 4. In the contractor example we introduced expected losses. Here we see that under certain mild conditions there is a utility, so that one should measure random outcomes by expected losses or gains in utility. Using this fact, we gain insight into when a statistician is interested in the mean, variance, median, or other parameters of a probability distribution.

CHAPTER 5.

(A) There are six basic factors that enter into a typical statistical problem of decision making. These are:

1. The possible actions a.
2. The possible states of nature θ.
3. The losses (consequences of acts) $l(\theta, a)$.
4. The experiment resulting in data $\mathbf{Z}$ with probability distribution $f(z \mid \theta)$.
5. The strategies s of how to react to information.
6. The expected losses or the risks (consequences of strategies) $L(\theta, s)$, $R(\theta, s)$.

(B) Various criteria have been proposed for selecting one of the available strategies:

1. Admissible strategies are those which are not dominated by other strategies.

2. Bayes strategies are those which minimize weighted averages of the expected losses (or of the risks).

3. Minimax expected loss — a conservative approach.

4. Minimax risk — a modification of the conservative approach.

(C) From the graphical point of view, certain results are almost obvious.

1. The set of all randomized strategies is represented by a convex set.

2. All admissible strategies are Bayes strategies for some *a priori* probabilities.

3 (a). The Bayes strategies corresponding to *positive a priori* probabilities are admissible.

3 (b). For any set of *a priori* probabilities there is at least one pure (nonrandomized) Bayes strategy.

4. It should not be surprising if the minimax expected loss strategy has equal expected losses for all states of nature.

CHAPTER 6. The fact that all admissible strategies are Bayes strategies for some *a priori* probabilities could be used to characterize the class of all admissible strategies if it were easy to compute Bayes strategies. The method suggested in Chapter 5, in principle, involves listing all possible pure (nonrandomized) strategies, evaluating the expected losses, and selecting the pure strategies which minimize the weighted average of these losses. This procedure may be completely impractical in problems where there is a large number of possible outcomes of the experiment, for then there are very many pure (nonrandomized) strategies.

The following is an alternative method of computing the Bayes strategies. As a piece of data is observed, "digest" it by replacing the *a priori* probabilities for the states of nature by the easily computed *a posteriori* probabilities. After each piece of data is observed, replace the preceding *a posteriori* probabilities by new ones. Finally, solve the no-data problem with the original *a priori* probabilities replaced by the final *a posteriori* probabilities.

Here there is no need to list all possible strategies. We can even evaluate the worth of the experiment before performing it and judge whether it is worth the cost, if any, to perform it.

Exercise 6.51. Suppose that in the contractor problem (Example 1.1), the *a priori* probabilities of θ_1, θ_2, and θ_3 are 0.3, 0.3, and 0.4 respectively.

(a) Compute the Bayes strategy.

(b) How much is the experiment worth?

(c) Compute the action probabilities and risks $R(\theta, s)$ for the Bayes strategy.

Exercise 6.52. In Exercise 1.5, the loss function given happened to be that of a settler who wanted to bring along a cello. He was unaware that his wife had resolved to bring along a sewing machine if they did not take an air conditioner. Her losses were given by

	a_1	a_2
θ_1	4	10
θ_2	4	4
θ_3	4	1

These states of nature have *a priori* probability 0.2, 0.3, and 0.5 respectively.

(a) Solve the no-data problems for the settler and for his wife.

(b) Mr. Clark comes back with the observation z_1. Find each of their Bayes actions.

(c) A couple of weeks later, Mr. Lewis arrives and announces that he had passed through East Phiggins and had observed z_3. Find each of the Bayes actions based on both observations. (Assume that the two observations are independent.)

SUGGESTED READINGS

See the readings suggested after Chapter 5.

CHAPTER 7

Introduction to Classical Statistics

1. INTRODUCTION

In the first six chapters, our examples were artificial or oversimplified so that the main ideas in decision making could be illustrated without being obscured by the complexities of real life. In common statistical practice, it is not necessary to simplify and idealize problems as much as we did. Nevertheless, there is still a need for considerable idealization to convert the problems to ones in which we can do the computing necessary to propose and evaluate reasonable strategies.

In this chapter we shall illustrate how the decision making ideas can be applied to several problems typical of the sort that occur in actual statistical practice. In our illustrations we shall forego the use of examples where the data consist of only one of a few possible observations and where there are only a few possible states of nature. Most problems in statistical practice are posed as problems in *testing hypotheses*, *estimation*, or *confidence intervals*. In this chapter we shall treat one or two simple examples of each type of problem.

2. AN EXAMPLE OF HYPOTHESIS TESTING

Example 7.1. Mr. Good, manufacturer of parachute cord, buys natural fibers from a supplier, Mr. Lacey, and fabricates them into cord. The strength of the cord is of critical importance to its salability. Consequently, the strength of the fibers from which it is made is also of critical importance. From time to time, he receives a batch of fibers from Mr. Lacey and, before beginning manufacture, he finds it necessary to ascertain whether the incoming fibers are sufficiently strong. If they are not, his own finished product will not be salable as parachute cord and must be marketed for some inferior use, such as binder twine, at a substantial economic loss.

There are two actions available to Mr. Good. These are: a_1—accept the batch and use the fibers in the manufacture of cord, and a_2—reject the batch and return the fibers to Mr. Lacey for a refund.

Mr. Good has been buying fibers from Mr. Lacey for many years and has learned from experience that in any batch, the fiber strength (force in grams required to tear a fiber) **X** of a fiber taken at random from the batch is approximately normally distributed with standard deviation $\sigma_X = 8$ grams. The mean fiber strength θ varies from batch to batch. The value of θ represents the unknown state of nature and determines the regrets incurred in accepting or rejecting the batch.

On the basis of his knowledge of the business, Mr. Good has approximated his regrets in Figure 7.1. This is based on the fact that, as θ decreases from 32, the number of parachute manufacturers who will buy his cord goes down considerably. When θ is 32, they will all be happy to buy his cord, and if $\theta > 32$, he will not only be able to sell all his cord, but his reputation will improve. When θ is 31, he will have so much difficulty selling his parachute cord that

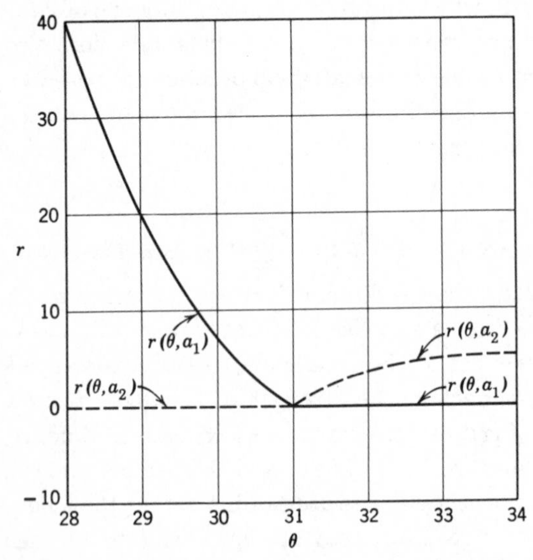

Figure 7.1. Regrets in Mr. Good's problem, Example 7.1.

he will do no better than by renting his factory facilities until another batch of fibers arrives.

To obtain relevant information about the unknown value of θ, it has always been Mr. Good's policy to perform the following experiment. Select 64 fibers at random from the batch and individually test them in the laboratory to determine their breaking strength. For his last batch, the laboratory reported the following list of strengths which are the observed values of $\mathbf{X}_1, \mathbf{X}_2, \cdots, \mathbf{X}_{64}$.

27.70	35.39	27.78	48.32	38.68	33.84	28.19	33.26	30.31	45.36
23.86	23.34	36.67	32.71	22.99	27.08	27.17	37.35	28.72	27.25
29.07	24.96	35.99	37.89	19.28	32.50	22.24	34.25	34.04	39.10
30.45	41.30	27.22	42.94	23.38	42.37	36.20	34.26	41.46	34.97
33.76	29.63	31.98	24.23	27.62	34.38	34.09	5.47	43.59	28.84
28.70	32.89	41.16	34.09	32.10	29.31	25.27	38.29	24.37	37.41
18.19	32.86	41.24	36.45						

A strategy s is a "recipe" which tells us what action to take for every possible sample $\mathbf{Z} = (\mathbf{X}_1, \mathbf{X}_2, \cdots, \mathbf{X}_{64})$. We must consider what are the reasonable strategies and select one of these on the basis of the risks corresponding to it. Once our strategy is selected, we shall apply it to see how we should react to the above data.

One type of strategy which suggests itself is to use $\overline{\mathbf{X}}$ as an indication of θ and to accept the batch if $\overline{\mathbf{X}}$ is "large" and to reject the batch if $\overline{\mathbf{X}}$ is "small." Three strategies of this type are the following:

$s_{31.0}$: Take action a_1 (accept batch) if $\overline{\mathbf{X}} \geq 31.0$ and action a_2 otherwise.
$s_{31.5}$: Take action a_1 (accept batch) if $\overline{\mathbf{X}} \geq 31.5$ and action a_2 otherwise.
$s_{32.0}$: Take action a_1 (accept batch) if $\overline{\mathbf{X}} \geq 32.0$ and action a_2 otherwise.

In general we use the designation:

s_c: Take action a_1 (accept batch) if $\overline{\mathbf{X}} \geq c$ and action a_2 otherwise.

Even though there are infinitely many strategies of the above type, there are many other kinds of strategies. For example, the contrary man may reject the batch if $\overline{\mathbf{X}} \geq 31.0$ and accept otherwise. The absent-minded man may accept the batch no matter what the data happen to be. The lazy man may accept the batch if $(\mathbf{X}_1 + \mathbf{X}_2)/2 \geq 31.0$ and reject otherwise. The involved man may accept the batch if $2\mathbf{X}_1 - \mathbf{X}_2 + 2\mathbf{X}_3 - \mathbf{X}_4 + \cdots + 2\mathbf{X}_{63} - \mathbf{X}_{64} \geq 1000$

and reject otherwise. However, we shall note in Chapter 9 that, in this problem, the class of all Bayes strategies (which contains all admissible strategies) is the class of strategies s_c, and, therefore, we need not consider these other varieties of strategies.[1] In Table 7.1 we illustrate the computation of the risks for $s_{31.5}$. We illustrate the computation of the row corresponding to $\theta = 30.5$. Referring to Figure 7.1, we have $r(\theta, a_1) = 3.55$ and $r(\theta, a_2) = 0$.

TABLE 7.1

COMPUTATION OF THE ACTION PROBABILITIES AND RISKS FOR STRATEGY $s_{31.5}$ IN EXAMPLE 7.1

State of Nature	Regrets		Action Probabilities		Risk
θ	a_1 $r(\theta, a_1)$	a_2 $r(\theta, a_2)$	a_1 $P\{\overline{X} \geq 31.5 \mid \theta\}$	a_2 $P\{\overline{X} < 31.5 \mid \theta\}$	$R(\theta, s_{31.5})$
28.00	40.00	0.00	0.000233	0.999767	0.00932
28.50	27.50	0.00	0.00135	0.99865	0.0371
29.00	19.40	0.00	0.00621	0.99379	0.120
29.25	16.20	0.00	0.0122	0.9878	0.198
29.50	13.15	0.00	0.0228	0.9772	0.300
29.75	10.50	0.00	0.0401	0.9599	0.421
30.00	8.00	0.00	0.0668	0.9332	0.534
30.25	5.60	0.00	0.1056	0.8944	0.591
30.50	3.55	0.00	0.1587	0.8413	0.563
30.75	1.60	0.00	0.2266	0.7734	0.363
31.00	0.00	0.00	0.3085	0.6915	0.000
31.25	0.00	1.20	0.4013	0.5987	0.718
31.50	0.00	2.20	0.5000	0.5000	1.100
31.75	0.00	2.95	0.5987	0.4013	1.184
32.00	0.00	3.55	0.6915	0.3085	1.095
32.25	0.00	4.00	0.7734	0.2266	0.906
32.50	0.00	4.40	0.8413	0.1587	0.698
33.00	0.00	5.00	0.9332	0.0668	0.334
33.50	0.00	5.35	0.9772	0.0228	0.122
34.00	0.00	5.50	0.99379	0.00621	0.0342

$$s_{31.5}\begin{cases}\text{Accept the batch } (a_1) \text{ if } \overline{X} \geq 31.5\\ \text{Reject the batch } (a_2) \text{ if } \overline{X} < 31.5\end{cases}$$

[1] Strictly speaking, the absent-minded man's strategy is also admissible. Our statement is correct only when we include this strategy (which we may label $s_{-\infty}$) and s_∞ in the class of strategies s_c.

To compute the action probabilities we note that $\bar{X}$ is normally distributed with mean θ and standard deviation $\sigma_X/\sqrt{64} = 8/8 = 1$ gram.[1] Then, applying $s_{31.5}$, we accept if $\bar{X} \geq 31.5$ and the probability of taking action a_1 is given by:

$$P\{\bar{X} \geq 31.5 \mid \theta = 30.5\} = 0.1587,$$

similarly

$$P\{\bar{X} < 31.5 \mid \theta = 30.5\} = 0.8413.$$

The risk is then obtained by taking the weighted average of the regrets:

$$R(30.5, s_{31.5}) = (0.1587)(3.55) + (0.8413)0 = 0.563.$$

In Figure 7.2 we present curves representing the *risk functions* for $s_{31.5}$ and for several other strategies. The risk function for $s_{31.5}$ seems to have a larger hump for $\theta > 31$ than for $\theta < 31$. If we wish to decrease the size of this hump, we must decrease the probability of taking the wrong action (a_2) when $\theta > 31$. To decrease the probability of taking action a_2, we must decrease the subscript of the strategy s_c. Thus s_{31} has a smaller hump for $\theta > 31$ than does $s_{31.5}$. Looking at Figure 7.2, we see that the peaks of the two humps can be equalized for some strategy "close" to $s_{31.25}$ yielding the minimax strategy. The minimax risk is close to 0.91.

There are several points which were brought out by the above discussion. Since $\bar{X}$ tends to be reasonably close to θ, and $\theta = 31.0$ is the break-even point where it does not matter which action is taken, $s_{31.0}$ is a reasonable strategy. Since the regrets $r(\theta, a_1)$ are larger (for $\theta < 31.0$) than $r(\theta, a_2)$ (for $\theta > 31.0$), it is advisable to decrease the probability of taking action a_1, which can be done by increasing the subscript of s_c.

Notice that the change called for in the strategy is quite small. As is indicated in Exercise 7.1, the change is smaller when the

[1] The "normal approximation" or "central limit" theorem states that $\bar{X}$ is approximately normally distributed. In this case where we assumed that the original observations are independent and normally distributed, we could state that $\bar{X}$ is actually, and not just approximately, normally distributed. As a matter of fact, the original observations in this example can be only *approximately* normally distributed, since they must be positive, and a normal random variable can take negative values.

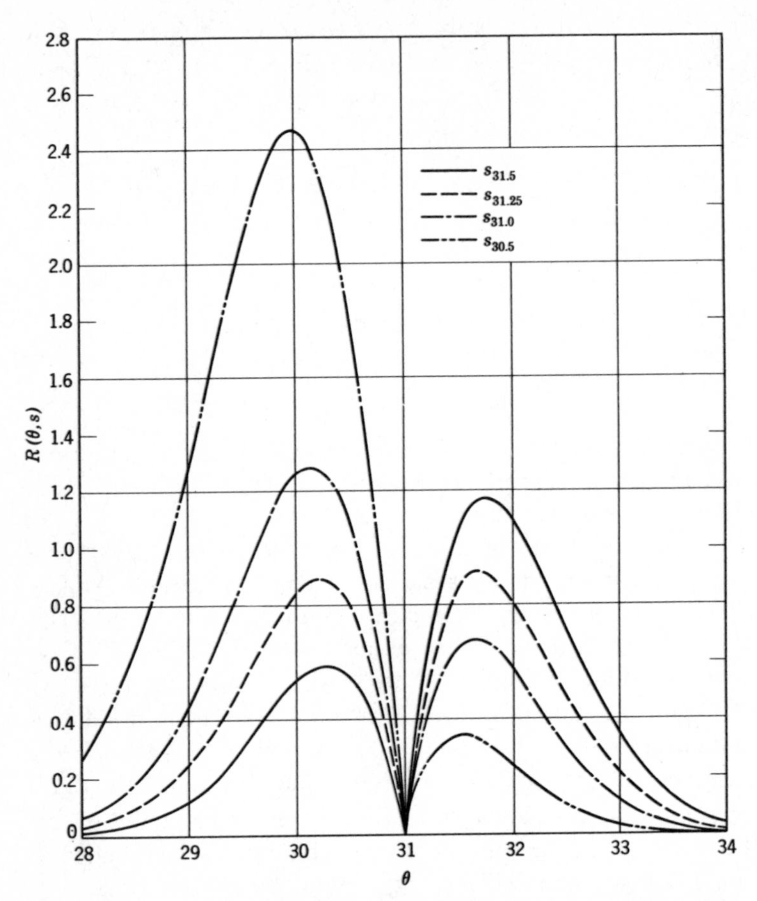

Figure 7.2. Risk functions for several strategies in Example 7.1.

sample size is larger. This fact illustrates the general remark that, although a good strategy depends on the nature of the regret function, it is relatively insensitive to small changes in the regrets if there are considerable data available.

**Exercise 7.1.* Mr. Good increases his sample size to 256. Let us designate by s_c^* the strategy which calls for action a_1 if $\overline{\mathbf{X}}_{256}$, the mean of 256 observations, exceeds c. Evaluate the risk functions for s_{31}^*, $s_{31.25}^*$. Guess the strategy which would be minimax.

2.1. Bayes Strategies

In the preceding section, the relative sizes of the regrets $r(\theta, a_1)$

and $r(\theta, a_2)$ made it advisable to modify strategy s_{31} to decrease the probability of taking action a_1. On the other hand, if there is *a priori* probability to believe that $\theta > 31.0$ (where a_1 is appropriate) is more likely than $\theta < 31.0$, one should tend to increase the probability of taking action a_1.

Let us suppose that Mr. Good has observed on the basis of past experience that the mean strength for batches of fibers behaves like a random variable $\boldsymbol{\theta}$ which has a distribution approximately given by the discrete density of Table 7.2.

TABLE 7.2

DISTRIBUTION OF MEAN FIBER STRENGTH FOR A RANDOM BATCH OF FIBERS RECEIVED FROM MR. LACEY (I.E., *A PRIORI* PROBABILITIES)

θ	28.0	28.5	29.0	29.5	30.0	30.5	31.0	31.5	32.0	32.5	33.0	33.5	34.0
$P\{\boldsymbol{\theta}=\theta\}$	0.00	0.01	0.02	0.03	0.05	0.07	0.12	0.15	0.20	0.15	0.10	0.07	0.03

The weighted average of risks corresponding to $s_{31.5}$ is

$$\mathscr{R}(s_{31.5}) = (0.00)(0.00932) + (0.01)(0.0371) + (0.02)(0.120) + (0.03)(0.300) + \cdots + (0.03)(0.0342) = 0.61$$

Similarly we obtain the weighted average of risks for other strategies and graph them in Figure 7.3. Consequently, the Bayes strategy is approximately $s_{30.85}$ with average risk 0.42. Note that the *a priori* probability distribution assigns much more weight to

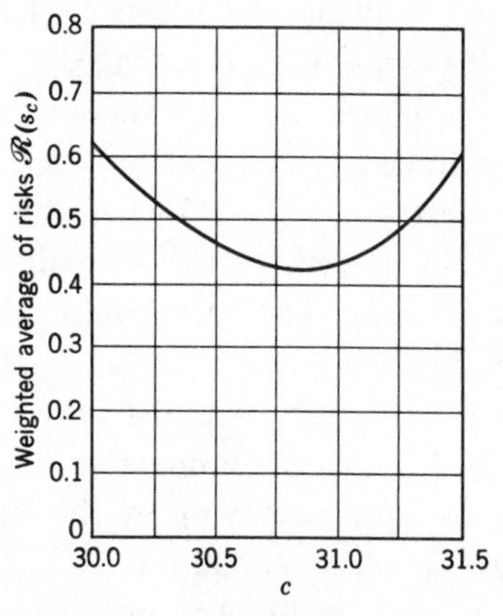

Figure 7.3. Weighted average of risks for strategy s_c in Example 7.1 with *a priori* probabilities given in Table 7.2.

$\theta > 31.0$ than to $\theta < 31.0$. It is reasonable to expect the risk for the Bayes strategy to have a considerably smaller hump for $\theta > 31.0$ than for $\theta < 31.0$, since the Bayes strategy is "tailored" to behave especially well for those states of nature most likely to occur (see Figure 7.2).

Exercise 7.2. How much should Mr. Good be willing to pay for the laboratory report on the 64 breaking strengths? (Refer to the *a priori* probabilities of Table 7.2.)

Exercise 7.3. What action does Mr. Good's Bayes strategy call for when $\bar{X} = 31.97$ (this is the sample mean of the 64 observations recorded)? What action does the minimax strategy call for?

Exercise 7.4 Using Table 7.2, evaluate the weighted average of risks for both strategies of Exercise 7.1. Guess at the Bayes strategy. How much is the large sample worth if s_{31}^* is used?

2.2. The Name "Testing Hypotheses"

Why is Mr. Good's example called an example of hypothesis testing? In this example there are only two actions. Action a_1 is appropriate for $\theta \geq 31.0$, and action a_2 is appropriate for $\theta < 31.0$. Taking action a_1 (accepting the batch) is equivalent to acting as though $\theta \geq 31.0$, or to *accepting the hypothesis* (*assumption*) *that* $\theta \geq 31.0$. Taking action a_2 is equivalent to *rejecting the hypothesis* $\theta \geq 31.0$ *in favor of the alternative hypothesis* $\theta < 31.0$. The fact that Mr. Good accepts the hypothesis or acts as though it were true does not imply that it *is* true, nor even that he is convinced that it is true. With some bad luck or lack of good data, he may be led to reject the hypothesis (take action a_2) when it is true ($\theta \geq 31.0$), or to accept the hypothesis (take action a_1) when it is false ($\theta < 31.0$).

The probabilities of making such errors are called the *error probabilities*. In Figure 7.4. we present (1) the action probabilities and (2) the error probabilities for s_{31}, $s_{31.25}$, and $s_{31.5}$. In presenting the action probabilities, we omitted the probability of taking action a_2 because that is simply 1 minus the probability of taking action a_1. The error probabilities $\varepsilon(\theta, s)$ are given by the difference between the probability of taking action a_1 and the ideal (1 or 0 depending on whether $\theta \geq 31.0$ or $\theta < 31.0$). Any one of the following three curves is completely descriptive of the action probabilities:

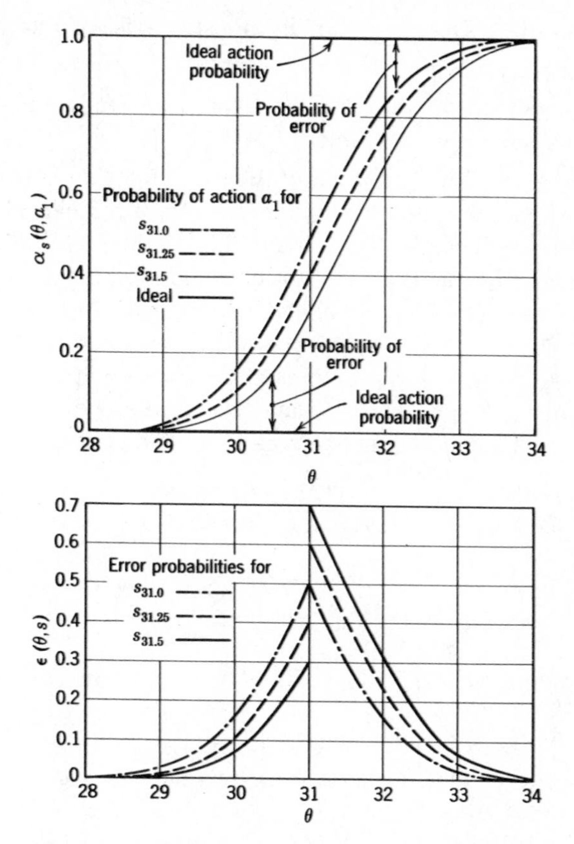

Figure 7.4. Action probabilities and error probabilities for several strategies in Example 7.1.

$\alpha_s(\theta, a_1)$—probability of taking action a_1;
$\alpha_s(\theta, a_2)$—probability of taking action a_2;
$\varepsilon(\theta, s)$ —probability of error.

The error probabilities are especially useful since, for two-action problems, the risk is given by

$$R(\theta, s) = \varepsilon(\theta, s)\, r(\theta)$$

where $r(\theta)$ is the regret due to taking the wrong action when θ is the state of the nature. Note that as the index c of s_c increases, the probability of taking action a_1 is diminished, increasing the error probabilities for $\theta \geq 31.0$ and decreasing them for $\theta < 31.0$.

Exercise 7.5. Plot $\varepsilon(\theta, s)$ for the two strategies of Exercise 7.1.

2.3. Another Example

It may help to fix the ideas of hypothesis testing if they are illustrated in another example.

Example 7.2. Mr. Baker has developed a new cake mix for retail consumption. He can either use it to replace his old mix (a_1) or stick to his old mix (a_2). He feels that it would pay to introduce the new mix if at least 60% of his customers prefer it to the old. His estimated regret function is given in Figure 7.5 where the unknown state of nature p represents the proportion of the customers who prefer the new mix. Let $\hat{\mathbf{p}}$ be the proportion of 100 customers selected at random who prefer the new mix. Let s_c be the strategy which consists of taking action a_1 if $\hat{\mathbf{p}} \geq c$ and action a_2 otherwise.

TABLE 7.3

ACTION PROBABILITIES AND RISKS FOR $s_{0.6}$ OF EXAMPLE 7.2

State of Nature	Regrets			Action Probablities		Risk
p	a_1 $r(p, a_1)$	a_2 $r(p, a_2)$	$\frac{0.6 - p}{\sqrt{p(1-p)/100}}$	a_1 $P\{\hat{\mathbf{p}} \geq 0.6 \mid p\}$	a_2 $P\{\hat{\mathbf{p}} < 0.6 \mid p\}$	$R(p, s_{0.6})$
0.450	0.342	0.000	3.015	0.0013	0.9987	0.0004
0.475	0.298	0.000	2.503	0.0062	0.9938	0.0018
0.500	0.241	0.000	2.000	0.0228	0.9772	0.0055
0.520	0.200	0.000	1.601	0.0547	0.9453	0.0109
0.540	0.149	0.000	1.204	0.1143	0.8857	0.0170
0.560	0.100	0.000	0.806	0.2102	0.7898	0.0210
0.580	0.052	0.000	0.405	0.3427	0.6573	0.0178
0.600	0.000	0.000	0.000	0.5000	0.5000	0.0000
0.620	0.000	0.048	−0.412	0.6598	0.3402	0.0163
0.640	0.000	0.100	−0.833	0.7977	0.2023	0.0202
0.660	0.000	0.145	−1.267	0.8974	0.1026	0.0149
0.680	0.000	0.200	−1.715	0.9568	0.0432	0.0086
0.700	0.000	0.258	−2.182	0.9854	0.0146	0.0038
0.725	0.000	0.332	−2.800	0.9974	0.0026	0.0009
0.750	0.000	0.418	−3.464	0.9997	0.0003	0.0001

$s_{0.6}$: take action a_1 if $\hat{\mathbf{p}} \geq 0.6$ and action a_2 otherwise.

We compute $\alpha_{s_{0.6}}(p, a_1) = P\{\hat{\mathbf{p}} \geq 0.6 \mid p\}$ by applying the normal cdf to the number of standard deviations between the mean of $\hat{\mathbf{p}}$ and 0.6. This is given by $(0.6 - p/\sqrt{p(1-p)/100}$.

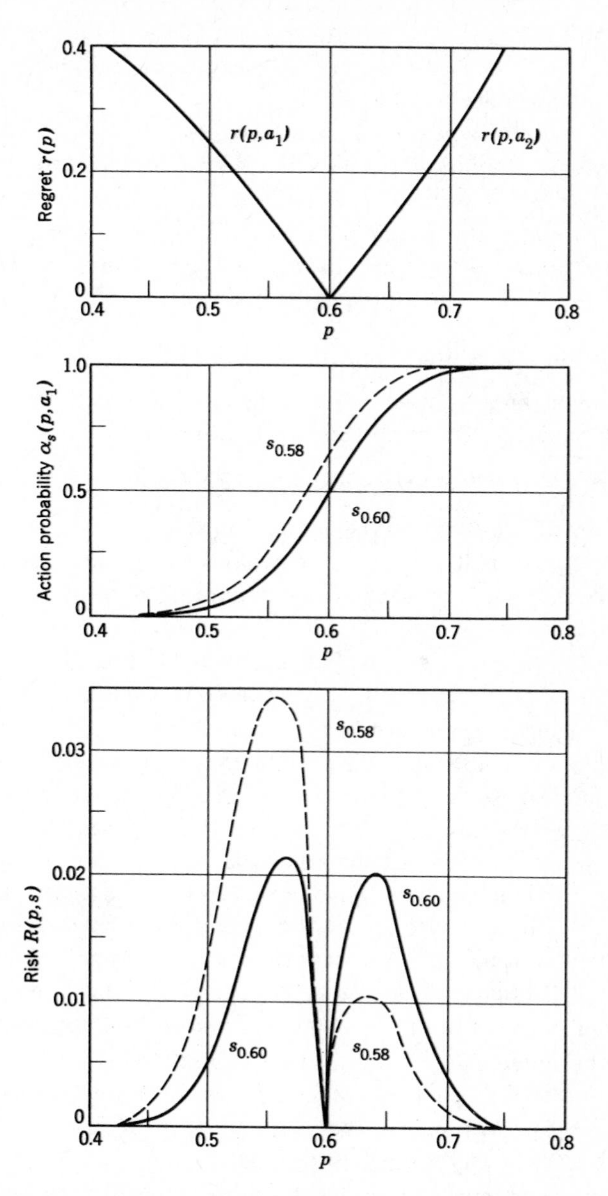

Figure 7.5. Regrets, action probabilities, and risks for Example 7.2.

In Chapter 9 it is shown that the admissible strategies are of the form s_c. In Table 7.3 the action probabilities and risks for the strategy $s_{0.6}$ are evaluated.

To compute the action probabilities, we make use of the fact that $\hat{\mathbf{p}}$ is approximately normally distributed with mean p and standard deviation $\sqrt{p(1-p)/100}$ (see Section 6, Chapter 6, and Exercise 6.45). We illustrate the computation for the row corresponding to $p = 0.66$ where $\hat{\mathbf{p}}$ has mean 0.66 and standard deviation $\sqrt{0.66(0.34)/100} = 0.0474$. Then

$$\alpha_{s_{0.6}}(p, a_1) = P\{\hat{\mathbf{p}} \geq 0.6 \mid p = 0.66\} .$$

Then the distance between 0.6 and the mean of $\hat{\mathbf{p}}$ is 0.06 which is 1.267 standard deviations. Thus $\alpha_{s_{0.6}}(p, a_1) = 0.8974$ and

$$R(p, s_{0.6}) = (0.8974)(0.000) + (0.1026)(0.145) = 0.0149 .$$

In Figure 7.5 we compare the action probabilities and the risks for $s_{0.6}$ and $s_{0.58}$. Thus it is easy to see that $s_{0.6}$ is approximately the minimax risk strategy and has maximum risk slightly larger than 0.02.

Exercise 7.6. Compute the action probabilities and risk function for strategy $s_{0.61}$ in Example 7.2. (For convenience, approximate $\sqrt{p(1-p)}$ by 1/2 for p between 0.25 and 0.75.)

Exercise 7.7. An election is being held for Governor of Phiggins. Mr. Smith, whose money is invested in stocks would (a_1) convert to bonds if he know that the incumbent would be defeated. He would (a_2) stick to stocks if he knew that the incumbent would be re-elected. Indicate what would seem a reasonable regret function for Mr. Smith. Indicate a reasonable strategy if he had available the results of a poll of 400 voters selected at random, and evaluate the risk function. (The regret functions for this problem are somewhat unusual in that it is not very important by how many votes the incumbent wins or loses the election. Assume that the population of Phiggindian voters is very large and approximate $\sqrt{p(1-p)}$ by 0.5 for p between 0.2 and 0.8.)

Exercise 7.8. The Bumble Seed Company has found an unidentified barrel of mustard seeds. If the seeds were fresh (θ_1), two-thirds of them would germinate and the Bumble Seed Company would desire to (a_1) market them under its own label. If they were one year old (θ_2), only half would germinate and the company would prefer to (a_2) market them under another label. The regrets are given by

	a_1	a_2
θ_1	0	12
θ_2	18	0

A random sample of 50 of the seeds are carefully planted and observed for germination.

(a) Indicate and evaluate risks for three reasonable strategies.
(b) Estimate the minimax risk strategy.

3. ESTIMATION

Example 7.3. When Mr. Good expanded his business, he began to supply cord not only to parachute manufacturers but also to all users of cord. Naturally, the value of the cord increases with the strength of the fibers used. In his new business the batches of fibers were no longer supplied to him at a fixed price. He had to make an offer for each batch. Mr. Good was no longer faced with the problem of accepting or rejecting the batch of fibers. His problem had become one of deciding how much to offer for the batch. The states of nature are described by θ, the mean fiber strength, as in Example 7.1. The available actions are now increased from two (accept or reject) to many. The action is the price offered for the batch. The regret corresponding to θ and the price offered are difficult to evaluate. If the price is too low for the quality, Mr. Lacey may sell the batch to another manufacturer. If the price is too high, Mr. Good will not make as much profit as he could. Let us assume that Mr. Good has made up a table or graph which represents the price he is willing to offer when he knows θ. (See Figure 7.6.) As θ increases, the price increases continuously. Thus, each action (price offered) corresponds to an estimate of θ (the mean fiber strength). For theoretical convenience, it is useful to label the action not by the price offered but by T, the corresponding value of θ. Now the regrets depend on θ and T. If our estimate T is equal to θ, we should have $r(\theta, T) = 0$. As T moves away from θ, $r(\theta, T)$ tends to increase.

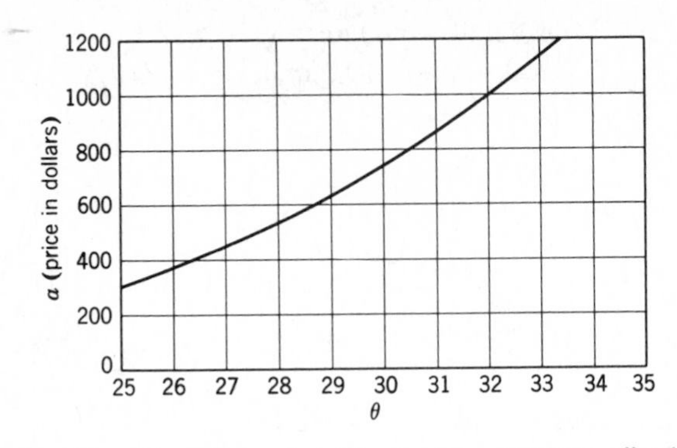

Figure 7.6. The price a that Mr. Good is willing to offer if he knows θ (Example 7.3).

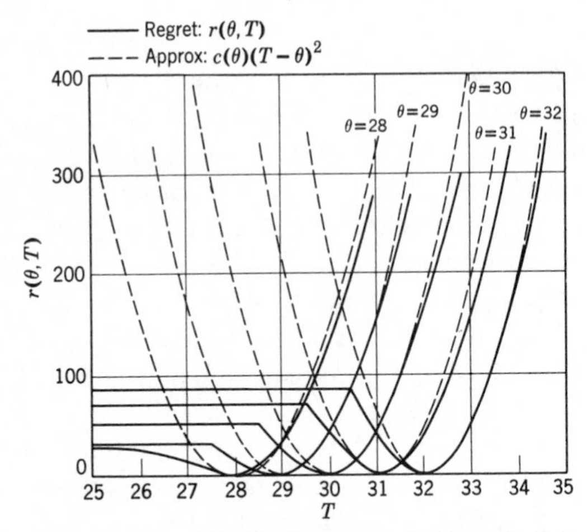

Figure 7.7. Mr. Good's regrets in Example 7.3 and approximating parabolas.

We indicate in Figure 7.7 Mr. Good's evaluation of $r(\theta, T)$ for several values of θ. We note that, for these curves, $r(\theta, T)$ is approximated by the following equations.[1]

$$r(\theta, T) = 47(T - \theta)^2 \qquad \text{for } \theta = 30 \text{ and } T \text{ close to } \theta$$

$$r(\theta, T) = 51(T - \theta)^2 \qquad \text{for } \theta = 31 \text{ and } T \text{ close to } \theta$$

$$r(\theta, T) = 55(T - \theta)^2 \qquad \text{for } \theta = 32 \text{ and } T \text{ close to } \theta.$$

In general, for T close to θ, $r(\theta, T)$ is approximated by Equation (7.1).

[1] See Exercise 7.9 to see how these equations are arrived at.

$$r(\theta, T) = c(\theta)(T - \theta)^2 \tag{7.1}$$

where $c(\theta)$ depends on θ. (See Figure 7.7.)

Mr. Good still performs the same experiment of measuring the breaking strengths of 64 fibers randomly selected from the batch.

Mr. Good's strategy must assign an action or estimate to each possible sample $\mathbf{Z} = (\mathbf{X}_1, \mathbf{X}_2, \cdots, \mathbf{X}_{64})$. We call such a strategy an *estimator* and denote it by t instead of the usual s. Since a strategy is a function on the set of possible data to the set of possible actions, an estimator is a function on the set of possible $\mathbf{Z}$ to the set of possible estimates of θ. The actual estimate $\mathbf{T}$ is denoted by:

$$\mathbf{T} = t(\mathbf{Z})\ . \tag{7.2}$$

Let us consider three simple estimators. These are t_1, t_2, and t_3, given by

$$t_1(\mathbf{Z}) = \mathbf{T}_1 = \overline{\mathbf{X}}$$
$$t_2(\mathbf{Z}) = \mathbf{T}_2 = \breve{\mathbf{X}} = \text{sample median}$$
$$t_3(\mathbf{Z}) = \mathbf{T}_3 = 30.$$

Let us examine the properties (action probabilities and risks) of these three estimators. Estimator t_1 yields an estimate $\overline{\mathbf{X}}$ which is normally distributed with mean θ and variance $\sigma^2_{\overline{\mathbf{X}}} = \sigma^2_{\mathbf{X}}/64 = 1$. The corresponding risk is given by

$$R(\theta, t_1) = E[c(\theta)(\overline{\mathbf{X}} - \theta)^2] = c(\theta)\,E[(\overline{\mathbf{X}} - \theta)^2] = c(\theta).$$

Estimator t_2 yields $\breve{\mathbf{X}}$ which mathematicians have shown to be approximately normally distributed with mean θ and variance $\sigma^2_{\breve{\mathbf{X}}} = 1.57\sigma^2_{\mathbf{X}}/64 = 1.57$. Then

$$R(\theta, t_2) = E[c(\theta)(\breve{\mathbf{X}} - \theta)^2] = c(\theta)\,E[(\breve{\mathbf{X}} - \theta)^2] = 1.57c(\theta).$$

Estimator t_3 yields $\mathbf{T}_3 = 30$ which is a fixed number. This estimator ignores the data and corresponds to a guess. Here we have

$$R(\theta, t_3) = c(\theta)(30 - \theta)^2.$$

These three risk functions are compared in Figure 7.8. More generally, we have for any estimator t yielding the estimate

$$\mathbf{T} = t(\mathbf{Z})$$

$$R(\theta, t) = c(\theta)\,E[(\mathbf{T} - \theta)^2]. \tag{7.3}$$

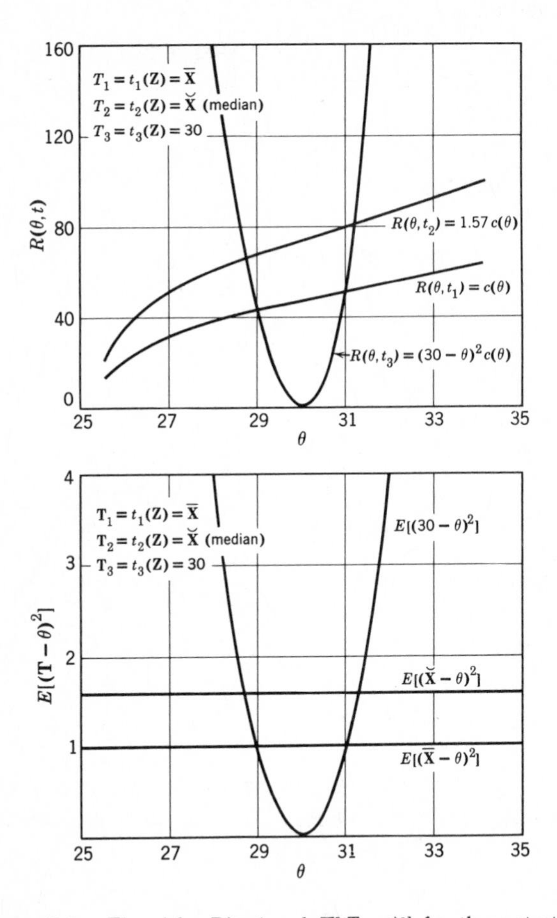

Figure 7.8. The risks $R(\theta, t)$ and $E[(\mathbf{T} - \theta)^2]$ for three strategies in Example 7.3. $R(\theta, t) = c(\theta)\, E[(\mathbf{T} - \theta)^2]$.

In this particular problem where the observations are assumed to be normally and independently distributed, t_1 (sample mean) has several good properties which make it desirable and commonly used. Notice that $R(\theta, t_1) < R(\theta, t_2)$ for all θ, and thus the sample median estimator is dominated. On the other hand, the rather "foolish" estimator t_3 is admissible, even though it ignores the data, because it cannot be improved if $\theta = 30$. However, statisticians usually discard this estimator, since there are other estimators that *are almost certain to give estimates close to θ if the sample size is large,* and this can *not* be said about t_3.

We remark that in many, and perhaps most, examples the regret function can be approximated by $c(\theta)(T-\theta)^2$. Then the most important aspects of the action probabilities of an estimator are summarized by

$$E[(\mathbf{T}-\theta)^2] = \sigma_{\mathbf{T}}^2 + [E(\mathbf{T})-\theta]^2. \tag{7.4}$$

In other words, for comparing two estimators, the values of $E[(\mathbf{T}-\theta)^2]$ for these two estimators are more important than a very precise knowledge of the actual regret function or of $c(\theta)$.

Exercise 7.9. This exercise is to illustrate how one could compute the regret function for Mr. Good's problem if it were very important to do so. We suppose that Mr. Good has a supplier of fibers whose material and price are very stable and such that Mr. Good can be assured of \$250 profit by using this supplier's fibers. The batch brought by Mr. Lacey might yield a greater profit. On the other hand, if he overpays, he might not do so well. Let us suppose if Mr. Good uses Mr. Lacey's batch, his profit will be $[f(\theta)-a]$ where

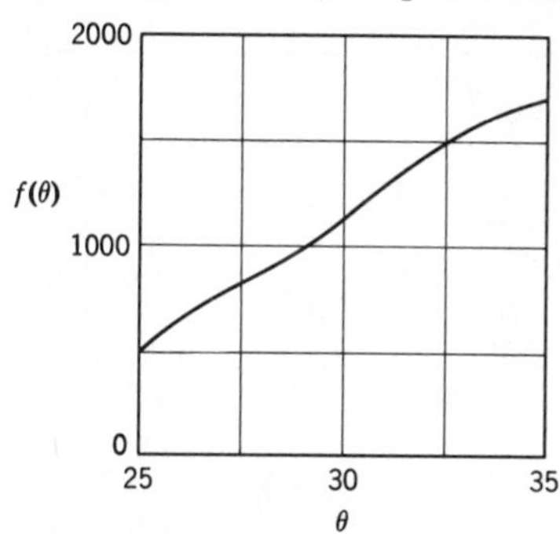

Figure 7.9. Gross income $f(\theta)$ for batch with mean θ in the cord problem (Example 7.3.). (Profit $= f(\theta) - a$, where a is the price paid for the batch.)

a is the price he pays for the batch (see Figure 7.9). If he offers a dollars, the probability that Mr. Lacey will agree to accept is $p(\theta, a)$, as given in Figure 7.10. Thus Mr. Good either (1) profits by $f(\theta)-a$ with probability $p(\theta, a)$ (if Mr. Lacey accepts) or (2) profits by \$250 with probability $1-p(\theta, a)$ (if Mr. Lacey refuses,

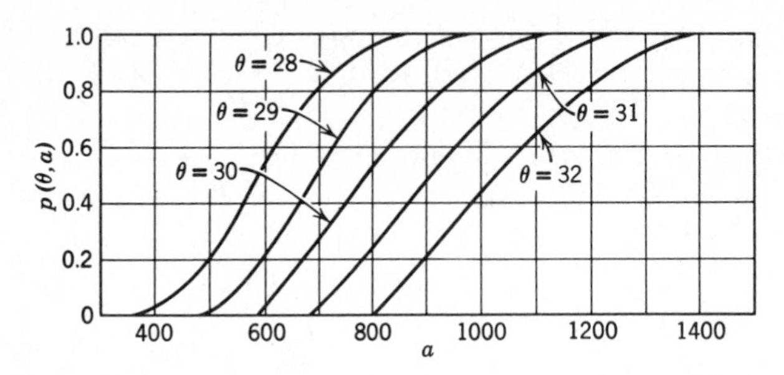

Figure 7.10. Probability $p(\theta, a)$ that Mr. Lacey will accept offer of a for the batch.

and Mr. Good has to revert to his other supplier). Since \$1000 is a relatively small amount in this big business, we shall assume that utility is proportional to dollars. In Figure 7.11, we graph the expected profit against the offered price for given values of θ. Check a few points on the curves of Figure 7.11. Check a few points on the curves of Figure 7.6 which yield the price that should be offered if θ is known. If T is an estimate of θ, we act as though T were θ and offer the price that would then be appropriate. Now check a few points on the curves representing $r(\theta, T)$.

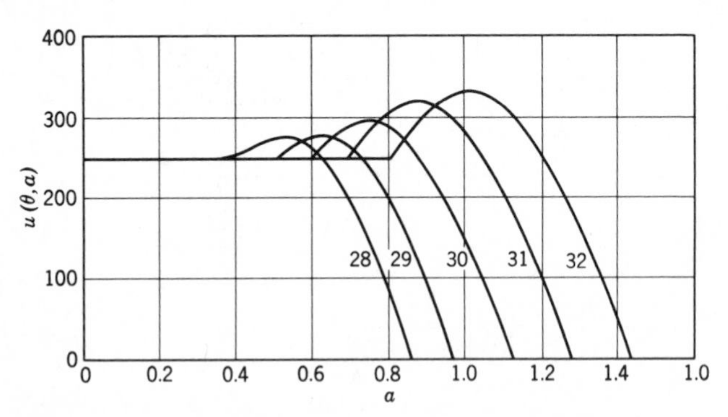

Figure 7.11. Expected profit $u(\theta, a)$ if price a is offered for batch. $u(\theta, a) = p(\theta, a)[f(\theta) - a] + [1 - p(\theta, a)](250)$.

Example 7.4. A market research firm is asked to determine what proportion p of New York families owning television sets heard a certain commercial. The firm has a random sample of 100 families owning television sets (the population of New York families owning television sets is very large). The proportion of the sample who heard the commercial is $\hat{\mathbf{p}}$. The market research firm uses $\hat{\mathbf{p}}$ as the estimate of p.

Thus the action is to report an estimate of p. The firm's strategy is to use $\hat{\mathbf{p}}$ as the value of the estimate. The action probabilities are the probabilities with which $\hat{\mathbf{p}}$ takes on the possible values 0.00, 0.01, 0.02, $\cdots$, 0.99, 1.00. The action probabilities can be summarized by the remark that $\hat{\mathbf{p}}$ is approximately normally distributed with mean p and variance $p(1-p)/100$.

Suppose the regret is given by $r(p, T) = (T-p)^2/p(1-p)$. Then, since $E[(\hat{\mathbf{p}}-p)^2] = p(1-p)/100$, we have risk

$$R(p) = \frac{p(1-p)/100}{p(1-p)} = 0.01.$$

It is interesting that, for the above regret function, $R(p)$ is constant.

Exercise 7.10. If, in Example 7.4, $r(p, T) = (T-p)^2$, evaluate and graph the risk function for using $\hat{\mathbf{p}}$ based on a sample of 400.

Exercise 7.11. For Example 7.4, we define the estimators t_1 and t_2 by $t_1(\mathbf{Z}) = \mathbf{T}_1 = 0.5\hat{\mathbf{p}} + 0.25$ and $t_2(\mathbf{Z}) = \mathbf{T}_2$ equal the proportion of the first 50 people sampled who heard the commercial. Indicate the action probabilities and compute and graph the risk functions for t_1 and t_2. Is there any value of p for which t_1 is better than t (which consists of using $\hat{\mathbf{p}}$)?

Exercise 7.12. The pulling strength of a randomly selected South Phiggindian is approximately normally distributed with mean θ and standard deviation 100 lb. A team of 100 randomly selected South Phiggindians barely manage to pull a weight $\mathbf{W}$. Estimate θ. What is the probability distribution of the estimate?

Exercise 7.13. It is desired to estimate the systematic error θ (at 10,000 yards) of an optical range finder. The regret is given by $14(T-\theta)^2$. Let $\overline{\mathbf{X}}$ be the average of N independent observations, each of which has mean $10{,}000 + \theta$ and variance 12.

(a) What is the risk based on the use of $\mathbf{T} = \overline{\mathbf{X}} - 10{,}000$?

(b) Graph this risk as a function of the sample size N.

(c) If each observation costs one unit, determine an optimal sample size by graphing risk plus cost of sampling as a function of N.

4. CONFIDENCE INTERVALS

In Chapters 1 through 6, we studied decision making by comparing strategies on the basis of their risks and without considering any intuitive reasons for preferring one strategy over others. Thus, in Example 1.1, the contrary man's strategy was treated without discrimination until it was seen to be dominated.

Should we not be able to look at the contrary man's strategy and reject it outright? Roughly speaking, what constitute reasonable strategies for the busy man who does not have time to formulate the problem in complete detail and compare all strategies? This question is especially important for scientists who must plan future experiments without the possibility of complete analysis of risks for all strategies. A rough answer is that reasonable strategies use the data to estimate the state of nature θ. If the estimate is very good, one acts as though the estimate were θ. If the estimate is rather poor, one tends to hedge accordingly.

Thus in Example 5.1, the strategy $s_6 = (a_1, a_2, a_3)$ may be interpreted as follows. The observation "fair" is strong evidence in favor of θ_1, and we take action a_1 (wear ordinary suit). Similarly, the observation "foul" is strong evidence in favor of θ_2, and we take action a_3 (wear complete rain outfit). But the observation "dubious" is inconclusive and, to prevent the disaster of guessing wrong, we take action a_2 which is not very bad (but not optimal) for either θ. In other words, we *hedge*. Thus it is important to estimate θ and to measure how good this estimate is.

In Example 7.3 we know that $\bar{\mathbf{X}}$ is normally distributed with mean θ and standard deviation one. Here the standard deviation of the estimate is a good measure of the inexactness of the estimate. But suppose we consider the modification of this example where $\sigma_{\mathbf{X}}$, the standard deviation of the $\mathbf{X}_i$, is not known. Then the standard deviation of $\bar{\mathbf{X}}$ is $\sigma_{\mathbf{X}}/\sqrt{n}$, which is not known. It is true that we can estimate $\sigma_{\mathbf{X}}$ also, but then we would have to worry about how good is our estimate of the standard deviation of our estimate. Although this problem is not at all serious for large samples, it can, for small samples, lead to estimates of standard deviations of estimates of standard deviation of estimates . . . of estimates, which is plainly a nuisance. This is especially painful when the estimate is not normally distributed.

A more convenient method of indicating the reliability of an estimate is that of *confidence intervals*. For large samples, this method does not yield any results which differ essentially from those obtained by a simple estimate of the standard deviation of the estimate. For small and medium size samples, this method is often mathematically elegant and convenient to apply. Although confidence intervals play a somewhat obscure role in decision making problems, it appears that scientists often find them useful as devices of description, or inference, in complicated problems where it is difficult to specify actions, states of nature, and regrets. *A confidence interval* $\mathbf{\Gamma} = \Gamma(\mathbf{Z})$ is a selected range of possible values for a parameter, which, the statistician asserts, actually does contain the parameter. The method of confidence intervals provides two things:

1. A "recipe" for constructing the "selected range" (interval) from any given set of data.
2. A value for the probability that the method leads to an interval which actually succeeds in bracketing (or covering, or containing) the unknown parameter.

The usual method of constructing such sets is closely associated with the problem of testing hypotheses.

We illustrate the confidence interval approach in several variations of the cord problem (see Examples 7.1 and 7.3).

CASE 1. $\sigma_{\mathbf{X}}$ known.

Here we have $\sigma_{\overline{\mathbf{X}}} = \sigma_{\mathbf{X}}/\sqrt{n}$. Thus

$$P\{-1.96\sigma_{\mathbf{X}}/\sqrt{n} < \overline{\mathbf{X}} - \theta < 1.96\sigma_{\mathbf{X}}/\sqrt{n}\} = 0.95.$$

That is to say that $\overline{\mathbf{X}}$ is within $1.96\sigma_{\mathbf{X}}/\sqrt{n}$ of θ with probability 0.95 or that the random interval[1]

$$\mathbf{\Gamma} = \left(\overline{\mathbf{X}} - \frac{1.96\sigma_{\mathbf{X}}}{\sqrt{n}}, \overline{\mathbf{X}} + \frac{1.96\sigma_{\mathbf{X}}}{\sqrt{n}}\right)$$

will contain θ with probability 0.95. For this particular outcome of the experiment, $\overline{\mathbf{X}} = 31.97$ and $\sigma_{\mathbf{X}}/\sqrt{n} = 1$, and Mr. Good may state: "*With confidence* 95%, *the interval* (30.01, 33.93) covers θ." The reader should note that no matter what the value of θ, the

[1] We often designate by (a, b) the interval from a to b, i.e., $\{x: a < x < b\}$.

probability that it will be covered by $\mathbf{\Gamma}$ is 0.95.

CASE 2. σ_X unknown, n large.

Here the interval $\mathbf{\Gamma}$ is closely approximated by

$$\mathbf{\Gamma}^* = \left(\bar{\mathbf{X}} - \frac{1.96\mathbf{s}_X}{\sqrt{n}}, \bar{\mathbf{X}} + \frac{1.96\mathbf{s}_X}{\sqrt{n}}\right)$$

since, for large n, $\mathbf{s}_X$ tends to be close to σ_X.

CASE 3. σ_X unknown, n not large (say 10).

If n is small, $P\{\mathbf{\Gamma}^* \text{ contains } \theta\}$ is no longer equal to or very close to 0.95. In Case 1, the 1.96 was derived from the fact that

$$P\left\{\frac{|\bar{\mathbf{X}} - \theta|}{\sigma_X/\sqrt{n}} < 1.96\right\} = 0.95.$$

If $(\bar{\mathbf{X}} - \theta)/(\sigma_X/\sqrt{n})$ is replaced by $(\bar{\mathbf{X}} - \theta)/(\mathbf{s}_X\sqrt{n})$, the probability distribution is modified. The fact is that the distribution of

$$\mathbf{t}_{n-1} = \frac{\bar{\mathbf{X}} - \theta}{\mathbf{s}_X/\sqrt{n}}$$

has been catalogued as the t distribution with $(n - 1)$ degrees of freedom (*S*ee Appendix D_4)[1]. Hence, if $n = 10$, we should apply

$$P\left\{-2.262 < \frac{\bar{\mathbf{X}} - \theta}{\mathbf{s}_X/\sqrt{10}} < 2.262\right\} = 0.95$$

which gives

$$\mathbf{\Gamma}^{**} = \left(\bar{\mathbf{X}} - 2.262\frac{\mathbf{s}_X}{\sqrt{10}}, \bar{\mathbf{X}} + 2.262\frac{\mathbf{s}_X}{\sqrt{10}}\right).$$

Note that the uncertainty about σ_X has led to replacing σ_X by $\mathbf{s}_X$ and to increasing the multiplier from 1.96 to 2.262. From Appendix D_4 we see that, as n increases, the multiplier of $\mathbf{s}_X/\sqrt{n}$ decreases to 1.96. In Chapter 10 we shall indicate a general method of obtaining confidence intervals.

To illustrate another important role of confidence intervals, suppose that in Example 7.1 Mr. Good observed $\bar{\mathbf{X}} = 65$. Then according to his strategy he would undoubtedly accept the batch. Presumably this would be the end of the problem. However, if Mr. Good were at all imaginative, he might note that a 95% confidence

[1] This fact depends on the normality of the observations.

interval for θ is (63.04, 66.96) which indicates fibers of extremely high strength. A reasonable manufacturer, scientist, or statistician would be very strongly tempted to reformulate the problem. Mr. Good might now wish to consider the possibility of (a_3), making twice as much cord from a single batch of fibers. The scientist might consider the possible action a_4 (investigate in what way the production of this batch differed from that of past batches). The statistician might even consider checking whether the laboratory technician had been intoxicated. A confidence interval is often useful in giving a good idea of the neighborhood in which θ lies so that the decision maker can formulate an appropriate problem without too complicated a set of actions.

Exercise 7.14. What is the 99% confidence interval for θ in Example 7.1 when $\sigma_{\mathbf{X}} = 8$ and $n = 64$?

Exercise 7.15. What is the 99% confidence interval for θ in Example 7.1 when $\sigma_{\mathbf{X}}$ is unknown, $\mathbf{s}_{\mathbf{X}} = 7$, and $n = 9$?

Exercise 7.16. The proportion of the population of prospective voters favoring candidate A is p. A polling company has taken a random sample of 10,000 of the large population of prospective voters and finds the proportion $\hat{\mathbf{p}}$, of the sample, favoring A. We know that $\hat{\mathbf{p}}$ is approximately normally distributed with mean p and standard deviation $\sigma_{\hat{\mathbf{p}}} = \sqrt{[p(1-p)]/n}$. What is an approximate 95% confidence interval for p? Evaluate the interval for $\hat{\mathbf{p}} = 0.3, 0.4, 0.5$. *Hint*: Assume n is large, as in Case 2, and approximate $\sigma_{\hat{\mathbf{p}}}$ by $\sqrt{[\hat{\mathbf{p}}(1-\hat{\mathbf{p}})]/n}$.

†5. SIGNIFICANCE TESTING

The most commonly used methods of statistics have a relatively short history of use. Nevertheless, most of them antedate the decision making formulation of statistics. At the time of their development, the formulation of the problem of hypothesis testing was incomplete, and considerable confusion resulted. However, the technique of *significance testing* which developed is well established, convenient, and also useful *so long as the underlying decision making problem is carefully kept in mind.*

In a test of significance, we typically consider a hypothesis specifying the value of a parameter. This hypothesis, often called the *null* hypothesis, is to be rejected if the sample turns out to be of a sort very unlikely to occur if the hypothesis is true, i.e., if the sample falls in a region, called the *rejection region*, which has small probability under the hypothesis. Then the data are considered to be "significant" evidence against the hypothesis.

In this formulation, the regrets and, incidentally, the alternative hypothesis, are not explicitly taken into account. These gaps have resulted in confusion concerning what constituted a reasonable rejection region. For example, consider the four following tests of the hypothesis H that a probability p is equal to 1/2. For each of these tests based on 10,000 observations, the probability of rejecting H if it is true (called the *significance level* of the test) is 0.05. These are: (1) Reject if $\hat{\mathbf{p}} > 0.5082$; (2) reject if $\hat{\mathbf{p}} < 0.4918$; (3) reject if $|\hat{\mathbf{p}} - 0.5| > 0.0098$; and (4) reject if $|\hat{\mathbf{p}} - 0.5| \leq 0.0003$. Which of these tests should be prefered?

This question can only be answered by considering the regrets and those states of nature for which it is important to reject the hypothesis. For example, Test 1 would be appropriate for investigating someone's claims of having extra sensory perception (ESP) and having probability $p > 0.5$ of predicting the color of a card selected at random. If, as he claimed, $p > 0.5$, $\hat{\mathbf{p}}$ would tend to be large, thus large values of $\hat{\mathbf{p}}$ should lead to rejecting the null hypothesis H: $p = 0.5$ (he has no ability to guess colors). On the other hand, Test 3 would be appropriate for Mr. Sharp who wants to know if a coin is sufficiently well balanced to be used in his gambling casino. Here, Mr. Sharp would like to reject H: $p = 1/2$ (the coin is well balanced) if p is much larger or much smaller than 0.5. Then both very small and very large values of $\hat{\mathbf{p}}$ may be taken as evidence of unbalance and lead to rejecting H. Similarly, Test 4 would be appropriate for the statistics instructor who suspects that his students have been falsifying the data so as to give the impression of a well-balanced coin.

Thus, we repeat, a rational use of the method of significance testing requires contemplation of the underlying decision making problem and, in particular, those states of nature for which rejection is highly desirable.

An important advantage of the method of significance testing is the following. Once we have decided on the nature of the rejection criterion (e.g., reject if $\hat{\mathbf{p}}$ is too large, or reject if $\hat{\mathbf{p}}$ is too far from

0.5), the test is easily derived for any specified significance level α. This test will depend only on H and α. For instance, the 0.05 significance level test which consists of rejecting H if $\hat{\mathbf{p}}$ is too large is derived as follows. Under the hypothesis H: $p = 1/2$, $\hat{\mathbf{p}}$ is approximately normally distributed with mean 0.5 and standard deviation $\sqrt{(0.5)(0.5)/10{,}000} = 0.005$. Thus we reject H if $\hat{\mathbf{p}}$ exceeds 0.5 by 1.645 standard deviations, i.e., if $\hat{\mathbf{p}} > 0.5082$.

However one difficulty remains. What is an appropriate significance level? Again, the answer depends upon the nature of the regrets.

We illustrate with the following two examples.

Example 7.5. A psychologist is studying Mr. Brown who claims to have ESP and, consequently, the ability to predict the color of a card selected at random from a deck with probability $p > 0.5$. If he had no ESP, p would be equal to 0.5. The experiment consists of 10,000 trials.

Example 7.6. Mr. Sharp wants to check whether a coin is sufficiently well balanced to be used as a gambling device in his casino. Because of favorable odds, he considers the coin as adequate so long as the probability of heads, p, lies between 0.48 and 0.52. The experiment consists of 10,000 tosses.

In both of these examples, the significance testing method calls for specifying the null hypothesis H: $p = 0.5$ although, in the second example, a more realistic formulation would involve the hypothesis H^*: $0.48 \leq p \leq 0.52$. The first, (Example 7.5), typifies the situation in scientific research where rejecting the hypothesis involves announcing a new "fact" or "effect" (e.g., existence of ESP). Such an announcement, if false, can lead to much waste of time and effort and, worse, may introduce confusion into the development of knowledge. Also, it is very embarrassing to the scientist who makes the announcement of an untrue "fact." The significance test controls the probability of such errors by setting up a *null* hypothesis which means "this effect does not exist." Only if the evidence is extremely unlikely under that hypothesis do we reject H and conclude that the effect *does* exist. A very strict (small) significance level is called for to give protection against the error of incorrectly rejecting the hypothesis—but at the cost of making it difficult to demonstrate new "facts" when they are true.

The emphasis in this kind of application is on the question: Is an effect *real*?

Example 7.6 is of the type where we are interested in whether an effect is *large*. By setting a strict significance level, one ensures that, if no effect is present, one is very likely to accept the null hypothesis of no effect—and if an effect is small, one is still likely to conclude "no effect." Only a large effect will give high probability of rejecting H. Thus, if one compares a new process with an established process, or an expensive material with a cheap material, one may set up the null hypothesis that the established (or cheap) material is as good as the other, and then set a strict (small) significance level. This has the effect of making it likely that the established (or cheap) material will be regarded as equal to its competitor unless the competitor is much better.

In Example 7.6, no one believes H—that a real coin is perfectly well balanced. On the basis of 10,000 observations, the 0.05 significance level test is very likely to reject H if $p = 0.485$, where it would be preferable to accept the coin. Because the sample size is large, a much stricter significance level is called for. In fact, for large samples, the procedure which accepts H so long as $0.48 \leq \hat{\mathbf{p}} \leq 0.52$ would be quite reasonable. Here a result $\hat{\mathbf{p}} = 0.485$ would sometimes be called *statistically significant* evidence against H: $p = 1/2$, but also evidence that the difference between p and $1/2$ is not of *practical significance*.

In review, significance testing is a convenient method often applied to problems where it is of interest to find out whether an effect is real or large. The null hypothesis denies (nullifies) this effect. The appropriate choice of α represents a compromise between the probability of rejecting H when it is true (or when the effect is small) and the probability of accepting H when it is false (or when the effect is large). If a large sample is available, a small α is called for. If very few observations are available, a large α may be necessary.

Exercise 7.17. A psychologist is interested in two problems. First he wants to know if Mr. Brown has E S P and, consequently, has probability $p > 1/2$ of guessing the color of a card selected at random from a deck of cards. Second, he wants to know if the proportion of women who prefer a white dress, when given a choice between white and black, exceeds 1/2. For each problem he has accumulated 1000 observations. In which case should he use a lower significance level? Why?

Exercise 7.18. A biologist is involved in two experiments concerning the lengths of mouse tails. In one he compares males and females. In the second he compares ordinary mice with mice whose parents' tails had been cut off. Specify the null hypotheses for the two problems in terms of mean tail lengths. Given the same number of observations in each experiment, which calls for a lower significance level? Why?

Exercise 7.19. Derive 0.01 significance level tests for Examples 7.5 and 7.6. Present a 20% significance level test for the teacher who suspects his students of falsifying the data.

Exercise 7.20. For Example 7.1, present a 0.05 significance level test of the hypothesis that the cord is satisfactory, i.e., state the null hypothesis H: $\theta = 32$ and reject if the evidence indicates that the cord is unsatisfactory.

Exercise 7.21. Mr. Brown does very poorly in a preliminary ESP experiment. He claims that on some days he has negative ESP and $p < 0.5$. How should the next day's experiment be analyzed?

Exercise 7.22. State a 0.01 significance level test of H: $p = 0.2$ based on 400 observations where rejection is desired if p is far from 0.2 in either direction.

The scientist who publishes the results of his experiments may feel that other interested scientists with different regret functions are entitled to more information than a statement that a 0.05 significance level test led to the rejection of the null hypothesis. In this case, we recommend the use of a confidence interval as a supplementary description which is often helpful.

†6. A DECISION MAKING PROBLEM WHERE COMMON STATISTICAL PROCEDURES DO NOT APPLY

Customarily, statisticians select a strategy by proposing reasonable strategies, comparing their risks, and selecting a strategy for which the risks are "small." Sometimes problems arise where it is impractical even to describe a strategy, simply because there is a tremendous variety of possible observations which cannot all be considered. If, in such cases, *a priori* probabilities are relevant and available, one should apply the Bayes strategy. This is possible since the action determined by the Bayes strategy depends only on the regrets, *a priori* probabilities, and the *data actually observed.* One need not consider the entire set of possible observations to obtain the appropriate action. In a manner of speaking, we need cross only the bridges we come to. We illustrate with Example 7.7.

Example 7.7. A butler is suspected of having murdered his employer, Mr. White, whose will left a good deal of money to the butler. Mr. White was found lying on his bed with his suit jacket on and a knife in the right part of his chest. From additional information, it is clear that either the butler murdered Mr. White or that Mr. White committed suicide. The famous criminologist, Mr. Black, explains that, among suicides, only 10% kill themselves with a knife. Only 20% commit suicide while wearing a jacket. Finally, among those who kill themselves with a knife, only 7% insert the knife in the right part of the chest. Hence, argues Mr. Black, the probability of suicide is only $(0.10)(0.20)(0.07) = 0.0014$ and the butler is surely guilty.

Mr. White's lawyer points out that Mr. Black has forgotten that the dead man weighed 235 lb and had red hair, and only 1/100 of 1% of all people weigh 235 lb and have red hair. With these somewhat irrelevent data, we could make the "probability of suicide" 0.00000014. We could, if we wanted, make this probability arbitrarily small. Similarly, we could make the "probability of murder" pretty small and prove that most likely Mr. White is not really dead.

What was wrong with Mr. Black's argument? First, he was not computing the probability of suicide. He was computing the *probability of obtaining these data* if Mr. White had committed suicide. Suicide is a possible state of nature or cause of death and not a possible observation here. Technically, this probability is called the *likelihood of suicide* (in view of the above data). Secondly, when he multiplied probabilities, he assumed that for suicides the use of a knife and keeping the jacket on are independent. Are they? In any case, we shall see that Mr. Black should compare the *likelihood of murder with that of suicide.* That is, he must also compute the probability of obtaining these data if Mr. White had been murdered. After a good deal of questioning, it is determined that, if Mr. White had been murdered by the butler (θ_1), the probability of his being stabbed, would be 1/5.

Let us abbreviate the relevant data as follows.[1]

E_1: Mr. White was stabbed to death.
E_2: There was exactly one knife wound.
E_3: The knife wound was in the right part of the chest.
E_4: The knife was left in the body.
E_5: Mr. White was wearing his jacket.
E_6: Mr. White was in his room.
E_7: Mr. White was lying on his bed.

Consider the hypothesis of murder:[2]

$$P\{E_1|\theta_1\} = 0.2, \qquad P\{E_2|E_1, \theta_1\} = 0.3, \qquad P\{E_3|E_1, E_2, \theta_1\} = 0.6,$$
$$P\{E_4|E_1, E_2, E_3, \theta_1\} = 0.5, \qquad P\{E_5|E_1, E_2, E_3, E_4, \theta_1\} = 0.7,$$
$$P\{E_6|E_1, E_2, E_3, E_4, E_5, \theta_1\} = 0.7,$$
$$P\{E_7|E_1, E_2, E_3, E_4, E_5, E_6, \theta_1\} = 0.2.$$

Under the hypothesis of suicide (θ_2), we have

$$P\{E_1|\theta_2\} = 0.1, \qquad P\{E_2|E_1, \theta_2\} = 0.9, \qquad P\{E_3|E_1, E_2, \theta_2\} = 0.07,$$
$$P\{E_4|E_1, E_2, E_3, \theta_2\} = 0.9, \qquad P\{E_5|E_1, E_2, E_3, E_4, \theta_2\} = 0.2,$$
$$P\{E_6|E_1, E_2, E_3, E_4, E_5, \theta_2\} = 0.9,$$
$$P\{E_7|E_1, E_2, E_3, E_4, E_5, E_6, \theta_2\} = 0.4.$$

Hence, if $\mathbf{Z}$ constitutes the entirety of all 6 pieces of information,

$$P\{\mathbf{Z}|\theta_1\} = 0.001764 \qquad \text{and} \qquad P\{\mathbf{Z}|\theta_2\} = 0.000408.$$

The jury has available two actions. These are a_1 (convict the butler) and a_2 (acquit the butler). From society's point of view, it is worse to convict an innocent man than to acquit a guilty one. Suppose that the regrets corresponding to the two errors are 4 and 1 respectively.

On the basis of the evidence concerning the motivations and circumstances of Mr. White's household, it seems reasonable to the court to assume that the *a priori* probability (prior to the consideration of $E_1, E_2, \cdots, E_6$) of H_2 is about 0.5.

[1] As the reader may be aware, the example and probabilities are somewhat fabricated.

[2] The notation $P\{E_3|E_1, E_2, \theta_1\}$ represents $P\{E_3|E_1$ and $E_2\}$ when θ_1 is the state of nature and is often read "the probability of E_3 given E_1, E_2, and θ_1."

We now compute the *a posteriori* probabilities and the appropriate Bayes action in Table 7.4.

TABLE 7.4

A POSTERIORI PROBABILITIES AND BAYES ACTION FOR EXAMPLE 7.7

	$f(\mathbf{Z} \mid \theta)$	$\bar{w}$
(Murder) θ_1	0.001764	0.5
(Suicide) θ_2	0.000408	0.5

$w_1 f(\mathbf{Z} \mid \theta_1)$	0.000882
$w_2 f(\mathbf{Z} \mid \theta_2)$	0.000204
$f(\mathbf{Z})$	0.001086

$r(\theta, a)$	a_1 (Convict)	a_2 (Acquit)
θ_1	0	1
θ_2	4	0
$B(\bar{\mathbf{w}}, a)$	0.752	0.812

$\mathbf{w}_1$	0.812
$\mathbf{w}_2$	0.188

Bayes action $= a_1$ (convict)
Bayes risk $= \mathscr{B}(\bar{\mathbf{w}}) = 0.752$

If this example were to be treated from a non-Bayesian point of view, it would be necessary to list various strategies for comparison. But it would be impossible to specify a strategy, for this would require considering *all possible circumstances of any kind which would have aroused the suspicion of the police.* Among other items, we would have to consider how we would react if Mr. White had been found dead of arsenic poisoning in his garden or drowned in his bathtub, etc. It is simply unmanageable to consider all possible varieties of circumstances and, worse, to compute the appropriate probabilities.

It should be admitted that in the above example the *a priori* probability of 1/2 was obtained in a somewhat mysterious fashion. In more classical statistical problems, it often occurs that the exact value of the original *a priori* probability is not too important, for usually there is a considerable amount of data available, and the conclusion would not depend strongly on $\bar{w}$. In the above example, the conclusion depended very heavily on the value of the *a priori* probability and the regrets. There $B(\bar{\mathbf{w}}, a_1)$ and $B(\bar{\mathbf{w}}, a_2)$ were very close to one another. Occasionally, in statistical problems, one finds that $\bar{\mathbf{w}}$ is "very close" to the boundary of the set leading to action

a_1, and one is inclined to worry about taking action a_1. As long as there are only two actions to take and no possibility of accumulating additional data, the statistician should take the appropriate action. In many cases, the situation is not quite so simple or desperate. For example, there may be other actions available. In the butler's case, one might consider conviction with a recommendation of a five-year jail sentence. Then the jury would be hedging against the possibility of making a bad mistake by taking an action which is not especially good if the butler is innocent or if he is guilty.

7. SUMMARY

In this chapter we have considered some very simple examples of the type of problem that statisticians ordinarily treat. There were examples in testing hypotheses. Characteristically, in such problems, there are two possible actions, each of which is appropriate for certain states of nature. In Example 7.1, taking action a_1 is equivalent to acting as though $\theta > 31.0$ and is, therefore, identified with *accepting* the hypothesis $\theta > 31.0$.

The risk function $R(\theta, s)$ is given by

$$R(\theta, s) = \alpha_s(\theta, a_1)\, r(\theta, a_1) + \alpha_s(\theta, a_2)\, r(\theta, a_2) = \varepsilon(\theta, s)\, r(\theta)$$

where $\alpha_s(\theta, a_1)$ and $\alpha_s(\theta, a_2)$ are the action probabilities for strategy s and add up to one; $r(\theta)$ is the regret corresponding to the wrong action for θ; and $\varepsilon(\theta, s)$, the error probability, is the corresponding action probability. Since $\alpha_s(\theta, a_1) + \alpha_s(\theta, a_2) = 1$ and $\varepsilon(\theta, a_1) = \alpha_s(\theta, a_1)$ or $\alpha_s(\theta, a_2)$, any one of $\alpha_s(\theta, a_1)$ or $\alpha_s(\theta, a_2)$ or $\varepsilon(\theta, s)$ can be used to describe the action probabilities for s.

Examples in estimation are characterized by a continuous range of possible actions. Here, again, each possible action is appropriate for some state of nature and can be identified with an estimate T of θ. If the regrets are approximated by

$$r(\theta, T) = c(\theta)(T - \theta)^2,$$

it is desirable to find an estimator t, yielding an estimate

$$\mathbf{T} = t(\mathbf{Z})$$

such that

$$R(\theta, t) = c(\theta)\, E(\mathbf{T} - \theta)^2$$

tends to be small.

For problems which are not very carefully specified, it is often desirable to estimate the state of nature θ and find some idea of how good this estimate is. An elegant approach is that of confidence intervals, where one constructs a random interval $\mathbf{\Gamma}$ based on the data such that, no matter what the value of θ may be,

$$P\{\mathbf{\Gamma} \text{ covers } \theta\} = \gamma.$$

$\mathbf{\Gamma}$ is called a γ confidence interval for θ. Commonly used values of γ include 0.95 and 0.99.

In this chapter we have given up the simplifying assumption of a finite number of possible states of nature. However, the ideas developed in Chapters 5 and 6 are still applicable. We have still neglected, except to a slight extent in our exercises, the problem of design of experiments.

To help fix some of the ideas of this chapter, we present the following problem.

Exercise 7.23. "Repel" is a new insect repellent. The manufacturers claim that it is highly effective against mosquitoes. A representative of the Bureau of Standards takes 100 men and selects an arm at random for each man. He coats this arm with Repel and the other arm with water, which smells like Repel but which is supposedly ineffective. To assure an honest report, the men are not told which arm is coated with Repel. They go out at dusk and report which arm is the first to be bitten by a mosquito. Let $\hat{\mathbf{p}}$ be the proportion of people for which the first arm bitten is the one coated with Repel. Put yourself in the position of a government employee in the Bureau of Standards. You can either announce that Repel is practically worthless or let its manufacturer continue advertising.

(a) What do your regret functions for the two actions look like?

(b) Indicate what seems a reasonable strategy and compute and graph its risk function. (For convenience, approximate $p(1-p)$ by 1/4 here.)

(c) If this strategy seems to bear improvement, indicate how you would improve it.

A scientist visiting the Bureau of Standards was mildly interested in the Repel experiment. When he observed that $\hat{\mathbf{p}}$ was 0.80, he constructed a 99% confidence interval for p, the probability that the mosquitoes would first bite the arm coated with Repel.

(d) Present a 99% confidence interval for p.

(e) If you were the scientist, how would you react to the above result?

(f) To what action would the government employee's strategy lead when he observes that $\hat{\mathbf{p}} = 0.80$?

CHAPTER 8

Models

1. INTRODUCTION

It is not uncommon for an architect designing a house to build a model representing his ideas of the house. Customarily the model is built to scale and is much smaller than the house. What are the main characteristics of the model that make it desirable? It is relatively cheap to construct and resembles the house in so many fundamental respects that the architect can visualize many faults that the house may have and correct them before putting up the main building.

We must note that even though the model is a useful representation of the house, it is far from identical to the house. It is usually much smaller and lacks many details that the architect considers relatively unimportant in his design problem.

There are several possible useful ways to represent the house. The prospective buyer may find a simple outline of the floor plan sufficiently descriptive to decide whether he will like the house. Thus the floor plan can also be considered as a model of the house. The builder finds neither the architect's model nor the floor plan sufficient. He will often require a detailed list of specifications. To the layman, this list of specifications may look only like a meaningless pile of paper and may in no way resemble the house. Still, for the builder, it is a more complete and useful representation of the house than the architect's model.

We shall regard a *model* as a *useful convenient simplified representation of the essentially important aspects of a real object or situation.* In this sense, the architect's model, the floor plan, and the builder's list of specifications are all models of the house. Not one of these is a perfect description, but each is adequate for its special purpose.

In real life situations, we frequently encounter problems that can be treated only after simplifying them by eliminating many apparently minor aspects. The parent who wishes to divide four apples equally between her two children may tacitly assume that the apples are equal and give two to each. The statement "all the apples are equal" is never precisely correct but it may be sufficiently close to correct to make a useful model.

In the application of statistical ideas to real problems, it is necessary to represent a problem in a simplified form or model, to solve the corresponding model by the appropriate computations, and to convert the solution into the action it represents. The form of the model we have used in this book is one in which there is is a set representing the possible actions, a set representing the possible states of nature, a loss or regret function, a set representing the possible observations resulting from the experiment, a probability distribution on this set for each possible state of nature, a set representing the possible strategies, a risk function, and, finally, some criterion for selecting among the strategies on the basis of the risk functions. Thus the model of the problem consists in part of models of various aspects (actions, losses, etc.) of the problem. The practical solution of the problem requires the construction of an adequate model, the computation of the solution of the model, and the transformation of this solution to real action.

How do we know when a model is adequate? Often there are clear-cut signs when the model is inadequate. Two such signs are ridiculous answers and failure of the solution. Sometimes there are no clear signs available before applying the solution. There does not seem to be a well-developed theory on this problem, and the task of constructing adequate workable models seems to be a major aspect of the "art" in the various sciences. In this chapter we shall discuss models of aspects of decision making. The objective is to give some ideas of effective model building and to acquaint the student with a few of the simpler and more useful models commonly used in the practice of statistics.

In the earlier chapters, highly artificial examples have often been used. It was thus possible to shun problems of realism in the models and to focus attention on the features illustrative of the topics under discussion.

2. MODELS OF PROBABILITY AND UTILITY

Throughout this book, probability and utility considerations play a primary role. Many examples of considerable depth and importance have been studied. However, we shall find it more convenient to illustrate our building of a probability model with a more trivial case that is not of great obvious practical importance.

We return to the problem where the experiment consists of tossing a nickel and a penny, and we are interested in the number of heads. To apply the probability model, we make the following simplifying assumptions about reality. First, we assume that there are only four possible outcomes; i.e., both coins fall heads, the nickel falls heads and the penny falls tails, etc. It is assumed that we need not worry about a coin falling in a crack and standing on end and that we are not interested in such extraneous aspects as the direction in which the face of the coin points. Second, we assume that, if the experiment were repeated under "similar circumstances" many many times, the frequencies of the four possible outcomes would tend to be close to certain specific numbers called "long-run relative frequencies."

The probability model corresponding to the above simplified assumptions is a mathematical entity. It is described by the set $\mathscr{X} = \{(H, H), (H, T), (T, H), (T, T)\}$ and a probability function P defined on the subsets E of $\mathscr{X}$. To qualify as a probability function, P must satisfy the five probability properties described in Section 3, Chapter 4. Notice that these properties involve only $\mathscr{X}$ and P and do not make any reference to reality.

The following is a trivial application of the model. First, the elements of $\mathscr{X}$ are related to the concrete possible outcomes in the obvious way, and P is related to the long-run relative frequency. Then, if it were known that two heads fall 20% of the time and two tails 40% of the time (in the long run), one may translate to the model as follows: $P\{(H, H)\} = 0.2$, $P\{(T, T)\} = 0.4$. Using the model, we easily derive $P\{(H, T), (T, H)\} = 0.4$. Returning to the concrete application, we may feel assured that the long-run relative frequency of exactly one head is 40%.

Nontrivial applications of probability such as the approximate normality of $\hat{p}$ may be obtained by using the notion of independence which is a mathematical term defined by reference to the probability model. Thus, two subsets of $\mathscr{X}$, A and B, are defined to be independent if $P\{A \text{ and } B\} = P\{A\}\,P\{B\}$. What assumptions about reality correspond to the independence of A and B? In the two-coin problem, it is difficult to conceive of some common causal factor which simultaneously affects and tends to relate the outcomes of both coins. If we assume that there is no such factor, we translate this assumption about "concrete reality" to our model by stating that $\{(H, H), (H, T)\}$ (which corresponds to the nickel falling heads) and $\{(H, H), (T, H)\}$ (which corresponds to the penny falling heads) are independent.

In applications of probability theory, it is very natural to use a given term to indicate both a concept in the model and one in "concrete reality." For example, we use probability to represent the function P and long-run relative frequency. Such double usage is common and convenient, but occasionally leads to confusion. At times it is important to separate the model from reality.

A careful analysis of utility requires a mathematical model translating the assumptions of Section 2.1, Chapter 4. A detailed translation would be quite tedious to indicate at this point. Such a translation would be necessary to permit a complete and formal derivation of the utility function properties. Such a derivation appears in *Theory of Games and Economic Behavior* by Von Neumann and Morgenstern but is beyond the scope of this book. Consequently, we shall not delve into the mathematical model of utility.

3. MODELS OF THE SET OF AVAILABLE ACTIONS

To illustrate how we build and use models of the set of available actions let us re-examine Example 7.1. Imagine that, when the batch of fibers arrives, Mr. Good can think of only the following actions:

1. Accept the batch at Mr. Lacey's usual price and make parachute cord as usual.
2. Reject the batch and rent out his factory for the week required to receive a new batch.
3. Offer Mr. Lacey a price lower than the usual price for the batch.

4. Accept the batch at the usual price and readjust the machines to make cord with more or fewer strands of fiber than usual in a given length of cord.

With a little thought Mr. Good decides that, if he tried to offer a lower price, Mr. Lacey would probably be upset and his entire business relationship with Mr. Lacey would be disturbed at great financial loss to himself. Mr. Good rejects this possibility.

Readjusting the machines to use fewer strands of fiber is a very delicate and expensive operation and not worth considering unless the strands were extremely strong ($\theta > 50$ or so). On the basis of previous experience, Mr. Good sees no reason to expect such strong strands.

Thus Mr. Good feels that only the two actions described in (1) and (2) need be considered, and he proposes to solve the model set up in Example 7.1 with only two available actions.

Notice how we built this model where the set of possible actions consisted of a_1 (also named "accept the batch") and a_2 (also named "reject the batch"). We essentially considered a decision making problem which was more complex with four types of actions. This complex problem was never completely formulated. It too was a *rough preliminary model* of the still more complicated real situation, for, obviously, one could conceive of still more actions. A rough analysis of this problem seemed to indicate that the simpler model was appropriate. An attempt to formulate this more complicated model in a clearer fashion and to solve it precisely would, of course, have been very difficult.

The reader may recall how in the discussion of confidence intervals we hypothesized the data $\overline{X} = 65$ yielding a 95% confidence interval (63.04, 66.96). Such a confidence interval would lead a *flexible* Mr. Good to reconsider the discussion of the fourth type of action (readjusting the machine).

It is important to note that statisticians are rightly suspicious of such reconsiderations. Without scrupulous care, it is easy for a statistician to get into a habit of carelessly re-evaluating his model in view of the data. Such a habit may reduce his behavior from objective decision making to subjective wishful thinking and vitiate completely his evaluation of risks.[1]

[1] This statement is partly due to the fact that any large set of data has certain

In Example 7.3, we used a model where each state of nature θ determined an appropriate action (the optimal price to offer) and each price was appropriate for some θ. Here our model of the set of available actions was the set of possible prices. For technical reasons,. we found it more convenient to translate this model to an essentially equivalent model where each price was labeled by the θ for which it was appropriate. In our translated model, the set of available actions is exactly the same as the set of possible states of nature, and each action is labeled by a number T called "the estimate of θ." (By taking action T, we mean that Mr. Good offers the price which would be optimal if θ were equal to T.) The technical advantage of this model is that we can often approximate $r(\theta, T)$ by a simple expression.

4. MODELS OF SETS OF POSSIBLE STATES OF NATURE

We illustrate with two models of the sets of possible states of nature. In Example 7.1 we represented the state of nature by the following model: Individual fibers in the batch have breaking strengths which are independently and normally distributed with mean θ and standard deviation 8. The set of possible states of nature corresponds to the set of positive numbers θ.

Clearly this model is not completely realistic. First of all, any normally distributed random variable can be negative with some probability. Thus breaking strengths, which are always positive, cannot be strictly normally distributed. The assumption of normality is relatively serious but, unfortunately, we cannot discuss the effects of deviations from normality here. Second, the assumption that σ is known to be 8 would be preposterous without a good deal of previous experience. The use of the estimate **s** in place of σ is a useful technique which is easy to apply. Our assumption that σ was known was made in the interest of simplifying the elementary discussion of Example 7.1.

A second model which is used in practice is obtained as follows: Suppose that before an election a public opinion research com-

---

peculiarities which one may not have expected. For example, in the first 11 pairs of random digits in the table of random digits (Appendix C_1), the second element of the pair is a 3 or 6, 8 times out of 11. This is a very peculiar phenomenon, but we should ignore it unless (1) we have some sound reason, based on our rough preliminary model, for suspicion that our table may tend to exhibit such a phenomenon or (2) considerable additional data exhibit this phenomenon.

pany wishes to determine who will win the election, the Whigs or the Populists. In order to do so, they assume that:

1. Each prospective voter has already determined how he will vote;
2. if questioned by the company's representatives, he will answer honestly; and
3. the party which is favored by the majority of prospective voters will win. (In other words we neglect the possibility that prospective voters who favor the Whigs may be more or less likely to abstain from voting.)

These assumptions yield the following model. A randomly selected prospective voter has probability p of responding that he favors the Whigs, where p is the proportion of all prospective voters who favor the Whigs. They will win if $p > 0.5$, and lose if $p < 0.5$.

Ordinarily this model is supplemented by assuming that:

4. The population of voters is very large, which leads to the following: Let $\mathbf{X}_i = 1$, if the ith in a random sample of n prospective voters claims he favors the Whig party and let $\mathbf{X}_i = 0$ otherwise. Then $\mathbf{X}_1, \mathbf{X}_2, \ldots, \mathbf{X}_n$ are independent random variables each of which is 0 with probability $1 - p$ and 1 with probability p.

5. MODELS OF REGRET FUNCTIONS

Throughout this book we have continually discussed the construction of numerous models for regret and utility functions. One typical model is for estimation problems where it is assumed that the regret corresponding to an estimate T of a parameter θ is given by

$$r(\theta, T) = c(\theta)(T - \theta)^2.$$

This model, an obvious approximation, seems to be apropriate in many cases. In these cases, it is generally true that the approximation is very good when T is close to θ. Example 4.3 (concerned with Mr. Sheppard's rifle) is another case where the squared error is reasonable. However, there are other problems where

$$r(\theta, T) = c(\theta) \mid T - \theta \mid$$

is a closer approximation to reality. A related example is the warehouse problem, Example 4.5.

In practical applications, it is customary for the rare statistician who asks his client to specify a regret function to receive a blank stare. It turns out that clients will find difficulty in specifying regret functions because (1) they are difficult to specify and (2) the clients are not accustomed to thinking in terms of regret functions.

The student may ask whether our elaborate structure for statistical problems serves any useful purpose when it is so difficult to specify a regret function. In reply we have stated, and repeat, that, *in most statistical problems with a reasonable amount of data available, small variations in the regret function have negligible effects on the strategies selected.*

However, gross variations in the regret function do have an effect, and a rough idea of the regret function should be made available if the client wants a reasonable procedure.

Later we shall consider problems where the statistician plays a role in the design of the experiment (i.e., the experiment has not yet been performed when our statistician is called in). Then it will be important to consider the cost of an experiment. Suppose that an experiment consists of determining the breaking strength of an unspecified number n of individual fibers. Here the experiment designer still has a choice of determining n. A good choice must balance the cost of acquiring additional data against the utility of having the extra information. How does $C(n)$, the cost of examining n fibers, depend on n? One typical model would be expressed by

$$C(n) = cn.$$

This model corresponds to assuming that each observation has a fixed cost c. Another would be given by

$$C(n) = a + cn.$$

This model corresponds to assuming a cost of a for setting up the equipment and a charge of c for each observation. These two models are frequently used and give reasonable results although, in most situations, more complicated functions would be more realistic.

6. MODELS OF EXPERIMENTS

In Chapter 1 we described the experiment in terms of (1) the set

$\mathscr{Z}$ of all possible outcomes z and (2) the response frequencies of the possible observations for each possible θ. After our discussion of probability in Chapters 3, 4, and 6 and our discussion of the probability model in Section 2, we may summarize as follows. The model of our experiment gives for each θ a probability distribution on the set $\mathscr{Z}$. In the cases where the distributions are continuous or discrete, the probability distributions can be concisely summarized by a probability density function $f(z \mid \theta)$. Then $f(z \mid \theta)$ completely specifies the model of the experiment.

In most statistical problems where the data are the observed values of numerical random variables, such a density function is appropriate. To be sure, there are situations where the data are not numerical (e.g., the sociologist who observes an angry stare or a smile in reply to a remark). In such situations, the mathematician can often transform the data to the values of numerical random variables. For example, in the polling problem, a Whig is assigned the value one and a Populist the value zero. For simplicity, we shall hereafter assume that each experiment can be described by a density $f(z \mid \theta)$.

The reader may note that in many problems the model of the state of nature more or less automatically determines $f(z \mid \theta)$. For example in the cord problem (Example 7.1), the state of nature describes the probability distribution of individual breaking strengths which determines f.

Now we shall proceed to discuss several kinds of models or parts of models of the experiments which find frequent application in statistical practice.

†6.1. Specifications of Distributions

It is not uncommon to assume that the data are normally distributed with mean or variance, or both, depending on the state of nature. In some problems, other distributions suggest themselves. For example, there is reason to believe that the lifetime (number of hours of use before failure) of a radio tube is a random variable with approximately an exponential distribution. The exponential distribution with mean a is given by

$$\begin{aligned} f(x) &= 0 && \text{for } x < 0 \\ f(x) &= \frac{1}{a} e^{-x/a} && \text{for } x \geq 0. \end{aligned}$$

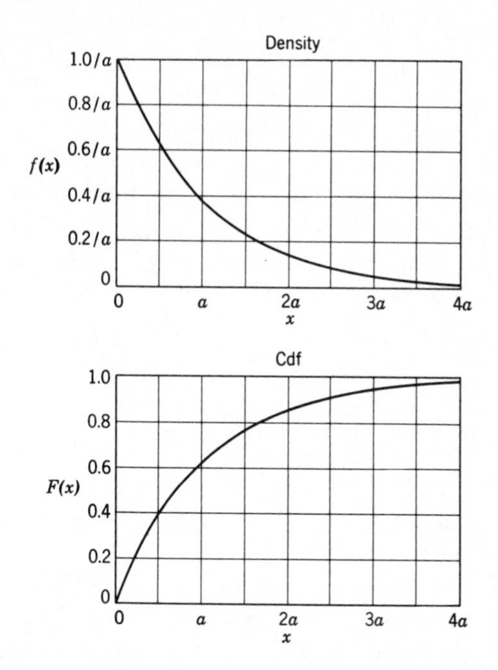

Figure 8.1. Density $f(x)$ and cdf for the exponential distribution $f(x) = (1/a)e^{-x/a}$ for $x > 0$.

Figure 8.1 represents the form of the density and cumulative distribution function. The cdf is tabulated in Appendix D_3.

The exponential distribution has many other uses. For example, if a specimen of radioactive material is put in a Geiger counter, the time between two successive counts is commonly assumed to have an exponential distribution. This is related to the additional fact that, if radioactivity R (as measured by the counting rate) is studied, it will decline exponentially in time, i.e., $R = ce^{-bt}$, where t is time.

Sometimes there is reason to use the model that a density is symmetric about its "center." This model can be expressed by the equation

$$f(\theta + x) = f(\theta - x)$$

where θ is the center. Two examples of symmetric distributions are the normal distribution and the distribution of **X**, the sum of the two faces on a pair of ideal dice.

Exercise 8.1. Does the outcome **X** of an ideal dial have a symmetric distribution? Is the exponential distribution symmetric? Give reasons.

Many continuous distributions have only a single mode. By this we mean that the density has only a single peak. Histograms of grades on an examination often seem to indicate distributions which are bimodal, i.e., having two peaks. Distributions which have a single mode are called "unimodal." Such a distribution can be expressed by the relations

$$f(x_1) \leq f(x_2) \qquad \text{if } x_1 < x_2 \leq \theta$$

and

$$f(x_3) \leq f(x_4) \qquad \text{if } x_3 > x_4 \geq \theta.$$

Here θ represents the mode.

Exercise 8.2. Is the normal distribution unimodal? Is the exponential distribution unimodal?

Exercise 8.3. Would you guess that the heights of American males over 21 years of age have a unimodal distribution? What would you say about the heights of American adults (combining men and women over 21)?

†6.2. Models Concerned with Relations Between Variables

Physics abounds with models of this type. For example, Boyle's law is an assertion about the mutual relationships amongst the pressure, temperature, and volume of a gas. Newton's laws of motion express relationship among variables such as force, mass, and acceleration. Often there is reason to suppose, or to inquire about, the existence of a functional relationship among variables. Perhaps the simplest type of functional relationship is such a one as Boyle's law, where the value of one variable can be expressed in terms of the values of others. However, where data are obtained with random errors too large to be ignored, models are less likely to be fruitful if they relate to the random variables themselves than if they relate to parameters of the probability distributions. For example, it is a matter of everyday experience that, in general, tall people are heavier than short people. It would be a mistake to try to determine the weight of a person from his height alone since there are many people of the same height who have different weights. On the other hand, it may be possible to specify the average weight of people of a given height. Let $\mathbf{X}$ and $\mathbf{Y}$ represent the height and weight respectively of an individual selected at random from the population. Let $E(\mathbf{Y} \mid x)$ represent the *average weight of individuals of height* x. The expression $E(\mathbf{Y} \mid x)$ is called the regression of $\mathbf{Y}$ on $\mathbf{X}$ (weight on height). We might be interested in how $E(\mathbf{Y} \mid x)$ depends on x. For example, it seems reasonable to propose the model

$$E(\mathbf{Y} \mid x) \text{ increases as } x \text{ increases}$$

(i.e., weight tends to increase with height).

This model is not a very interesting one for height and weight. If it were an assertion about honesty and intelligence, or income and hours of sleep, it might be more exciting.

Sometimes we can expect that a more definitely prescribed model is appropriate. If $\mathbf{Y}$ is the yield of penicillin in a certain production setup where the only unspecified variable is the temperature t of the mix, we might consider many possible models; three such possibilities are:

1. $E(\mathbf{Y} \mid t) = A + Bt$;
2. $E(\mathbf{Y} \mid t) = A + Bt + Ct^2$;
3. $E(\mathbf{Y} \mid t) = A + B \log t$.

Notice that, if $B > 0$, then it follows in Models 1 and 3 that $E(\mathbf{Y} \mid t)$ increases as t increases. Notice that, if $C = 0$ in Model 2, it becomes Model 1. Notice that we would not seriously expect any of these three models to hold over a wide range of temperatures t since, at high temperatures, the penicillin organism would be killed, and below some temperature the yield should be exactly zero.

Problems concerned with relations between variables are notorious for the frequency with which fallacies arise from inadequacies in the model. For example, one can show beyond a shadow of a doubt that among children in elementary school the swifter runners are on the average better spellers. (Among children, both running and spelling ability increase with age.) Any model which fails to take into account one or more important variables runs the risk of generating nonsense.

Exercise 8.4. For each of the three models above, write the conditions on A, B, and C so that $E(\mathbf{Y} \mid t)$ does not depend on t.

One important kind of relation between variables involves variables of classification. Recall Mr. Sharp's observation that customers at the restaurant could be classified as "Big Eaters" or "Not Big Eaters" and as "Affluent" or "Not Affluent." Here there were four possible outcomes for the random variable. The distribution of such a random outcome (customer) can be specified as follows.[1]

	B	Not B	
A	p_{11}	p_{12}	$p_{1.}$
Not A	p_{21}	p_{22}	$p_{2.}$
	$p_{.1}$	$p_{.2}$	1

If now A stands for "Affluent" and B for "Big Eater," we have Mr. Sharp's situation. If on the other hand A stands for "Anemic" and B stands for "Vitamin B Deficiency," we have another situation. If A stands for "Native of the State of Residence" and B stands for "High School Graduate," we have still another situation. Often relevant questions can be posed in terms of this model. Is $p_{11} > 1/2$? Is $p_{.1} > 1/2$? An especially important question may be: Are A and B independent, i.e., is

$$p_{11} = p_{1.}p_{.1}?$$

Exercise 8.5. For which of the following cross classifications would you expect independence?

People classified by: sex; over 150 lb in weight or not.
People classified by: sex; blue eyes or not.

7. MODELS OF THE SET OF AVAILABLE STRATEGIES

In Chapters 1 and 5, our model of the decision making problem involved a set of possible actions, a set of possible states of nature, a regret function, an experiment, a set of available strategies, and a risk function. Implicit in this model is the assumption that the experiment has been determined. This model is especially appropriate for problems where the statistician is called in after all the

[1] In Chapter 6 we used $P\{A \text{ and } B\}$ instead of p_{11}, $P\{A\}$ instead of $p_{1.}$, etc. The subscript notation is frequently more convenient.

data have been collected. Here the strategy consists of a rule or recipe for reacting to the data, i.e., *the set of available strategies is the set of all possible functions s on the set $\mathscr{Z}$ of possible data to the set of possible actions.* A typical strategy s will lead to a random action given by

$$\mathbf{A} = s(\mathbf{Z})$$

where $\mathbf{Z}$ is the outcome of the experiment.

For simplicity we have used the above structure for the model of the decision making problem. However, in many applications, the decision maker can exert some influence on the choice of an experiment. Then, roughly speaking, his object is to select an experiment which will be highly informative and not very costly. Furthermore, after the experiment, he must face the question of whether to take one of the available actions or to do more experimentation. These intermediate choices or decisions which finally culminate in terminating experimentation and taking an action are properly all part of an over-all strategy.

From this point of view, the proper structure of a decision making problem should be as follows: first we have actions, states of nature, and a regret function. Then we must select a strategy; *the strategy is a rule which tells how to react to data as they become available.* At each point the strategy determines which experiment to perform next or dictates a final action from the set of available actions. Thus, the design of experiments is one part of the selection of the strategy.

Finally, we compute the risk function $R(\theta, s)$ which is the expected regret and includes the regret due to taking the final action and due to the cost of experimentation.

It is customary and convenient but not necessary to assume that

$$R(\theta, s) = E[r(\theta, \mathbf{A})] + E(\mathbf{C})$$

where $\mathbf{A}$ is the final action and $\mathbf{C}$ is the cost of experimentation. Generally, both $\mathbf{A}$ and $\mathbf{C}$ are random depending upon the random outcomes of experiments.

Although our notions concerning strategies have been broadened considerably by this discussion, the main ideas of Chapters 5 and 6 dealing with the theory of decision making are still valid.

Furthermore, we still denote our final action by $\mathbf{A} = s(\mathbf{Z})$ even though s does considerably more than merely select $\mathbf{A}$.

8. THE MODELS FOR THE PROBLEMS OF TESTING AND ESTIMATION

Now we are prepared to discuss concisely the character of problems of testing hypotheses and estimation. We shall call a problem one of testing hypotheses if (1) there are two possible actions a_1 and a_2 and (2) the set of states of nature can be decomposed into two nonoverlapping sets $\mathscr{N}_1$ and $\mathscr{N}_2$ such that a_1 is optimal for $\theta \in \mathscr{N}_1$ and a_2 for $\theta \in \mathscr{N}_2$, i.e.,

$$r(\theta, a_1) = 0 \qquad \text{for } \theta \in \mathscr{N}_1$$

and

$$r(\theta, a_2) = 0 \qquad \text{for } \theta \in \mathscr{N}_2 .$$

Traditionally, one talks of the hypotheses H_i that $\theta \in \mathscr{N}_i$. Customarily, one equates taking action a_i with accepting the hypothesis H_i. The action probabilities are given by

$$\alpha_s(\theta, a_1) = P\{s(\mathbf{Z}) = a_1 \mid \theta\}$$

and

$$\alpha_s(\theta, a_2) = P\{s(\mathbf{Z}) = a_2 \mid \theta\} .$$

The error probabilities are given by

$$\varepsilon(\theta, s) = P\{s(\mathbf{Z}) = a_2 \mid \theta\} \qquad \text{for } \theta \in \mathscr{N}_1$$

and

$$\varepsilon(\theta, s) = P\{s(\mathbf{Z}) = a_1 \mid \theta\} \qquad \text{for } \theta \in \mathscr{N}_2 .$$

The risk function is given by

$$R(\theta, s) = \alpha_s(\theta, a_1)\, r(\theta, a_1) + \alpha_s(\theta, a_2)\, r(\theta, a_2) + E(\mathbf{C}) .$$

Note that there are no hedging actions since each action is optimal for some states of nature.

The test of significance (see Section 5, Chapter 7) is a hypothesis testing procedure which is formally treated without giving explicit consideration to $\mathscr{N}_2$, $r(\theta, a_i)$, or $\alpha_s(\theta, a_1)$. The method thus deals with an incompletely formulated model of the statistical problem.

Some real life statistical problems are so complicated as to make the use of significance tests convenient. However, a rational use of this model must necessarily involve consideration of the underlying decision problem.

The estimation problem is characterized by the following model. The set of possible actions (called estimates) coincides with the set of possible states of nature. The risk function $r(\theta, T)$ satisfies the conditions

$$r(\theta, T) = 0 \qquad \text{for } T = \theta$$

$$r(\theta, T) > 0 \qquad \text{for } T \neq \theta.$$

Here the strategy (at least that aspect leading to an action) is called an estimator and is denoted by t.

Note that there are no hedging actions since each action T is optimal for $\theta = T$. We may point out that a problem with two states of nature and two actions could be called either a hypothesis testing problem or an estimation problem. Here the actions a_1 and a_2 may either be called accepting H_1: $\theta = \theta_1$ and accepting H_2: $\theta = \theta_2$ or estimating θ as θ_1 and estimating θ as θ_2. Often problems with a finite number of available actions are called testing problems. We usually reserve the estimation title for cases where there is a continuous range of possible estimates.

9. SUMMARY

In this chapter we discussed the concept of a model as a possibly simplified representation of reality which permits one to come to conclusions which can be translated into useful decisions about reality. In discussing models of sets of available actions, we illustrated the need and usefulness of simplification, of rough preliminary models, and some of the art of model building. The fact that models of the regret function must usually be oversimplified because of difficulty in specifying regrets precisely is not a fundamental difficulty. The reason for this is that, in problems where considerable data are available, strategies are negligibly affected by small variations in the regrets.

The model of the experiment is the probability distribution of the experiment for each state of nature. Usually this is given by a density $f(z \mid \theta)$. Certain characteristics of the distribution are often of interest. These include symmetry and unimodality. Models involving relations between variables have wide applicability, and some were discussed in Section 6.2.

In Chapters 1 and 5 our concept of strategy was based on the assumption that the experiment was determined. In Section 7 we indicated that the role of the strategy can be widened to include the design of experiments.

Finally, the models for testing hypotheses and estimation are given. Both models fail to allow for the possibility of hedging actions.

The art of statistics seems to lie in the construction of good models, for there seems to be no substantial theory for this topic.

CHAPTER 9

Testing Hypotheses

1. INTRODUCTION

A problem in which one of two actions must be chosen is said to be a problem in hypothesis testing. The trial of the butler in Example 7.7 is of this type (acquit or convict being the two actions); also Example 7.1 (accept the lot of fibers or reject the lot being the two actions). In each of these examples one action was appropriate to certain possible states of nature, and the other action appropriate to the other possible states of nature. The least complicated hypothesis-testing problems are those where there are only two possible states of nature, θ_1 and θ_2. Although nonartificial problems of this special character rarely occur (see Exercise 7.8), study of this two-state, two-action type of problem increases insight into the often encountered problems where there are two actions and many states of nature which naturally divide into two classes. (Observe that in Example 7.1 "good fibers" correspond to the *many* possible states where the value of θ is a number greater than 31).

2. NOTATION

Suppose that we have a two-action problem where action a_1 is optimal for any θ in $\mathscr{N}_1$ and a_2 is optimal for any θ in $\mathscr{N}_2$, and where θ is known to be in one of the two nonoverlapping sets $\mathscr{N}_1$, $\mathscr{N}_2$. We denote the hypotheses by H_1: $\theta \in \mathscr{N}_1$ and $H_2 : \theta \in \mathscr{N}_2$.

We say that the hypothesis H: $\theta \in \mathscr{N}$ is *simple* if $\mathscr{N}$ has only one element. If $\mathscr{N}$ has more than one element, we say that H is *composite*. Obviously the problem of testing a simple hypothesis versus a simple hypothesis is a two-action, two-state problem. For problems in testing hypotheses, a strategy is called a *test*. A pure strategy or test can be described by the set A_1 of possible outcomes which lead to action a_1 (called accepting H_1 because a_1 is the action one would take if one believed H_1). It can also be described by the set $A_2 = \tilde{A}_1$ of possible outcomes which lead to action a_2 (called *rejecting* H_1 or *accepting* H_2). The statistician can make one of two kinds of errors. He can reject the hypothesis H_1 (i.e., take action a_2) when it is true, or he can accept the hypothesis when it is false. These are sometimes called errors of Type I and Type II respectively.

In a testing problem, we have

$$R(\theta, s) = \alpha_s(\theta, a_1)\, r(\theta, a_1) + \alpha_s(\theta, a_2)\, r(\theta, a_2) \tag{9.1}$$

where $\alpha_s(\theta, a_1)$ and $\alpha_s(\theta, a_2)$ are the action probabilities for strategy s. But if θ were known, one of the two actions would be appropriate, and the corresponding regret would be zero. Then we can call the probability of taking the wrong action the *error probability* $\varepsilon(\theta, s)$, and the *regret* due to the error of taking the wrong action we call $r(\theta)$. Then

$$R(\theta, s) = \varepsilon(\theta, s)\, r(\theta). \tag{9.1a}$$

3. SIMPLE HYPOTHESIS VERSUS SIMPLE HYPOTHESIS (TWO STATES OF NATURE)

We shall illustrate our ideas with the following example.

Example 9.1. Mr. Jones is faced with the necessity of choosing between the two actions: (a_1) bet a large sum that a coin will fall heads and (a_2) bet it on tails. It is known that the probability p of falling heads is either given by H_1: $p = p_1 = 3/5$ or H_2: $p = p_2 = 1/3$. Suppose the regrets due to the *errors* of taking action a_2 when H_1 is correct, and action a_1 when H_2 is correct, are 3 and 5 respectively. He is allowed to toss the coin three times before betting.

We tabulate the regret function and the probability distribution of the data in Tables 9.1 and 9.2. In the latter table, we include the *likelihood ratio*

$$\lambda(z) = \frac{f(z \mid \theta_1)}{f(z \mid \theta_2)}.$$

In Figure 9.1 we present the set of risk points $(R_1, R_2) = (R(\theta_1, s), R(\theta_2, s))$ corresponding to the set of all strategies (pure and randomized). We also present the set of error points $(\varepsilon_1, \varepsilon_2) = (\varepsilon(\theta_1, s), \varepsilon(\theta_2, s))$ corresponding to the set of all strategies. Since $R(\theta, s) = r(\theta)\, \varepsilon(\theta, s)$, the set of risk points is obtained from the set

TABLE 9.1

REGRET DUE TO TAKING WRONG ACTION: $r(\theta)$

θ	$r(\theta)$
θ_1	$r(\theta_1)$
θ_2	$r(\theta_2)$

	$r(p)$
$p = 3/5$	3
$p = 1/3$	5

TABLE 9.2

PROBABILITY DISTRIBUTION OF DATA $f(z \mid \theta)$ AND LIKELIHOOD RATIO $\lambda(z)$

	z_1	z_2	$\cdots$	z_k
θ_1	$f(z_1 \mid \theta_1)$	$f(z_2 \mid \theta_1)$	$\cdots$	$f(z_k \mid \theta_1)$
θ_2	$f(z_1 \mid \theta_2)$	$f(z_2 \mid \theta_2)$	$\cdots$	$f(z_k \mid \theta_2)$
$\lambda(z)$	$\frac{f(z_1 \mid \theta_1)}{f(z_1 \mid \theta_2)}$	$\frac{f(z_2 \mid \theta_1)}{f(z_2 \mid \theta_2)}$	$\cdots$	$\frac{f(z_k \mid \theta_1)}{f(z_k \mid \theta_2)}$

	z_1 *HHH*	z_2 *HHT*	z_3 *HTH*	z_4 *THH*	z_5 *HTT*	z_6 *THT*	z_7 *THH*	z_8 *TTT*
$p = 3/5$	0.216	0.144	0.144	0.144	0.096	0.096	0.096	0.064
$p = 1/3$	0.037	0.074	0.074	0.074	0.148	0.148	0.148	0.297
$\lambda(z)$	5.838	1.946	1.946	1.946	0.649	0.649	0.649	0.215

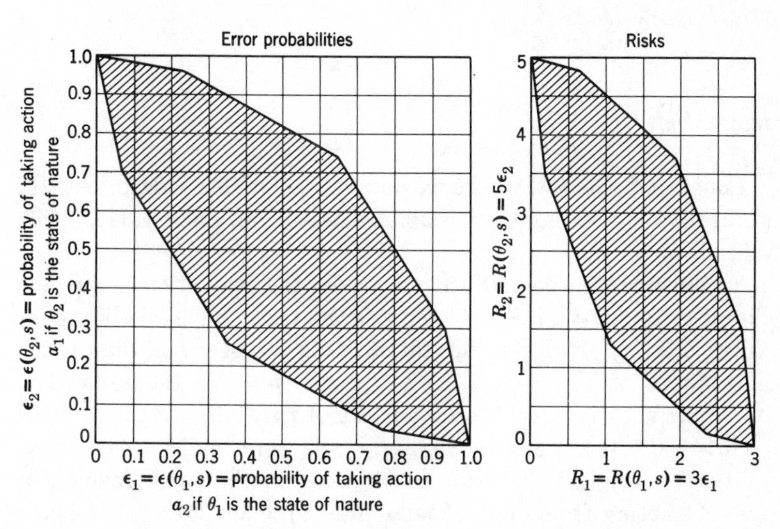

Figure 9.1. Error probabilities and risks for all strategies of Example 9.1.

of error points by multiplying the scales by $r(\theta_1)$ and $r(\theta_2)$ respectively.[1] It is clear that both sets are convex and that the admissible strategies correspond to the admissible part of the boundary on each set. The admissible part of the boundary of the set of error points is called the *error curve.*

In Example 9.1 there are 256 pure strategies, and it would be quite a task to obtain Figure 9.1 by listing and evaluating them all. Is there a simple method of characterizing the boundary of the figure or at least the admissible strategies? The answer to this question is "yes." We know that each admissible strategy is the Bayes strategy for some *a priori* probabilities. In Appendix E_{11} we use the methods of Chapter 6 to show that *every Bayes strategy is a likelihood-ratio test.* We define the likelihood-ratio tests as follows. If the data $\mathbf{Z}$ are observed, $f(\mathbf{Z}\,|\,\theta)$ is called the *likelihood* of θ. Note that the likelihood represents the probability of observing $\mathbf{Z}$ when θ is the state of nature. The *likelihood ratio* is then given by

$$\lambda(\mathbf{Z}) = \frac{f(\mathbf{Z}\,|\,\theta_1)}{f(\mathbf{Z}\,|\,\theta_2)}.$$

A test is said to be a *likelihood-ratio test if there is a number k such that this test leads to*

$$\begin{array}{lll} & a_1 & \text{if } \lambda(\mathbf{Z}) > k \\ & a_2 & \text{if } \lambda(\mathbf{Z}) < k \\ \text{and either} & a_1 \text{ or } a_2 & \text{if } \lambda(\mathbf{Z}) = k. \end{array}$$

Consider Table 9.3, where we obtain the likelihood-ratio tests for values of k bracketing the possible values of $\lambda(z)$, 5.838, 1.946, 0.649, and 0.215.

To illustrate the computation, consider the evaluation of the likelihood-ratio test with $k = 1$. This is the test s where action a_1 is taken if $\lambda(\mathbf{Z}) > 1$. Referring to Table 9.2, we see then that action a_1 is taken if $\mathbf{Z} \in A_1 = \{z_1, z_2, z_3, z_4\}$. For θ_1 ($p = 3/5$), the probability of taking action a_1 is $\alpha_s(\theta_1, a_1) = P\{\mathbf{Z} \in A_1\,|\,p = 3/5\} = 0.216 + 0.144 + 0.144 + 0.144 = 0.648$. Similarly $\alpha_s(\theta_2, a_1) = 0.259$. The probability of taking action a_2 can be obtained in the same fashion or by subtracting from one. Finally, $R(\theta_1, s) = 3(0.352) = 1.056$ and

[1] It is sometimes convenient to consider the error points as risk points for a problem where $r(\theta_1) = r(\theta_2) = 1$.

TABLE 9.3

SOME LIKELIHOOD-RATIO TESTS FOR EXAMPLE 9.1[1]

k	A_1	θ	Regrets a_1	Regrets a_2	Action Probabilities a_1	Action Probabilities a_2	Risks $R(\theta, s)$
7	ϕ	θ_1	0	3	0.000	1.000	3.000
		θ_2	5	0	0.000	1.000	0.000
5	$\{z_1\}$	θ_1	0	3	0.216	0.784	2.352
		θ_2	5	0	0.037	0.963	0.185
1	$\{z_1, z_2, z_3, z_4\}$	θ_1	0	3	0.648	0.352	1.056
		θ_2	5	0	0.259	0.741	1.295
0.5	$\{z_1, z_2, \cdots, z_7\}$	θ_1	0	3	0.936	0.064	0.192
		θ_2	5	0	0.703	0.297	3.515
0.2	$\{z_1, z_2, \cdots, z_8\}$	θ_1	0	3	1.000	0.000	0.000
		θ_2	5	0	1.000	0.000	5.000

$R(\theta_2, s) = 5(0.259) = 1.295$. Note that the error probabilities are $\varepsilon(\theta_1, s) = \alpha_s(\theta_1, a_2) = 0.352$, $\varepsilon(\theta_2, s) = \alpha_s(\theta_2, a_1) = 0.259$.

All other values of k were deliberately omitted in Table 9.3. First of all nothing new is obtained if we use more than one value of k between adjacent possible values of the likelihood ratio. Specifically, referring to Table 9.2, the likelihood-ratio test for $k = 1.5$ yields $A_1 = \{z_1, z_2, z_3, z_4\}$ which is the same as for $k = 1$. Secondly, the values of k we used yield the vertices of the admissible boundary. If we use for k a possible value[2] of λ, we obtain points on the line segments connecting the vertices. We illustrate with four of the eight pure strategies corresponding to $k = 1.946$.

A_1	θ	Regrets a_1	Regrets a_2	Action Probabilities a_1	Action Probabilities a_2	Risk $R(\theta, s)$
$\{z_1\}$	θ_1	0	3	0.216	0.784	2.352
	θ_2	5	0	0.037	0.963	0.185
$\{z_1, z_2\}$	θ_1	0	3	0.360	0.640	1.920
	θ_2	5	0	0.111	0.889	0.555
$\{z_1, z_3, z_4\}$	θ_1	0	3	0.504	0.496	1.488
	θ_2	5	0	0.185	0.815	0.925
$\{z_1, z_2, z_3, z_4\}$	θ_1	0	3	0.648	0.352	1.056
	θ_2	5	0	0.259	0.741	1.295

[1] This title is more modest than necessary. Using this table, the entire boundary of the set of risk points can be obtained.

[2] More precisely, by a possible value of λ, we mean a number k such that $P\{\lambda(\mathbf{Z}) = k\} > 0$ for some θ.

All of these strategies lie on the line segment connecting the vertices corresponding to $k = 5$ and $k = 1$ and so are equivalent to mixtures of these two strategies.

Finally, suppose that we wish to obtain the rest of the boundary (dominated part) of the sets of risk points and error points. It can be shown that the vertices of the inadmissible part of the boundary are obtained by replacing the A_1 sets in Table 9.3 by their complements. This yields the foolish tests of accepting H_1 when the likelihood ratio is small.

To summarize, for tests of a simple hypothesis versus a simple alternative, every admissible strategy is a Bayes strategy and every Bayes strategy is a likelihood-ratio test. This result may be reasonably interpreted to mean that high values of the likelihood ratio

$$\lambda(\mathbf{Z}) = \frac{f(\mathbf{Z} | \theta_1)}{f(\mathbf{Z} | \theta_2)}$$

tend to support $H_1 : \theta = \theta_1$ and low values tend to support the alternative hypothesis $H_2 : \theta = \theta_2$.

Exercise 9.1. Derive and graph the error curve for the strategies in the problem described by $f(z | \theta)$ in Table 9.4.

TABLE 9.4

$f(z | \theta)$

	z_1	z_2	z_3	z_4	z_5	z_6
θ_1	0.04	0.10	0.16	0.20	0.30	0.20
θ_2	0.32	0.20	0.16	0.15	0.12	0.05

Remarks

1. In the two-state testing problem, the class of admissible strategies corresponds to the error curve (admissible part of the set of error points). But the error curve does not involve the particular regrets $r(\theta_1)$ and $r(\theta_2)$. In any testing problem, admissibility does not depend on the particular values of $r(\theta)$. Of course, if $r(\theta_1)$ is much larger than $r(\theta_2)$, one would be inclined to select an admissible strategy for which $\varepsilon_1 = \varepsilon(\theta_1, s)$ is relatively small, i.e., to use a likelihood-ratio test with a small k so that action a_1 is very likely to be used and ε_1 to be small. In fact, Appendix E_{11} shows that the Bayes strategy is the likelihood-ratio test for $k = w\, r(\theta_2)/(1 - w)\, r(\theta_1)$

and, as $r(\theta_1)/r(\theta_2)$ increases, k decreases, making it easier to take action a_1. Similarly if $w_1/w_2 = (1 - w)/w$ is large, one prefers to believe in H_1 and to take action a_1.

2. Our proof that the Bayes, and therefore admissible, strategies are likelihood-ratio tests involved the assumption of discrete distributions. This assumption is not necessary and the result also holds when the data have a continuous distribution.

We present two important examples applying the likelihood-ratio test. In one of these examples we generalize Example 9.1. When a coin, which has probability p of falling heads, is tossed n times, the number of heads, $\mathbf{m}$, is said to have a *binomial distribution*. Of course, the result indicated below for the coin-tossing problem applies whenever the data have a binomial distribution with unknown p. For example, it applies to polling problems.

Example 9.2. Binomial Distribution. We generalize Example 9.1 for any p_1 and p_2 with $p_1 > p_2$, for any $r(\theta_1)$ and $r(\theta_2)$, and for the experiment consisting of n tosses of the coin for any n.

In Appendix E_{12} we show that the likelihood-ratio tests of H_1: $p = p_1$ versus H_2: $p = p_2$ can be reduced to the following form:

$$\begin{array}{rl} \text{Accept } H_1 & \text{if } \hat{\mathbf{p}} > k' \\ \text{Reject } H_1 & \text{if } \hat{\mathbf{p}} < k' \\ \text{Take either action} & \text{if } \hat{\mathbf{p}} = k' \end{array}$$

where $\hat{\mathbf{p}}$ is the proportion of heads observed.

This result applies to Example 9.1 since $\hat{\mathbf{p}}$ takes on the values 1, 2/3, 2/3, 2/3, 1/3, 1/3, 1/3, 0 for $\mathbf{Z} = z_1, z_2, \cdots, z_8$. Notice that the class of likelihood-ratio tests does not involve the particular values of p_1 and p_2. In Figure 9.2 we present the error curves for $p_1 = 3/5$, $p_2 = 1/3$, and $n = 3$, 10, 50, and 100.

Exercise 9.2. Use the fact that, when n is large, $\hat{\mathbf{p}}$ is approximately normally distributed with mean p and standard deviation $\sqrt{p(1-p)/n}$ to verify some points on the error curve for $n = 100$ in Figure 9.2.

Example 9.3. Find the sample size required to test H_1: $p = p_1 = 3/5$ versus H_2: $p = p_2 = 1/3$ so that $\varepsilon(p_1, s) = 0.05$ and $\varepsilon(p_2, s) = 0.01$.

In Figure 9.3, we draw typical (approximate) densities of $\hat{\mathbf{p}}$ under H_1 and H_2 and locate a hypothetical k' between p_1 and p_2. The error

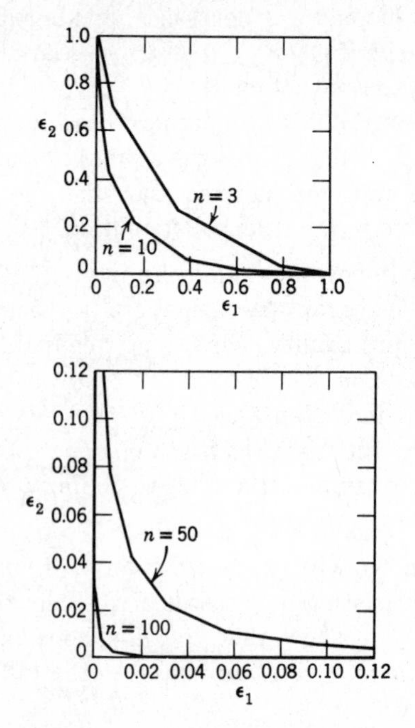

Figure 9.2. Error curve for testing $p = 3/5$ versus $p = 1/3$ (Example 9.2) for sample sizes $n = 3$, 10, 50, and 100.

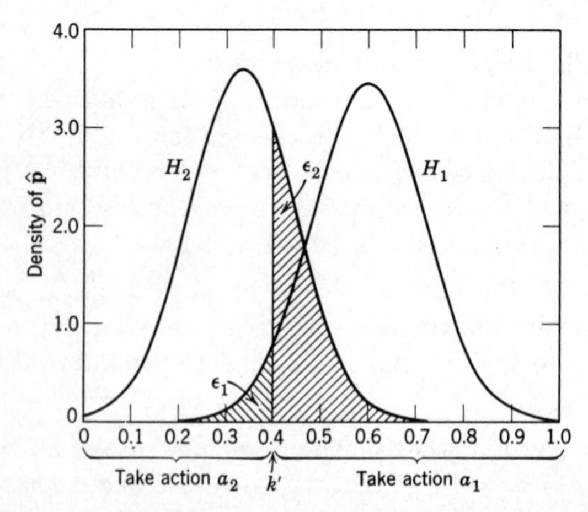

Figure 9.3. Approximate densities of $\hat{\mathbf{p}}$ under H_1 and H_2. $\varepsilon_1 = \varepsilon(p_1, s) = P\{\hat{\mathbf{p}} < k' \mid p_1\}$, $\varepsilon_2 = \varepsilon(p_2, s) = P\{\hat{\mathbf{p}} \geq k' \mid p_2\}$ if s consists of accepting $p = p_1$ when $\hat{\mathbf{p}} \geq k'$.

probabilities are given by the shaded areas. Using the normal cdf, for the appropriate k' and n, the shaded area on the left should be $\varepsilon_1 = 0.05$. Thus k' should be 1.645 standard deviations from the mean p_1. That is

$$p_1 - k' = 1.645\sqrt{p_1(1 - p_1)/n}.$$

Similarly, the shaded area on the right should be $\varepsilon_2 = 0.01$ and thus k' should be 2.326 standard deviations from the mean of $\hat{\mathbf{p}}$ (under the hypothesis $p = p_2$). Thus

$$k' - p_2 = 2.326\sqrt{p_2(1 - p_2)/n}.$$

Solving for n and k' (the first step is to divide one equation into the other, eliminating n), we obtain $n = 50.89$ and $k' = 0.487$. Thus for $n = 51$ and $k' = 0.487$, we can improve slightly on $\varepsilon_1 = 0.05$ and $\varepsilon_2 = 0.01$.

Exercise 9.3. Find the sample size required to test $H_1: p = p_1 = 0.55$ versus $H_2: p = p_2 = 0.45$ so that $\varepsilon(p_1, s) = \varepsilon(p_2, s) = 0.05$.

Exercise 9.4. Let $p_1 = 3/5$, $p_2 = 1/3$, $r(p_1) = 3$, $r(p_2) = 5$, and $n = 100$. Compute $R(p_1, s)$ and $R(p_2, s)$ for several likelihood-ratio tests, i.e., several values of k'. Graph $R(p_1, s)$ against k' and $R(p_2, s)$ against k'. What is the minimax risk strategy? What is the corresponding risk?

Exercise 9.5. Compute $R(p_1, s)$ and $R(p_2, s)$ as in Exercise 9.4. For each strategy graph $(1/4)R(p_1, s) + (3/4)R(p_2, s)$ against k'. What is the Bayes strategy for *a priori* probabilities 1/4 and 3/4 for $p = p_1$ and $p = p_2$ respectively.[1]

Example 9.4. Normal Distribution. Consider the two-state testing problem where the experiment yields n independent observations $\mathbf{X}_1, \mathbf{X}_2, \cdots, \mathbf{X}_n$ from a normal population with unknown mean μ and known standard deviation σ. The hypotheses are $H_1: \mu = \mu_1$ and $H_2: \mu = \mu_2$ (where $\mu_1 > \mu_2$).

In Appendix E_{12} we show that the likelihood-ratio tests of H_1 versus H_2 are of the following form:

[1] It is easy to modify the argument of Appendix E_{12} to find the relationship between k' and $k = [w\,r(\theta_2)]/[(1 - w)\,r(\theta_1)]$. Then there would be a more direct method of finding the Bayes strategy. However the two-state testing problem is more of a theoretical tool than a practical problem and there is little call for Bayes solutions of two-state testing problems in actual practice.

$$\begin{aligned} &\text{Accept } H_1 && \text{if } \bar{\mathbf{X}} > k' \\ &\text{Reject } H_1 && \text{if } \bar{\mathbf{X}} < k' \\ &\text{Take either action} && \text{if } \bar{\mathbf{X}} = k'. \end{aligned}$$

Since the probability that $\bar{\mathbf{X}} = k'$ for some specified k' is zero, the choice of the action taken when $\bar{\mathbf{X}} = k'$ has no effect on the error probabilities. Notice that in this example also the class of likelihood-ratio tests does not involve the particular values of the parameters μ_1 and μ_2. This property plays an important role in extending the results for the two-state binomial and normal problems to the many-state case.

In Figure 9.4 we indicate the error curves for Example 9.4 with $\mu_1 = 32$, $\mu_2 = 30$, $\sigma = 8$, and $n = 16$, 64, and 256.

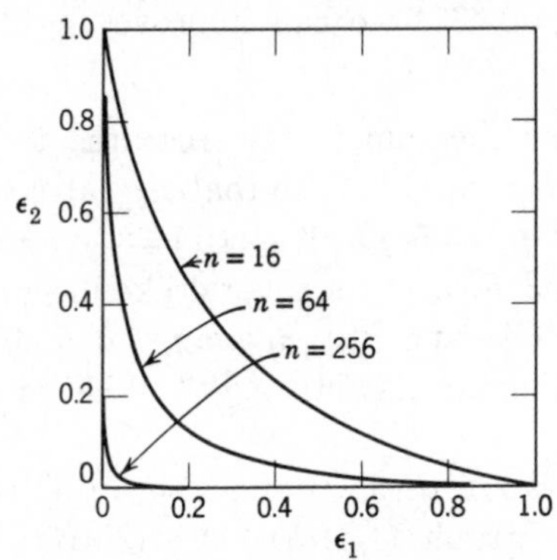

Figure 9.4. Error curves for Example 9.4 with $\mu_1 = 32$, $\mu_2 = 30$, $\sigma = 8$, and sample sizes $n = 16$, 64, and 256.

Exercise 9.6. Verify some points of Figure 9.4.

Exercise 9.7. Find the sample size required to test $H_1 : \mu = \mu_1 = 32$ versus $H_2 : \mu = \mu_2 = 30$ so that $\varepsilon(\mu_1, s) = \varepsilon(\mu_2, s) = 0.05$. (Assume normal distribution with $\sigma = 8$.)

Exercise 9.8. Show that the error curves of Figure 9.4 depend only on $\sqrt{n}\,(\mu_1 - \mu_2)/\sigma$. When this number increases, the error curve *improves.*

It may be remarked that for both examples discussed in this section, the admissible tests were exactly as would be expected. This is not always the case. In Appendix E_{12} we present an example where the likelihood-ratio tests would strike the uninitiated as quite strange.

°*Exercise 9.9.* Derive a simple form of the likelihood-ratio test for $H_1: \sigma = \sigma_1$ versus $H_2: \sigma = \sigma_2$, where σ represents the standard deviation of a normally distributed random variable with known mean μ and $\sigma_1 > \sigma_2$.

4. COMPOSITE HYPOTHESES INVOLVING ONE PARAMETER

The two-state testing problem is artificial but, as we shall see, instructive. Practical problems in hypothesis testing usually involve many possible states of nature. One case of a testing problem with many states of nature is the Example 7.1. Here the problem consists of testing $H_1: \theta \geq 31$ versus $H_2: \theta < 31$. It is not uncommon to have such tests where the state of nature is specified by a single number θ and to be interested in whether θ is large or small. Another example is the following generalization of the coin-tossing problem (Example 9.1).

Example 9.5. Mr. Jones is given a choice between the two actions: (a_1) bet that a coin will fall heads and (a_2) bet on tails. The unknown probability of a head is p between 0 and 1. When $p>1/2$, he should bet heads and when $p<1/2$ he should bet tails. Suppose that his regrets due to taking the wrong action are given by

$$r(p) = 30|p - 1/2|$$

and that his data consist of the result of 100 tosses of the coin.

This problem is readily seen to be a test of the composite hypothesis $H_1: p \geq 1/2$ versus $H_2: p < 1/2$ and is also of the form described above. In general, we may consider the problem of testing $H_1: \theta \geq \theta_0$ versus $H_2: \theta < \theta_0$, where the regrets are specified by $r(\theta)$ and the data have the distribution $f(z|\theta)$.

Before we proceed with the discussion of the general problem, let us consider the two special problems of Example 7.1 and Example 9.5. We already claimed that in Example 7.1 the admissible tests consist of accepting H_1 if $\bar{\mathbf{X}}$ is large enough. In the general normal problem with unknown mean μ and known variance σ^2, the admissible tests of $H_1: \mu \geq \mu_0$ versus $H_2: \mu < \mu_0$ are of this form. Similarly, in the general binomial problem the admissible tests of $H_1: p \geq p_0$ versus $H_2: p < p_0$ consist of accepting H_1 if $\hat{\mathbf{p}}$ is large enough. (This result applies to Example 9.5, of course.) These statements are proved in Appendix E_{13}. The proofs are based on the facts that: (1) for the two-state normal problem of Section 3 with known variance, the class of admissible tests is not affected by the particular values of μ_1 and μ_2, and (2) for the two-state binomial problem, the class of admissible tests is not affected by the particular values of p_1 and p_2.

Exercise 9.10. Evaluate and graph $\varepsilon(p, s)$ and $R(p, s)$ for Example 9.5, where s consists of accepting H_1 if $\hat{\mathbf{p}} \geq 0.5$ and H_2 if $\hat{\mathbf{p}} < 0.5$, and $n = 100$.

Thus, in these two composite hypothesis testing problems of rather general interest, the class of admissible tests is easily characterized. However, in the general problem of testing $H_1: \theta > \theta_0$ versus $H_2: \theta < \theta_0$, this statement cannot be made. In many problems, a simple characterization of the admissible strategies is mathematically difficult or impossible to achieve. Frequently, the statistician will compromise by considering strategies which are not necessarily optimal or admissible. If his strategy, *when evaluated*, seems not to be too far from what he feels a good strategy should yield, the statistician may be quite content to use it.[1]

Thus we shall frequently find ourselves in the position of proposing strategies which seem reasonable. In the one-parameter problem of testing $H_1: \theta \geq \theta_0$ versus $H_2: \theta < \theta_0$, several approaches have been suggested for generating reasonable strategies.

The first of these is a generalization of the likelihood-ratio test. Except for the following brief description, we shall not refer to it again. The *generalized likelihood ratio* for testing $H_1: \theta \in \mathcal{N}_1$ versus $H_2: \theta \in \mathcal{N}_2$ is defined by

$$\lambda(\mathbf{Z}) = \frac{\max_{\theta \in \mathcal{N}_1} f(\mathbf{Z} \mid \theta)}{\max_{\theta \in \mathcal{N}_2} f(\mathbf{Z} \mid \theta)}$$

and the *generalized likelihood-ratio test* consists of accepting H_1 if $\lambda(\mathbf{Z})$ is large enough. Roughly speaking, this method consists of comparing the likelihoods of the most likely states in $\mathcal{N}_1$ and $\mathcal{N}_2$ respectively. The method is often difficult to apply and generally involves the methods of calculus. On the other hand, it can be

[1] It is for this purpose that statisticians frequently study properties of optimal strategies which are too complicated to use in practice.

applied to the general two-action problem and not only to the one-parameter case. Furthermore, under many circumstances, this method has been shown to yield strategies which are very good when the sample size is large.

A second method which is specifically designed for the one-parameter case is the *indifference zone* approach. According to this approach, the statistician selects an interval (θ_2, θ_1) about θ_0 within which the regrets seem to be small, and such that the regrets are sizable outside this interval. This interval is called the indifference zone, and the statistician is not seriously concerned with the error probabilities in the indifference zone. On the other hand, he wants the error probabilities to be small for θ not in this zone. Accordingly, he applies a likelihood-ratio test of $H_1^*: \theta = \theta_1$ versus $H_2^*: \theta = \theta_2$ to test $H_1: \theta \geq \theta_0$ versus $H_2: \theta < \theta_0$. That is, he accepts H_1 if

$$\lambda^*(\mathbf{Z}) = \frac{f(\mathbf{Z} \mid \theta_1)}{f(\mathbf{Z} \mid \theta_2)}$$

is large enough. This test is relatively simple to apply. If this test gives small error probabilities for $\theta = \theta_1$ and $\theta = \theta_2$, one can expect, with a reasonable amount of luck, that it will give small error probabilities outside the indifference zone. (Of course one actually checks on this question by computing some error probabilities outside the indifference zone.)

In Example 7.1 Mr. Good might feel that he is not much concerned with what happens if $30.7 < \theta < 31.5$ for the regrets there are less than 2, whereas they are much larger outside the interval (30.7, 31.5). Similarly in Example 9.5, one might select $\{p : 0.45 < p < 0.55\}$ as an indifference zone.

How does one select an indifference zone? Why is (0.45, 0.55) a more reasonable choice than (0.499, 0.501), or is it? Strictly speaking, the indifference zone is merely a device which permits us to compute reasonable strategies readily. From this point of view, one indifference zone is more appropriate than another only if the strategies it yields have better risk functions than the other. Accordingly, one might compare the risks for strategies derived from several indifference zones.

Exercise 9.11. The Whig party has hired Mr. Turner to determine who will win an election. Mr. Turner will either (a_1) predict that the Whigs will win or (a_2) predict that the Populists will win.

He will base his action on a sample of n random voters. He is interested in being correct mainly to increase his prestige, and he considers his regret to be specified as in Figure 9.5, where p is the

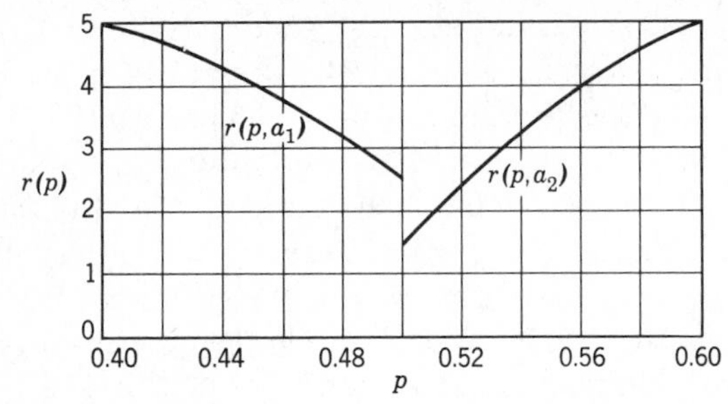

Figure 9.5. Mr. Turner's regret function for Exercise 9.11.

proportion of all voters who prefer the Whigs. This curve is based partly on his assumption that his employers will be annoyed with him if he guesses wrong on a very close election, and that they will be more disappointed in him if he guesses wrong in their favor than otherwise. Evaluate the risk functions for $s_{0.5}$, where $s_{k'}$ consists of taking action a_1 if $\hat{\mathbf{p}} \geq k'$ and the sample size $n = 100$.

Exercise 9.12. In the binomial problem of testing $H_1 : p \geq 0.5$ versus $H_2 : p < 0.5$, find the smallest sample size and the appropriate strategy for which $\varepsilon(0.45, s) = \varepsilon(0.60, s) = 0.01$.

5. COMPOSITE HYPOTHESES INVOLVING ONE PARAMETER: TWO-TAILED TESTS

Example 9.6. Mr. Sharp runs a gambling establishment where, for a small charge, people can bet against the house on the flip of a coin. It is important that the coin be rather well balanced, otherwise people will notice whether it favors heads or tails and bet accordingly, much to Mr. Sharp's regret, until he replaces the coin. His available actions are (a_1) to use the coin or (a_2) to discard it and search for another. Mr. Sharp represents his regret function by the graph of Figure 9.6. Notice that, if $0.48 \leq p \leq 0.52$, the coin is sufficiently well balanced so that the advantage gained by someone who knows of its bias does not outweigh the house's charge for betting.

This problem is that of testing H_1: $0.48 \leq p \leq 0.52$ against the alternative H_2: $p < 0.48$ or $p > 0.52$, where p is the probability of

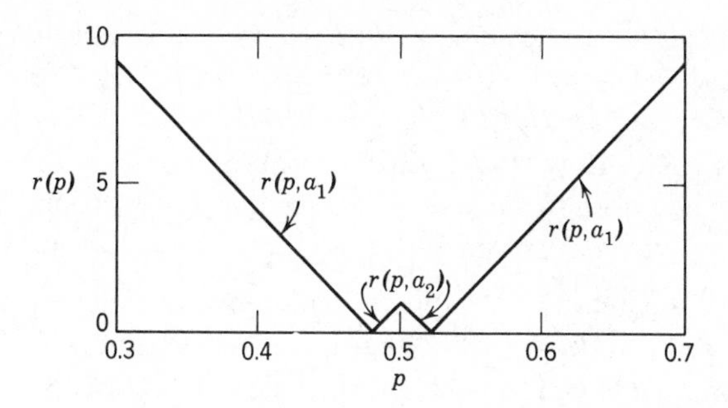

Figure 9.6. Mr. Sharp's regrets for Example 9.6.

the coin falling heads. If Mr. Sharp tosses his coin 400 times and $\hat{\mathbf{p}}$ is the observed proportion of heads, it is clear that a reasonable test would be to accept H_1 for $\hat{\mathbf{p}}$ sufficiently close to 1/2. Thus accepting H_1 if $0.48 \leq \hat{\mathbf{p}} \leq 0.52$ is one such test, accepting H_1 if $0.45 \leq \hat{\mathbf{p}} \leq 0.55$ might be another, and accepting H_1 if $0.45 \leq \hat{\mathbf{p}} \leq 0.51$ might be reasonable if Mr. Sharp's previous experience led him to believe that most biased coins favor heads. Let us consider the second of these tests. The probability of rejecting H_1 is the probability of $\hat{\mathbf{p}} < 0.45$ plus the probability of $\hat{\mathbf{p}} > 0.55$ and is represented by the areas in the two tails of the distribution of $\hat{\mathbf{p}}$ (see Figure 9.7). For this reason, the three tests indicated above and

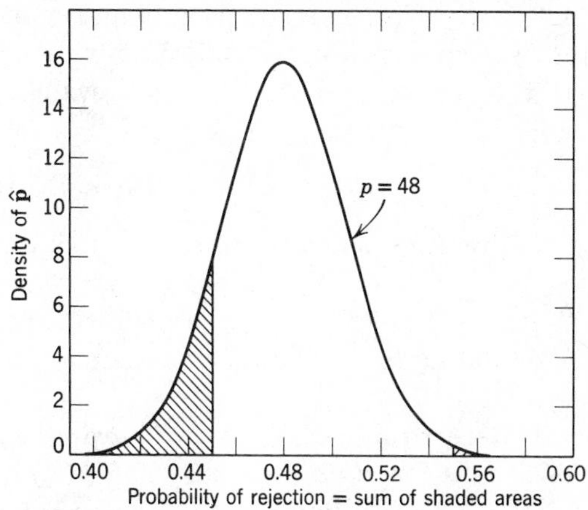

Figure 9.7. Representation of the action probability for a two-tailed test.

all tests which lead to rejection of H_1, if either $\hat{\mathbf{p}} < k_2$ or $\hat{\mathbf{p}} > k_1$, are called *two-tailed tests.*

Exercise 9.13. Indicate the ideal action probabilities for Mr. Sharp. Evaluate the error probabilities and risks for s which consists of accepting H_1 if $0.45 \leq \hat{\mathbf{p}} \leq 0.55$.

Exercise 9.14. How would you improve on the test of Exercise 9.13? If p had a known *a priori* probability distribution, describe roughly how you would use it.

Example 9.7. Let us suppose that in Mr. Jones' problem (Example 9.5) the size of the bet has increased considerably but that he is no longer obliged to bet. On the basis of 100 tosses of a coin, he can decide on one of three actions. These are: (a_1) do not bet; (a_2) bet on heads; and (a_3) bet on tails. Assuming that the bet, if made, is a very large one at even odds, he would actually have a loss of utility if he bet and p were close to 1/2.[1]

This problem of Mr. Jones resembles that of Mr. Sharp in Example 9.6. While Mr. Sharp is interested in testing whether the coin is too unbalanced to use, Mr. Jones wants to know if it is sufficiently unbalanced to take advantage of, and, if so, whether to bet on heads or tails. The two problems differ in one important respect. Mr. Sharp has two actions available (keep the coin or discard it) while Mr. Jones must decide among three actions. These are: (a_1) do not bet; (a_2) bet on heads; and (a_3) bet on tails. Thus we may call Mr. Jones' problem a trilemma to distinguish it from the dilemma or two-action choice, which is usually called a testing problem. In the trilemma problem, we must evaluate $\alpha_s(\theta, a_1)$, $\alpha_s(\theta, a_2)$, and $\alpha_s(\theta, a_3)$ to compute the risk

$$R(\theta, s) = \alpha_s(\theta, a_1)\, r(\theta, a_1) + \alpha_s(\theta, a_2)\, r(\theta, a_2) + \alpha_s(\theta, a_3)\, r(\theta, a_3).$$

Exercise 9.15. Present the graphs of a version of a reasonable regret function for Mr. Jones. You must consider $r(\theta, a_1)$, $r(\theta, a_2)$, and $r(\theta, a_3)$. What are the ideal action probabilities? Express this problem as one of testing hypotheses where there are three alternative hypotheses.

Exercise 9.16. Referring to Exercise 9.15, evaluate and graph $R(\theta, s)$ for s which calls for (a_1) if $0.45 \leq \hat{\mathbf{p}} \leq 0.55$, ($a_2$) if $\hat{\mathbf{p}} > 0.55$, and (a_3) if $\hat{\mathbf{p}} < 0.45$ where $n = 100$. In doing so, tabulate $r(\theta, a_1)$, $r(\theta, a_2)$, $r(\theta, a_3)$, $\alpha_s(\theta, a_1)$, $\alpha_s(\theta, a_2)$, and $\alpha_s(\theta, a_3)$ for each θ considered.

Characteristically the two-tailed testing problem involves

[1] This is true for anyone whose utility for money is concave.

testing a hypothesis $H_1: \theta_2 \leq \theta \leq \theta_1$ versus the alternative $H_2: \theta < \theta_2$ or $\theta > \theta_1$. The trilemma can be expressed as the selection of one of the three alternative hypotheses, $H_1: \theta_2 \leq \theta \leq \theta_1$, $H_2: \theta > \theta_1$, and $H_3: \theta < \theta_2$.

Now let us analyze Example 9.6 further. It is suggested that the reader review Section 5, Chapter 7, before proceeding. Strictly speaking, this problem is one of testing $H_1: 0.48 \leq p \leq 0.52$. In common statistical practice, Mr. Sharp's problem is usually called one of testing as to whether the coin is perfectly well balanced. Thus Mr. Sharp's problem is confused with that of testing $H_1^*: p = 1/2$ versus $H_2^*: p \neq 1/2$. If a scientist wanted to test whether this coin had a p within 0.001 of 1/2, his problem would also customarily be confused with the above testing problem. This is a special case of the test of a simple hypothesis $H_1^*: \theta = \theta_0$ versus the composite alternative $\theta \neq \theta_0$. The fact that the scientist and Mr. Sharp have radically different regrets should not be overlooked. (Consider their regrets for deciding "H_1 is true" when, in fact, $p = 0.51$.)

What should a statistician do if a customer comes to him for a test of $\theta = \theta_0$ versus $\theta \neq \theta_0$? In such a case, it is important to ask whether the customer would not prefer to accept H_1 (i.e., take the action that is appropriate for $\theta = \theta_0$) when θ is not necessarily θ_0 but very close to θ_0. Questions of this type help to specify what values of θ are to be considered *practically equal* to θ_0. Such a question would be unnecessary if the customer provided the statistician with his regret function. In practice, it is impossible to find customers who will specify a precise regret function. They are usually reluctant even to give a rough idea of their loss function. Although slight deviations in the regret function have little effect in problems involving considerable data, a rough idea of $r(\theta, a)$ is absolutely necessary to yield an intelligent analysis.

How should a statistician analyze a problem if he does not have an accurate picture of the customer's regret function? To compute the risk function, one needs the action probabilities and the regrets. If the regrets, which depend on the customer, are not specified, the statistician can indicate the action probabilities of the various strategies. Customarily, the statistician presents his customer with various strategies and their action probabilities.

The customer selects one of these strategies. Presumably, at least two kinds of considerations enter into making this choice:

1. *The relative regrets due to the various actions.* If incorrectly taking one action is far more costly than incorrectly taking the other, the customer may favor a strategy which holds the corresponding error probabilities low at the expense of raising the other error probabilities.

2. *Intuition, opinion, or knowledge as to which states of nature are likely and which are unlikely.* Thus if Mr. Good knew that bad lots were rarely sent to him by Mr. Lacey, he might choose a strategy giving a high probability of accepting a bad lot because it would at the same time give a high probability of accepting a good lot; that is, his *a priori* information would make him prefer that one of the error probabilities be small.

If we consider a problem of testing H_1^*: $\theta = \theta_0$ versus H_2^*: $\theta \neq \theta_0$, whether or not it would be more adequately phrased as one of testing H_1: $\theta_2 < \theta < \theta_1$, it is customary to call $\alpha_s(\theta_0, a_2) = \varepsilon(\theta_0, s)$ the *significance level of the test* s of H_1^*.

In the absence of *a priori* knowledge concerning θ, a relatively low (strict) significance level is called for if (1) rejecting H_1^* when it is true is very expensive, (2) it is desirable to accept H_1^* unless θ is far from θ_0, or (3) the sample size is large.

Exercise 9.17. On the basis of 100 observations, indicate two different two-sided tests of H_1: $p = 1/4$ versus H_2: $p \neq 1/4$ for which the significance level is 0.05. Graph the action probability $\alpha_s(\theta, a_1)$ for each test. *Hint*: You can compensate for increasing one tail by decreasing the other.

Exercise 9.18. Professor Barker wonders whether the graduate students who take his course are typical of the population of graduate students. Assuming the model that his students have normally distributed grades with mean μ and $\sigma = 100$, he wishes to test H_1: $\mu = 500$ versus H_2: $\mu \neq 500$. Indicate a test based on a sample of 100 students with a significance level of 0.05.

6. SEVERAL PARAMETERS

To this point we have discussed testing a simple hypothesis against a simple alternative, and testing possibly composite hypotheses against one another. In the former case, only two particular distributions come under consideration. In the composite cases discussed, the two alternative sets of states of nature were completely described in terms of two sets of possible values for a single parameter (e.g., p or μ).

Many (perhaps most) composite hypothesis-testing problems are more complex and involve two or more parameters. Only by assuming σ to have the known value 8 did we force the cord problem (Example 7.1) to be a one-parameter problem involving only μ.

Similarly, if it were desired to test whether one drug were better than another for curing a disease, two parameters would be involved. The state of nature could be represented by $\theta = (p_1, p_2)$, where p_i is the probability that drug i will cure a patient. Then we must test H_1: $p_2 \leq p_1$ versus H_2: $p_2 > p_1$. More specifically, we test H_1: $\theta \in \mathscr{N}_1$ versus H_2: $\theta \in \mathscr{N}_2$, where

$$\mathscr{N}_1 = \{(p_1, p_2): 0 \leq p_2 \leq p_1 \leq 1\}$$

and

$$\mathscr{N}_2 = \{(p_1, p_2): 0 \leq p_1 < p_2 \leq 1\}.$$

(If one of these drugs is a very well established drug and the other a relatively new one, then we might be inclined to modify H_1 to H_1^*: $p_2 \leq p_1 + 0.1$ and H_2 to H_2^*: $p_2 > p_1 + 0.1$.)

Although problems involving two or more parameters are more complex than one-parameter problems, most of the ideas developed in the one-parameter case are still useful; some new notions are needed in addition.

Some problems cannot be described in terms of distributions which are completely known except for the values of a few parameters. Instead, the distributions may be entirely unknown, or so vaguely known that expression in terms of a few parameters is not possible. For example, suppose that it is desired to test whether female students who receive a grade A in freshman chemistry tend to weigh more than female students who receive a grade of B. Here we wish to know whether one cdf is to the right of the other without any knowledge about the cdf's. The subject of nonparametric inference involves tests which would be appropriate for such problems.

A two-parameter problem of classical type is presented as Exercise 9.19.

Exercise 9.19. Let $\mathbf{X}_1, \mathbf{X}_2, \cdots, \mathbf{X}_{10}$ be n observations on a normally distributed random variable with mean μ and unknown variance σ^2. Construct a test of H_1: $\mu = \mu_0$ versus H_2: $\mu \neq \mu_0$ at a significance level of 0.05. *Hint*: If σ were known, a reasonable test would consist of accepting H_1 if $|\bar{\mathbf{X}} - \mu_0| < 1.96\sigma/\sqrt{n}$, i.e., if

$$\frac{|\bar{\mathbf{X}} - \mu_0|}{\sigma/\sqrt{n}} < 1.96 .$$

If σ is unknown, one might consider replacing σ by $\mathbf{s_X}$. However, the distribution of

$$\mathbf{t} = \frac{\bar{\mathbf{X}} - \mu_0}{\mathbf{s_X}/\sqrt{n}}$$

is not normal. In fact $\mathbf{t}$ has the t distribution with $n - 1$ degrees of freedom if $\mu = \mu_0$ (see Appendix D_4).

7. DESIGN OF EXPERIMENTS

There are two important aspects to the problem of design of experiments. One is to decide which experiment to perform and the other is to decide how many observations to take. The following example illustrates the first aspect.

Example 9.8. Let us return to Example 7.1. Mr. Good's vice president suggests that measuring the breaking strength of each of 64 strands is a tedious and expensive experiment. It would be half as expensive to subject each strand to a 30-gram load and to see whether it breaks or not. Would it be wiser to determine 64 breaking strengths or, for the same cost, to determine how many of 128 strands have breaking strength of over 30 grams?

Assuming that the strand strengths are normally distributed with mean μ and standard deviation 8, the probability p that a strand will not break under a 30-gram load is

$$p = P\{\mathbf{X} > 30\} = P\left\{\frac{\mathbf{X} - \mu}{\sigma} > \frac{30 - \mu}{8}\right\}.$$

For each value of μ there is a corresponding value of p. If $\mu = 31$, $p = 0.550$. Thus, if we decide to use the alternative experiment, our problem becomes one of testing H_1: $p \geq 0.550$ versus H_2: $p < 0.550$. In Figure 9.8 we compare the risks for the minimax strategies for

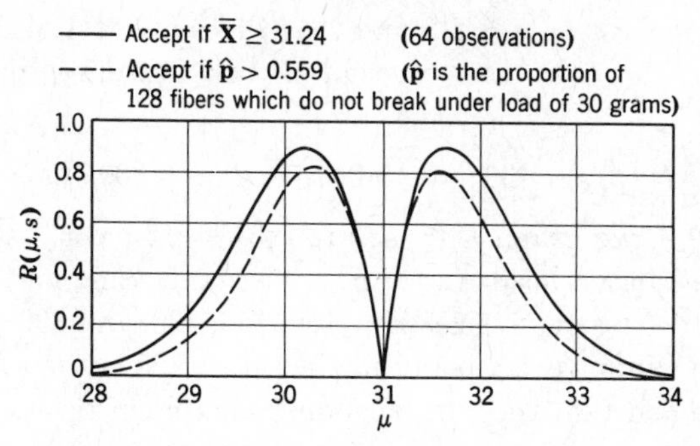

Figure 9.8. Risks for minimax strategies for two designs in Example 9.8.

both designs. It is clear that the new design is an improvement. One might also investigate whether a different load would yield a still better risk function.

To illustrate the aspect of experimental design which involves the number of observations to take, we extend Example 9.5.

Example 9.9. Assuming that the regret due to taking the wrong

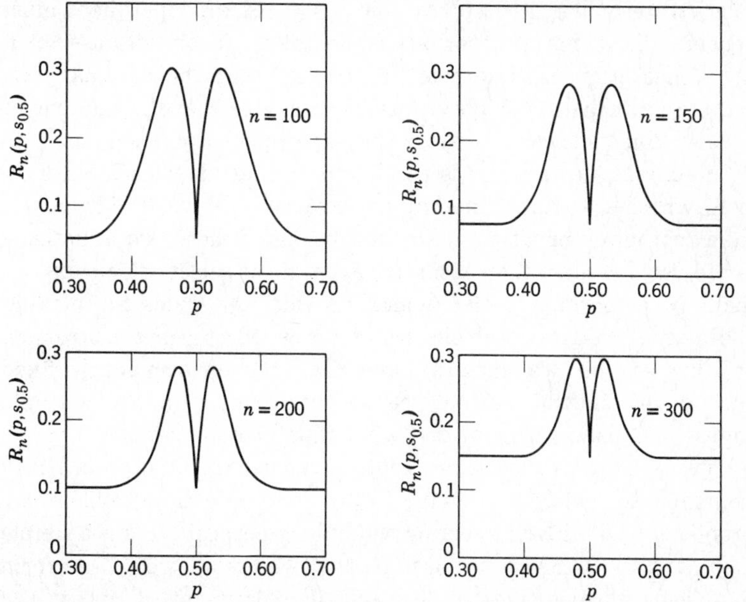

Figure 9.9. Risk functions for Example 9.9 for $s_{0.5}$ where the cost per observation is 0.0005 and for sample sizes $n = 100, 150, 200, 300$.

action is $r(p) = 30\,|\,p - 1/2\,|$, and that there is a cost of 0.0005 per toss of the coin, the risk associated with a strategy involving n tosses is

$$R_n(p, s) = 30\,|\,p - 1/2\,|\,\varepsilon_n(p, s) + 0.0005n.$$

In Figure 9.9, we graph $R_n(p, s_{0.5})$ for the strategy $s_{0.5}$ which consists of betting on heads (a_1) if $\hat{\mathbf{p}} \geq 0.5$ and betting on tails (a_2) if $\hat{\mathbf{p}} < 0.5$, and for sample sizes $n = 100, 150, 200$, and 300. It seems that from a minimax risk point of view, $n = 225$ is the appropriate sample size and that the corresponding maximum risk is 0.277.

Exercise 9.20. Referring to Exercise 9.11, evaluate the risk functions for $s_{0.5}$ and $s_{0.51}$ for $n = 100, 225$, and 400 where the cost per observation is 0.001. (This problem involves six risk function evaluations, each of which could appropriately be assigned to a portion of the class.) Comparing these risk functions, estimate the sample size and strategy which minimize the maximum risk.

8. SEQUENTIAL ANALYSIS

A strategy which involves the design of an experiment must consider how many observations to take. In our discussion of Mr. Jones' problem, Example 9.9, we saw that the minimax risk strategy involved 225 observations, provided that the number of observations is specified before the experiment is carried out.

In general, however, it is possible to decide, after each observation, whether to take another observation. If we decide not to take another observation, we must take one of the available actions. If we decide on another observation we must select which experiment to perform. A plan which provides the rules for making these decisions after each observation is called a *sequential strategy*. In this section we shall assume that the experiment is fixed throughout and that our only choices concern whether to continue repeating this experiment and what (final) action to take.

Furthermore, let us assume that we can use the indifference zone approach to replace our composite hypothesis testing problems by problems which involve testing a simple hypothesis versus a simple alternative. In short, we assume that we are allowed to perform independent repetitions of a specified experiment at a cost of c (units of utility) per repetition to test H_1: $\theta = \theta_1$ versus H_2: $\theta = \theta_2$.

The risk corresponding to a test is given by

$$R(\theta, s) = r(\theta)\, \varepsilon(\theta, s) + c\, E(\mathbf{N} \mid \theta)$$

where $\mathbf{N}$ is the number of repetitions performed. We use $\mathbf{N}$ to represent the number of observations finally taken because it is a random variable.

Let $\mathbf{X}_i$, the outcome of the ith experiment, have density $f(x \mid \theta)$. We define the likelihood ratio based on the first n observations $\mathbf{Z}_n = (\mathbf{X}_1, \mathbf{X}_2, \cdots, \mathbf{X}_n)$ by

$$\lambda_n(\mathbf{Z}_n) = \frac{f(\mathbf{X}_1 \mid \theta_1) f(\mathbf{X}_2 \mid \theta_1) \cdots f(\mathbf{X}_n \mid \theta_1)}{f(\mathbf{X}_1 \mid \theta_2) f(\mathbf{X}_2 \mid \theta_2) \cdots f(\mathbf{X}_n \mid \theta_2)}.$$

A test is said to be a *sequential likelihood-ratio test* if there are two numbers k_1 and k_2 and if, after the nth observation, the test calls for:

Another observation	when $k_2 < \lambda_n(\mathbf{Z}_n) < k_1$
Accepting H_1	when $\lambda_n(\mathbf{Z}) \geq k_1$
Rejecting H_1	when $\lambda_n(\mathbf{Z}_n) \leq k_2$.

In Appendix E_{14}, we show that *every admissible test* is a *sequential likelihood-ratio test*. The derivations of the error probabilities and expected sample size are beyond the scope of this book.

Example 9.10. Binomial Distribution. We extend Example 9.2 to allow for sequential testing. Let $\hat{\mathbf{p}}_n$ be the observed proportion of heads after n observations. In Appendix E_{15}, we show that, after the nth observation, a sequential likelihood-ratio test calls for:

Another observation	if	$a - \dfrac{b_2}{n} < \hat{\mathbf{p}}_n < a + \dfrac{b_1}{n}$
Accepting H_1	if	$\hat{\mathbf{p}}_n \geq a + \dfrac{b_1}{n}$
Rejecting H_1	if	$\hat{\mathbf{p}}_n \leq a - \dfrac{b_2}{n}$

where appropriate values of a, b_1, and b_2 may be determined in terms of the regrets, cost of sampling, and *a priori* probabilities. To specify appropriate values of b_1 and b_2 is too difficult for this course but more details about this example are given in Section 8.1. It suffices to indicate the intuitively obvious fact that increasing b_1 alone decreases the probability of accepting H_1 and tends to increase the expected sample size. Similarly, increasing b_2 alone will decrease the probability of rejecting H_1 and increase the expected sample size. Thus, if the cost of sampling is small, b_1 and b_2 should be large. If $r(\theta_1)$ is much larger then $r(\theta_2)$, we will tend toward accepting H_1, and b_2 will be much larger than b_1.

Example 9.11. Normal Distribution With Known Variance. We extend Example 9.4 to allow for sequential testing. Let $\overline{\mathbf{X}}_n$ be the sample mean of the first n observations. In Appendix E_{15} we show that after the nth observation, a sequential likelihood-ratio test calls for:

Another observation if $a - \dfrac{b_2}{n} < \overline{\mathbf{X}}_n < a + \dfrac{b_1}{n}$

Accepting H_1 if $\overline{\mathbf{X}}_n \geq a + \dfrac{b_1}{n}$

Rejecting H_1 if $\overline{\mathbf{X}}_n \leq a - \dfrac{b_2}{n}$

where

$$a = \frac{\mu_1 + \mu_2}{2}.$$

The remarks made concerning b_1 and b_2 for the binomial example are also appropriate here. More details will be given in Section 8.2.

In a testing problem it is customary to consider in advance the values of ε_1 and ε_2 which are tolerable. Then a sample size N^* can be chosen to yield these error probabilities. Alternatively, a

TABLE 9.5

AVERAGE PERCENTAGE SAVING IN SIZE OF SAMPLE WITH SEQUENTIAL ANALYSIS AS COMPARED WITH THE NONSEQUENTIAL TEST FOR TESTING THE MEAN OF A NORMALLY DISTRIBUTED VARIABLE

	When $\mu=\mu_1$					When $\mu=\mu_2$				
ε_1 \ ε_2	0.01	0.02	0.03	0.04	0.05	0.01	0.02	0.03	0.04	0.05
0.01	58	60	61	62	63	58	54	51	49	47
0.02	54	56	57	58	59	60	56	53	50	49
0.03	51	53	54	55	55	61	57	54	51	50
0.04	49	50	51	52	53	62	58	55	52	50
0.05	47	49	50	50	51	63	59	55	53	51

sequential test may be constructed to obtain these values; in this case, the sample size is random, but the expected sample size can be computed and is less than N^*. Table 9.5 shows the savings in sample size for various ε_1 and ε_2 obtained by sequentially testing H_1: $\mu = \mu_1$ versus H_2: $\mu = \mu_2$ for a normally distributed random variable with known variance.

†8.1. Details in the Binomial Example

For reference purposes, we present the following equations which approximate a, b_1, b_2, action probabilities, and expected sample size for a sequential likelihood-ratio test with specified ε_1 and ε_2. We also illustrate with a numerical example.

In Appendix E_{15}, we see that the value of a is given exactly by

$$a = g \log\left(\frac{1 - p_2}{1 - p_1}\right)$$

where

$$g = \frac{1}{\log \{[p_1/1 - p_1)]/[p_2/(1 - p_2)]\}}$$

To obtain a desired ε_1 and ε_2, we require (approximately)

$$b_1 = g \log\left(\frac{1 - \varepsilon_1}{\varepsilon_2}\right)$$

and

$$b_2 = g \log\left(\frac{1 - \varepsilon_2}{\varepsilon_1}\right).$$

Then the action probabilities are defined for all p and approximated by

$$\alpha(p, a_1) = \frac{1 - [\varepsilon_1/(1 - \varepsilon_2)]^h}{[(1 - \varepsilon_1)/\varepsilon_2]^h - [\varepsilon_1/(1 - \varepsilon_2)]^h}$$

where $h = -1$ if $p = p_1$, $h = 1$ if $p = p_2$, and, in general, h is related to p by

$$p = \frac{1 - [(1 - p_1)/(1 - p_2)]^h}{(p_1/p_2)^h - [(1 - p_1)/(1 - p_2)]^h}.$$

Finally

$$E(\mathbf{N} \mid p) = \frac{\alpha(p, a_1) \log [(1 - \varepsilon_1)/\varepsilon_2] + \alpha(p, a_2) \log [\varepsilon_1/(1 - \varepsilon_2)]}{p \log (p_1/p_2) + (1 - p) \log [(1 - p_1)/(1 - p_2)]}.$$

For comparison purposes, in Figure 9.10 we compare the minimax fixed sample size solution of Mr. Jones' problem (see Example 9.5) with the sequential test which yields the same error probabilities at $p_1 = 0.53$ and $p_2 = 0.47$. Here $a = 0.5$, $b_1 = b_2 = 6.2181$, and $\varepsilon_1 = \varepsilon_2 = 0.1833$.

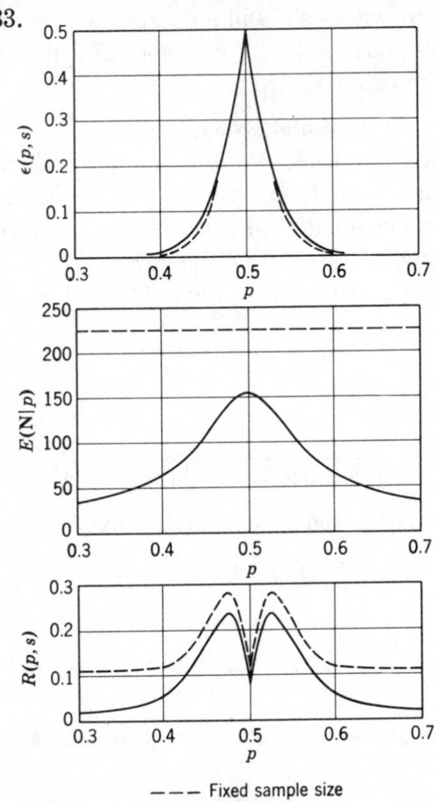

Figure 9.10. Comparison of minimax fixed sample size test with the sequential test which yields the same error probabilities at $p_1 = 0.53$ and $p_2 = 0.47$ for Example 9.5.

Sequential test: Continue sampling if $(0.5)n - 6.2181 < \mathbf{m}_n < (0.5)n + 6.2181$.

Nonsequential test: Sample size is 225. Accept if $\hat{\mathbf{p}} \geq 0.5$.

To illustrate how to apply such a test, we shall apply the sequential likelihood-ratio test with $a = 0.6$, $b_1 = 1$, and $b_2 = 2$ to the data $H, H, T, H, H, H, T, H, T, T, H, T, T, T, T, H, T, H, T, H, H, H, H, T, H, T, T, H, H, T, \cdots$.

It is more convenient to replace the inequality

$$a - \frac{b_2}{n} < \hat{\mathbf{p}}_n < a + \frac{b_1}{n}$$

by the equivalent one

$$na - b_2 < \mathbf{m}_n < na + b_1$$

where $\mathbf{m}_n$ is the number of heads in the first n tosses. In Table 9.6 we tabulate $na - b_2$, $\mathbf{m}_n$, and $na + b_1$.

TABLE 9.6

APPLICATION OF SEQUENTIAL LIKELIHOOD-RATIO TEST FOR THE BINOMIAL PROBLEM WITH $a = 0.6$, $b_1 = 1$, and $b_2 = 2$

n	$na - b_2$	$\mathbf{m}_n$	$na + b_1$
1	−1.4	1	1.6
2	−0.8	2	2.2
3	−0.2	2	2.8
4	0.4	3	3.4
5	1.0	4	4.0
6	1.6		4.6
7	2.2		5.2
8	2.8		5.8
9	3.4		6.4
10	4.0		7.0

Therefore we accept H_1 after the 5th observation.

Exercise 9.21. Use the table of random numbers (Table C_1) to simulate an experiment with $p = 0.4$. Apply Mr. Jones' sequential strategy with $a = 0.5$ and $b_1 = b_2 = 6.2181$. Does the strategy lead to error in this case? How many observations did you take?

†8.2. Details in the Normal Example with Known Variance

Here again we have the useful approximations for b_1 and b_2. These are[1]

$$b_1 = \frac{2.303\sigma^2}{(\mu_1 - \mu_2)} \log\left(\frac{1 - \varepsilon_1}{\varepsilon_2}\right)$$

and

$$b_2 = \frac{2.303\sigma^2}{(\mu_1 - \mu_2)} \log\left(\frac{1 - \varepsilon_2}{\varepsilon_1}\right)$$

The action probabilities (sometimes called the *operating characteristic*) and expected sample size are approximated by

[1] The factor 2.303 represents the reciprocal of the logarithm of e. If natural logarithms were used, this factor would not be present.

$$\alpha(\mu, a_1) = \frac{1 - [\varepsilon_1/(1 - \varepsilon_2)]^h}{[(1 - \varepsilon_1)/\varepsilon_2]^h - [\varepsilon_1/(1 - \varepsilon_2)]^h}$$

where

$$h = (\mu_1 + \mu_2 - 2\mu)/(\mu_1 - \mu_2)$$

and

$$E(\mathbf{N} \mid \mu) = [(b_1 + b_2)\alpha(\mu, a_1) - b_2]\Big/\left(\mu - \frac{\mu_1 + \mu_2}{2}\right).$$

Exercise 9.22. Consider Exercise 9.7. Specify the fixed-sample-size test and the sequential likelihood-ratio test which yield $\varepsilon_1 = 0.05$ and $\varepsilon_2 = 0.05$. Graph $\alpha(\mu, a_1)$ for each test and compare the fixed sample size with the graph of the expected sample size for the sequential test.

9. SUMMARY

In this chapter we discussed problems in testing simple and composite hypotheses where there are two actions, one being optimal for some states of nature, the other for the rest. For the two-state, two-action problem, admissible strategies are Bayes strategies. If the experiment (with sample size) is specified, the Bayes strategy for testing H_1: $\theta = \theta_1$ versus H_2: $\theta = \theta_2$ when the *a priori* probability of H_2 is w, is the *likelihood-ratio test* which consists of :

Accepting H_1	if $\lambda(\mathbf{Z}) > k = \dfrac{w\, r(\theta_2)}{(1 - w)\, r(\theta_1)}$
Rejecting H_1	if $\lambda(\mathbf{Z}) < k$
Taking either action	if $\lambda(\mathbf{Z}) = k$

where $\lambda(\mathbf{Z}) = f(\mathbf{Z} \mid \theta_1)/f(\mathbf{Z} \mid \theta_2)$ is the likelihood ratio. Thus the class of Bayes strategies is the class of likelihood-ratio tests. This *class* does not depend on the regrets $r(\theta_1) = r(\theta_1, a_2)$ and $r(\theta_2) = r(\theta_2, a_1)$, although cognizance should be taken of $r(\theta_1)$ and $r(\theta_2)$ and of the error curve in selecting a strategy from this class. The error curve is the part of the boundary of the set of $(\varepsilon_1, \varepsilon_2)$ for all strategies, where $\varepsilon_1 = \alpha_s(\theta_1, a_2)$, $\varepsilon_2 = \alpha_s(\theta_2, a_1)$, and $\alpha_s(\theta, a)$ is the probability that the strategy s leads to action a when θ is the set of nature.

For composite hypotheses, one can on occasion, apply the results for tests of simple hypotheses versus simple alternatives to obtain the admissible strategies. However, this is not always possible. Often one must be satisfied with strategies which seem reasonable and which yield reasonable risks. Here one must attempt to keep the error probabilities small. These are $\varepsilon(\theta, s) = \alpha_s(\theta, a_2)$ for those θ where a_1 is preferable, and $\varepsilon(\theta, s) = \alpha_s(\theta, a_1)$ for those θ where a_2 is preferable. Two approaches often used are the following:

1. *Indifference Zones.* If θ_1 and θ_2 are the end points of an indifference zone in which it is not especially urgent to take the best action, apply a likelihood-ratio test for testing $H_1^*: \theta = \theta_1$ versus $H_2^*: \theta = \theta_2$.

2. *The Generalized Likelihood-Ratio Test.* For composite hypotheses, the generalized likelihood ratio is defined by

$$\lambda(\mathbf{Z}) = \frac{\max\limits_{\theta \in \mathscr{N}_1} f(\mathbf{Z} \mid \theta)}{\max\limits_{\theta \in \mathscr{N}_2} f(\mathbf{Z} \mid \theta)}$$

where $\mathscr{N}_1$ and $\mathscr{N}_2$ are the sets of θ on which actions a_1 and a_2 are preferred respectively. The generalized likelihood-ratio procedure consists of:

Accepting H_1	if	$\lambda(\mathbf{Z}) > k$
Rejecting H_1	if	$\lambda(\mathbf{Z}) < k$
Taking either action	if	$\lambda(\mathbf{Z}) = k$.

The trilemma resembles the two-action problem (dilemma). It is somewhat more complex in that, to evaluate a strategy, three action probabilities $\alpha_s(\theta, a_1)$, $\alpha_s(\theta, a_2)$, and $\alpha_s(\theta, a_3)$ are involved in computing the risk function $R(\theta, s)$.

When faced with the problem of testing $H_1: \theta = \theta_0$ versus the alternative $H_2: \theta \neq \theta_0$, it is customary to call $\varepsilon(\theta_0, s) = \alpha_s(\theta_0, a_2)$ the *significance level* of test s.

When we consider the possibility of modifying the sample size in a testing problem, we must incorporate the cost of sampling into the risk function. A good procedure will balance the cost of sampling against the error probabilities. Extending our strategies, it is possible to consider sequential strategies which, after each observation, permit a choice between action a_1, action a_2, and taking another observation. The class of admissible sequential strategies for the test of a simple hypothesis versus a simple alternative with a fixed cost per observation is part of the class of sequential likelihood-ratio procedures which are described in Section 8. These procedures minimize the expected sample size for the error probabilities attained.

Throughout this chapter we have ignored Bayes strategies for testing problems except as a theoretical device to characterize the admissible strategies. The reason is that, for problems involving composite hypotheses, Bayes strategies are cumbersome to apply, and involve the giving of explicit loss functions, and, still more restrictively, require the existence and knowledge of the *a priori* probability.

In most applications, it is extremely difficult to know the regrets with any precision. Therefore, in statistical practice, considerable attention is paid to the action probabilities (*operating characteristics*) of tests. The individual using statistical techniques must select among the available tests by comparing the action probabilities, using his vague ideas about the regrets and about the *a priori* probabilities. Fortunately, what constitutes a good strategy is not very sensitive to small fluctuations in the regret function or *a priori* probabilities when considerable data are available.

SUGGESTED READINGS

[1] Blackwell, David, and M. A. Girschick, *Theory of Games and Statistical Decisions*, John Wiley and Sons, New York, 1950, Dover Publications, Inc., New York, 1979.

[2] Hoel, P.G., *Introduction to Mathematical Statistics*, second edition, John Wiley and Sons, 1954.

[3] Mood, A.M., *Introduction to the Theory of Statistics*, McGraw-Hill Book Co., New York, 1950.

References [2] and [3] are postcalculus books in mathematical statistics. Considerable space is devoted to the derivation of probability distributions which play an important role in statistics in general and in testing hypotheses in particular. Reference [1] applies decision theory to the problem of testing hypotheses.

CHAPTER 10

Estimation and Confidence Intervals

1. INTRODUCTION

Estimation problems are concerned with identifying the state of nature or certain aspects of it. The action taken is the one which would be best if the estimate were the true value of the parameter. Often, the action is to assert a value for the parameter. The loss involved in taking the wrong action depends, among other things, on how "far" the action taken is from the best action. In Example 7.3 the action was a price to offer, and if too much or too little were offered, utility would be lost.

In many examples, the state of nature θ can be represented by a number which is some parameter of the relevant probability distributions. In the cord problem (Example 7.3), θ was the mean fiber strength of the batch of fibers. The statistician will estimate θ by some number $\mathbf{T}$ derived from the data, i.e., he will act as though the state of nature were $\mathbf{T}$. For convenience, this action can be labeled $\mathbf{T}$.

In many real problems it is highly desirable to supplement an estimate with some information concerning how reliable the estimate tends to be. Frequently, the standard deviation of the estimate can be used for this purpose. If the standard deviation of the estimate is itself unknown but must also be estimated, it may be convenient to use the method of confidence intervals. Here a random interval $\mathbf{\Gamma}$ depending on the data $\mathbf{Z}$ is specified in such a fashion that for all θ

$$P\{\mathbf{\Gamma} \text{ contains } \theta\} = \gamma$$

where γ is some specified confidence level.

2. FORMAL STRUCTURE OF THE ESTIMATION PROBLEM: ONE-PARAMETER CASE

We assume that the state of nature is specified by a number θ.

The set of available actions (called estimates) is the same as the set of all possible θ. The regret due to the use of the estimate T when θ is the state of nature is $r(\theta, T)$. The strategy, denoted by t, is called an estimator. For each possible outcome $\mathbf{Z}$ of the experiment, the estimator yields an estimate

$$\mathbf{T} = t(\mathbf{Z}). \tag{10.1}$$

We shall assume that $r(\theta, T)$ increases as T moves further away from θ and $r(\theta, T) = 0$ when $T = \theta$. In particular, one possible regret function is

$$r(\theta, T) = c(\theta)(T - \theta)^2. \tag{10.2}$$

Then the risk function for an estimator t which yields estimates $\mathbf{T}$ is given by[1]

$$R(\theta, t) = c(\theta)\, E_\theta[(\mathbf{T} - \theta)^2]. \tag{10.3}$$

Other regret functions such as $c(\theta)|T - \theta|$, are sometimes appropriate and considered, but the "squared error" regret function of Equation (10.2) is the one with which most statistical theory of estimation is concerned. Briefly, there are three major reasons. First, many smooth regret functions which vanish at $T = \theta$ are well approximated by the "squared loss" function, especially for T close to θ. Second, the mathematics involved in using this regret function is relatively simple compared to other regret functions. Third, for large samples, many reasonable estimators yield estimates whose probability distributions are approximated by normal distributions with mean θ. Such distributions are completely specified by the variance of $\mathbf{T}$, and the smaller the variance the better the estimator, no matter what the regret function may be (so long as $r(\theta, T)$ increases as T moves away from θ).

In general

$$R(\theta, t) = E_\theta[r(\theta, \mathbf{T})] = E_\theta[r(\theta, t(\mathbf{Z}))] \tag{10.4}$$

but we shall devote most of our attention to the squared-error regret function of Equation (10.2). Since $c(\theta)\, E_\theta[(\mathbf{T} - \theta)^2]$ is small when $E_\theta[(\mathbf{T} - \theta)^2]$ is small, we shall be interested in

$$E_\theta[(\mathbf{T} - \theta)^2] = \sigma_{\mathbf{T}}^2 + [E_\theta(\mathbf{T}) - \theta]^2. \tag{10.5}$$

[1] The symbol E_θ represents expectation when θ is the state of nature.

Equation (10.5) is proved in deriving Consequence 11 of Appendix E_4.

3. METHODS OF ESTIMATION

In this section we shall describe three major methods of obtaining estimators. The general properties of these methods will be described in subsequent sections.

3.1. The Analogue Method or Method of Moments

Suppose that it is desired to estimate $\theta = E(\mathbf{X}^3)$ when the data consist of n independent observations $\mathbf{X}_1, \mathbf{X}_2, \cdots, \mathbf{X}_n$ on $\mathbf{X}$. Since θ is the long-run average of $\mathbf{X}^3$, it seems reasonable to estimate θ by the average of the $\mathbf{X}_i^3$ for the sample. This is a special case of the *analogue method* which consists of *estimating a parameter or property of a probability distribution by the same property of the sample.*

Thus, using the analogue method, we estimate $\mu = E(\mathbf{X})$ by $\overline{\mathbf{X}}$. In fact we estimate

$$\mu_k = E(\mathbf{X}^k)$$

which is called the kth *moment of* $\mathbf{X}$ (or the kth-order moment of $\mathbf{X}$) by

$$\mathbf{m}_k = \frac{1}{n}\sum_{i=1}^{n}\mathbf{X}_i^k$$

which is called the kth *moment of the sample.* Similarly, we estimate the variance of $\mathbf{X}$

$$\sigma_{\mathbf{X}}^2 = E[(\mathbf{X} - \mu_{\mathbf{X}})^2]$$

by the sample variance[1]

$$\mathbf{d}_{\mathbf{X}}^2 = \frac{1}{n}\sum_{i=1}^{n}(\mathbf{X}_i - \overline{\mathbf{X}})^2.$$

The sample median $\breve{\mathbf{X}}$ furnishes an estimate of the population median $\nu_{\mathbf{X}}$, and $\hat{\mathbf{p}}$ is used as an estimate of p.

Suppose that it is desired to estimate the mean of a distribution known to be exponential. It is a fact that the mean and standard

[1] We call $E[(\mathbf{X} - \mu_{\mathbf{X}})^k]$ the kth population moment about the mean and $\frac{1}{n}\sum_{i=1}^{n}(\mathbf{X}_i - \overline{\mathbf{X}})^k$ the kth sample moment about the mean. Thus $\sigma_{\mathbf{X}}^2$ is the second population moment about the mean.

deviation are equal in exponential distributions (see Section 6.1, Chapter 8). Then, should we use $\overline{\mathbf{X}}$, $\mathbf{d}_\mathbf{X}$, or $(1/2)(\overline{\mathbf{X}} + \mathbf{d}_\mathbf{X})$ as the estimate? The analogue method is not specific on this point. However, it is frequently called the *method of moments* because it is understood that, whenever possible, the parameter should be estimated by using moments and the lowest order moments that are convenient. Thus $\overline{\mathbf{X}}$ would be the appropriate estimate of the mean of an exponential distribution.

Exercise 10.1. Use the analogue method to estimate $\mu_\mathbf{X} + 3\sigma_\mathbf{X}$ when (a) nothing is assumed about **X** and when (b) **X** is assumed to have an exponential distribution.

Exercise 10.2. (a) Estimate the 25th percentile of a distribution on the basis of the 60 observations on Wearwell tires (see Table 2.4). (b) Assuming that the distribution is normal, revise the estimate.

3.2. The Maximum-Likelihood Method

Suppose that the sample is denoted by **Z** which has probability density function $f(z \mid \theta)$. If **Z** is observed, $f(\mathbf{Z} \mid \theta)$ is called the likelihood of θ. The maximum-likelihood method consists of estimating θ by that number $\hat{\theta}$ which maximizes the likelihood function, i.e., $\hat{\theta}$ is that value of the parameter with the largest likelihood. We illustrate with a simplified example, namely, an estimation version of Mr. Nelson's rain problem (Example 5.1).

Example 10.1. There are two possible states of nature, θ_1 and θ_2, and $f(z \mid \theta)$ is given by the following table.

	z_1 (fair)	z_2 (dubious)	z_3 (foul)
θ_1	0.60	0.25	0.15
θ_2	0.20	0.30	0.50
Value of $\hat{\theta}$	θ_1	θ_2	θ_2

If $\mathbf{Z} = z_1$, the likelihoods of θ_1 and θ_2 are 0.6 and 0.2, and $\hat{\theta} = \theta_1$. If $\mathbf{Z} = z_2$, $\hat{\theta} = \theta_2$, and if $\mathbf{Z} = z_3$, $\hat{\theta} = \theta_2$ since $0.30 > 0.25$ and $0.50 > 0.15$.

This simple example differs from most estimation problems in that it has only two possible states of nature while most estimation problems have infinitely many possible values of θ. The following examples are more typical.

Example 10.2. Normal Distribution With Known Variance. The maximum-likelihood estimate of μ, the mean of a normal distribution with known variance σ^2, based on n independent observations, is $\overline{\mathbf{X}}$. To show this we write the likelihood

$$f(\mathbf{Z} \mid \mu) = \left[\frac{1}{\sqrt{2\pi\sigma^2}} \exp\left(-\frac{1}{2\sigma^2}(\mathbf{X}_1 - \mu)^2\right)\right]$$

$$\cdot\left[\frac{1}{\sqrt{2\pi\sigma^2}} \exp\left(-\frac{1}{2\sigma^2}(\mathbf{X}_2 - \mu)^2\right)\right]\cdots\left[\frac{1}{\sqrt{2\pi\sigma^2}} \exp\left(-\frac{1}{2\sigma^2}(\mathbf{X}_2 - \mu)^2\right)\right]$$

$$f(\mathbf{Z} \mid \mu) = \frac{1}{(2\pi\sigma^2)^{n/2}} \exp\left(-\frac{1}{2\sigma^2}\sum_{i=1}^{n}(\mathbf{X}_i - \mu)^2\right).$$

This expression is largest when $\sum_{i=1}^{n}(\mathbf{X}_i - \mu)^2$ is smallest. But in Exercise 2.18 we saw that

$$\sum_{i=1}^{n}(\mathbf{X}_i - \mu)^2 = \sum_{i=1}^{n}(\mathbf{X}_i - \overline{\mathbf{X}})^2 + n(\overline{\mathbf{X}} - \mu)^2$$

which is minimized when μ is $\overline{\mathbf{X}}$. In other words $\hat{\mu} = \overline{\mathbf{X}}$.

Example 10.3. Binomial Distribution. Let p be the probability that a coin will fall heads. If $\mathbf{m}$ heads are observed in n tosses of the coin the maximum-likelihood estimate is $\hat{\mathbf{p}} = \mathbf{m}/n$. Elementary calculus is required to prove this.

Exercise 10.3. Suppose that three possible outcomes have probabilities p^*, $p^*(1 - p^*)$, and $(1 - p^*)^2$ respectively. Present the likelihood based on n observations in which these outcomes occur with frequencies $\mathbf{m}_1$, $\mathbf{m}_2$, and $\mathbf{m}_3$ respectively. Apply the result of Example 10.3 to obtain the maximum-likelihood estimate of p^*.

Example 10.4. Rectangular Distribution. See Mr. Sharp's dial problem (Example 3.2). Suppose that $\mathbf{X}$ has density f given by

$$f(x) = 1/2\mu \qquad \text{for } 0 \le x \le 2\mu$$
$$f(x) = 0 \qquad \text{elsewhere.}$$

Then μ is the mean of the distribution. While the method of moments would give $\overline{\mathbf{X}}$ as the estimate based on n independent observations, the maximum-likelihood estimate can be shown to be $(\max_{1 \le i \le n} \mathbf{X}_i)/2$, that is, one half the largest observation.

Exercise 10.4. The data 4.4, 7.3, 9.1, 6.5, 3.6 are obtained in five observations on a double-exponential distribution whose density is given by

$$f(x \mid \mu) = \frac{1}{4} \exp\left(-\frac{|x - \mu|}{2}\right).$$

Compute the maximum-likelihood estimate of μ. (*Hint*: $\sum_i |\mathbf{X}_i - a|$ is minimized by $a = \breve{\mathbf{X}}$, the sample median.)

3.3. Bayes Strategies

Given an *a priori* distribution $\bar{w}$ on the set of possible θ, the data $\mathbf{Z}$ induce an *a posteriori* probability distribution $\bar{\mathbf{w}}$ ($\bar{\mathbf{w}}$ depends on $\mathbf{Z}$). Thus the state of nature is regarded as a random variable, and $r(\boldsymbol{\theta}, T)$ may be thought of as a random quantity with a distribution which depends upon $\bar{\mathbf{w}}$. There is a corresponding value of T which minimizes the value of

$$B(\bar{\mathbf{w}}, T) = E^*[r(\boldsymbol{\theta}, T)].$$

(In this expression, E^* represents expectation with respect to the *a posteriori* distribution of $\boldsymbol{\theta}$.) Of course, the value of T which

TABLE 10.1

COMPUTATION OF BAYES ESTIMATE FOR EXAMPLE 10.1

$f(z \mid \theta)$

	z_1	z_2	z_3	$\bar{w}$
θ_1	0.60	0.25	0.15	0.6
θ_2	0.20	0.30	0.50	0.4

$r(\theta, T)$

$\theta \backslash T$	θ_1	θ_2
θ_1	0	4
θ_2	3	0

	z_1		z_2		z_3	
$w_1 f(z \mid \theta_1)$	0.36		0.15		0.09	
$w_2 f(z \mid \theta_2)$	0.08		0.12		0.20	
$f(z)$	0.44		0.27		0.29	
$\mathbf{w}_1$	$\frac{0.36}{0.44}$		$\frac{0.15}{0.27}$		$\frac{0.09}{0.29}$	
$\mathbf{w}_2$	$\frac{0.08}{0.44}$		$\frac{0.12}{0.27}$		$\frac{0.20}{0.29}$	
T	θ_1	θ_2	θ_1	θ_2	θ_1	θ_2
$B(\bar{\mathbf{w}}, T)$	$\frac{0.24}{0.44}$	$\frac{1.44}{0.44}$	$\frac{0.36}{0.27}$	$\frac{0.60}{0.27}$	$\frac{0.60}{0.29}$	$\frac{0.36}{0.29}$
Value of T	θ_1		θ_1		θ_2	

Note. We have modified the regret table from the natural estimation version of the rain problem (Example 5.1) so that $r(\theta_1, \theta_2) = 4$ instead of 3 in order to help the student trace the computations.

gives this minimum will depend upon $\bar{\mathbf{w}}$ and, thus, upon **Z**. In fact, we may write

$$\mathbf{T} = t(\mathbf{Z})$$

in which **T** denotes the minimizing value of T corresponding to the data **Z**. In these terms we have defined t as a *Bayes estimator*.

The above definition is quite abstract out of context and so we illustrate with Example 10.1. See Table 10.1. There we see that the Bayes estimates corresponding to z_1, z_2, z_3 are θ_1, θ_1, and θ_2. With different *a priori* probabilities or regrets, the Bayes estimator could change.

Let us study the behavior of the *a posteriori* distribution as the sample size increases. Start off with an arbitrary *a priori* probability distribution which does not neglect any possible θ. After each observation, compute the *a posteriori* probability distribution. As more and more data are collected, the *a posteriori* distribution tends to concentrate all of its probability very close to the true value θ. We illustrate with the following example.

Example 10.5. Suppose that the probability of a coin falling heads is $p = 0.4$. We simulate 20 tosses of the coin with the table of random numbers, obtaining, *H, H, T, T, T, H, T, H, T, T, H, T, T, T, T, T, H, T, T, H.*

Now suppose that someone gave us the *a priori* probability distribution of p, given by

$$\bar{w}(p) = 1 \qquad \text{for } 0 \le p \le 1$$
$$\bar{w}(p) = 0 \qquad \text{elsewhere}$$

Then, using the continuous analogue of the equations of Section 3, Chapter 6, the *a posteriori* probability for $0 \le p \le 1$ is given by

$$\bar{\mathbf{w}}(p) = \frac{\bar{w}(p) f(\mathbf{Z} \mid p)}{f(\mathbf{Z})} = \frac{1 \cdot p^{\mathbf{m}}(1 - p)^{n-\mathbf{m}}}{f(\mathbf{Z})}.$$

The denominator $f(\mathbf{Z})$ must be such that the area under the curve representing $\bar{\mathbf{w}}$ is one. It can be proved that

$$f(\mathbf{Z}) = [(\mathbf{m})!\,(n - \mathbf{m})!]/[(n + 1)!],$$

where $n! = 1 \cdot 2 \cdot 3 \cdots n$ represents the product of the first n integers. In Figure 10.1 we compare the *a priori* probability distribution with the *a posteriori* probability distriubtions for $n = 1, 2,$

3, 10, and 20. Note that the maximum-likelihood estimate when $n = 20$ is $\hat{\mathbf{p}} = 0.35$ but that the Bayes estimate is not defined until the regrets $r(p, T)$ are given.

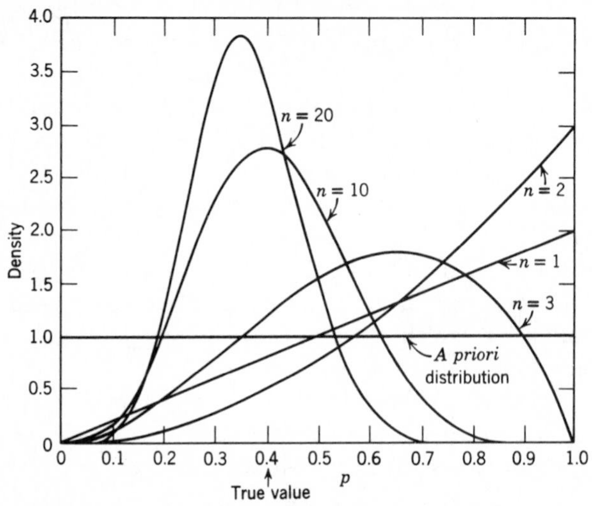

Figure 10.1. *A priori* distribution and *a posteriori* distributions of p based on the data in Example 10.5. For $0 \leqslant p \leqslant 1$

$$\bar{w}(p) = 1$$

$$\bar{\mathbf{w}}(p) = \frac{(n+1)!}{\mathbf{m}!\,(n-\mathbf{m})!}\, p^{\mathbf{m}}(1-p)^{n-\mathbf{m}}$$

3.4. Convenience of the Various Methods of Estimation

The properties of the methods of estimation will be considered in some detail in Section 4. Here we merely remark on the convenience of applying these methods. The analogue method is by far the easiest to apply. Furthermore, no knowledge of $f(z \mid \theta)$ is needed to apply this method. To apply the maximum-likelihood method or the Bayes method, $f(z \mid \theta)$ must be specified. For the Bayes method, some *a priori* probability distribution $\bar{w}$ and the regret function $r(\theta, T)$ must also be given. Ordinarily, calculus is required to derive maximum-likelihood estimates. The Bayes method is usually the clumsiest to apply because of the information required. Whereas the Bayes method is mainly used as a theoretical tool, the maximum-likelihood method is very widely applied in practical examples.

4. LARGE-SAMPLE PROPERTIES OF ESTIMATORS

A good estimator t is one for which $\mathbf{T} = t(\mathbf{Z})$ tends to be close to θ. To fix ideas, we repeat our assumption that the regret is approximated by

$$r(\theta, T) = c(\theta)(T - \theta)^2 \tag{10.2}$$

and

$$R(\theta, t) = c(\theta)\, E_\theta[(\mathbf{T} - \theta)^2]. \tag{10.3}$$

Consequently, a good estimator would be one for which

$$E_\theta[(\mathbf{T} - \theta)^2] = \sigma_{\mathbf{T}}^2 + [E(\mathbf{T}) - \theta]^2 \tag{10.5}$$

is small. For concreteness let us refer to the cord estimation problem (Example 7.3) and the three estimators of the mean breaking strength θ. To conform with the usual convention, we shall change our notation slightly to represent the mean of the breaking strength by μ instead of θ. Recall that in this problem, an observation $\mathbf{X}$ is normally distributed with unknown mean μ and known standard deviation $\sigma = 8$. In Figure 7.8, we compared the three estimators t_1, t_2, and t_3 where $t_1(\mathbf{Z}) = \overline{\mathbf{X}}$, $t_2(\mathbf{Z}) = \breve{\mathbf{X}}$ = sample median, and $t_3(\mathbf{Z}) = 30$ irrespective of the data. For each estimator t, both $R(\mu, t)$ and $E_\mu[(\mathbf{T} - \mu)^2]$ are functions of μ. It is interesting that $E_\mu[(\overline{\mathbf{X}} - \mu)^2] = 1$, $E_\mu[(\breve{\mathbf{X}} - \mu)^2] = 1.57$, and $E_\mu[(30 - \mu)^2] = (30 - \mu)^2$, so that t_1 dominates t_2 (i.e., $R(\mu, t_1) < R(\mu, t_2)$ for all μ) but t_1 does not dominate the ridiculous estimator t_3. In fact, *for μ close to* 30, $R(\mu, t_3) = c(\mu)(30 - \mu)^2 < R(\mu, t_1)$. In Chapter 7 we remarked that, even though t_3 is admissible, it should be discarded. This estimator is nothing more than a guess and is disastrous if the guess is wrong. Unless there is very strong *a priori* evidence concerning μ, a reasonable estimator should yield an estimate $\mathbf{T}$ which is close to μ, no matter what μ is. The estimators t_1 and t_2 behave in this fashion, thus even t_2 should be preferred to t_3 in spite of the fact that there is a somewhat better strategy than t_2.

To emphasize the point, let us consider the $E_\mu[(\mathbf{T} - \mu)^2]$ for these estimators when the sample size n is large. Then it can be shown that

$$E_\mu[(\overline{\mathbf{X}} - \mu)^2] = \sigma^2/n$$

$$E_\mu[(\breve{\mathbf{X}} - \mu)^2] = 1.57\sigma^2/n$$

and

$$E_\mu[(30 - \mu)^2] = (30 - \mu)^2.$$

Note that for *very large n*, the $E_\mu[(\mathbf{T} - \mu)^2]$ and, therefore, the associated risks for $\bar{\mathbf{X}}$, and $\breve{\mathbf{X}}$ are *very small for all* μ. This cannot be said for the guess 30. In fact, for both $\bar{\mathbf{X}}$ and $\breve{\mathbf{X}}$, the estimate tends to be close to μ, no matter what μ is, if the sample size is large. To formalize these properties of $\bar{\mathbf{X}}$ and $\breve{\mathbf{X}}$, we shall discuss the concept of the limit of a sequence.[1] Consider the sequences

$$\{a_n\} = \{a_1, a_2, \cdots, a_n, \cdots\}$$
$$= \{1, 1/2, 3/4, 7/8, 15/16, \cdots, 1 - 1/2^n, \cdots\}$$

and

$$\{b_n\} = \{b_1, b_2, \cdots, b_n, \cdots\}$$
$$= \{1/2, 3/4, 1/4, 7/8, 1/8, 15/16, 1/16, \cdots\}.$$

The first sequence is said to have the limit one because a_n is very close to one when n is very large. We write $\lim_{n\to\infty} a_n = 1$ or $a_n \to 1$. Graphically, if we plot $a_1, a_2, \cdots$ on a line, the points would tend to cluster near the point one. If you take a small interval about one, infinitely many terms of the sequence lie in this interval. With $\{b_n\}$, there are two points near which the b_n tend to cluster. They are 0 and 1. Then $\lim_{n\to\infty} b_n$ does not exist. On the other hand, we can talk about upper limits and lower limits. In fact the sequence $\{b_n\}$ has the upper limit one and lower limit zero. We write $\overline{\lim}_{n\to\infty} b_n = 1$, $\underline{\lim}_{n\to\infty} b_n = 0$. Essentially $\overline{\lim}_{n\to\infty} b_n$ is the greatest (rightmost) cluster point and $\underline{\lim}_{n\to\infty} b_n$ is the least. The limit of a sequence exists when the upper and lower limits coincide. For $\{a_n\}$, $\overline{\lim}_{n\to\infty} a_n = \underline{\lim}_{n\to\infty} a_n = 1$ and hence $\lim_{n\to\infty} a_n = 1$. The sequence $\{c_n\} = \{1, 1/2, 1, 1/4, 1, 1/8, \cdots\}$ has zero and one as its lower and upper limits.

Exercise 10.5. What are the lower and upper limits of $\{1, 0, -1, 1, 0, -1, 1, 0, -1, \cdots\}$?

Exercise 10.6. What are the lower and upper limits of $\{1, 1/2, 1/4, 1, 1/4, 1/8, 1, 1/8, 1/16, \cdots\}$?

[1] A sequence is a function defined on the set of positive integers. It is usually denoted by $\{a_n\} = \{a_1, a_2, \cdots, a_n, \cdots\}$, where a_n is the value of the function for the integer n and is called the nth term of the sequence. For example, $\{2/n\} = \{2, 1, 2/3, \cdots, 2/n, \cdots\}$ is a sequence.

Exercise 10.7. Find the upper and lower limits for the following sequences.

(a) $\{1/n\}$ (d) $\{(-1)^n/n\}$
(b) $\{1/3^n\}$ (e) $\{(-1)^n + 3/n\}$
(c) $\{(-1)^n\}$

In what follows *we regard an estimator t as a sequence of estimators, one for each possible sample size n.* For example, t_1 gives $\bar{\mathbf{X}}$ which is $(\mathbf{X}_1 + \mathbf{X}_2)/2$ for $n = 2$ and $(\mathbf{X}_1 + \mathbf{X}_2 + \mathbf{X}_3)/3$ for $n = 3$. Another example, which is slightly pathological, is t_4 which gives $\bar{\mathbf{X}}$ when n is even and $\breve{\mathbf{X}}$ when n is odd. Still another estimator is t_5, which gives $\mathbf{X}_1$ for $n = 1$ and $(\mathbf{X}_1 + \mathbf{X}_2)/2$ for $n \geq 2$. That is, t_5 ignores all but the first two observations. When it is important to regard an estimator as defined for one specified sample size only, we shall so indicate. Note that

$$E_\mu[(\mathbf{T}_4 - \mu)^2] = \sigma^2/n \qquad \text{if } n \text{ is even}$$

and

$$E_\mu[(\mathbf{T}_4 - \mu)^2] = 1.57\sigma^2/n \qquad \text{if } n \text{ is odd.}$$

Also

$$E_\mu[(\mathbf{T}_5 - \mu)^2] = \sigma^2 \qquad \text{if } n = 1$$

and

$$E_\mu[(\mathbf{T}_5 - \mu)^2] = \sigma^2/2 \qquad \text{if } n > 1.$$

DEFINITION 10.1. An estimator t is said to be a *consistent estimator* of θ if [1]

$$\lim_{n\to\infty} E_\theta[(\mathbf{T} - \theta)^2] = 0 \qquad \text{for all } \theta. \tag{10.6}$$

It is clear that the estimators t_1, t_2, and t_4 are consistent estimators of μ whereas t_3 and t_5 are not. Since consistent estimators are readily obtainable in most practical problems, we shall consider consistency as a fundamental property and restrict our attention to consistent estimators when sample sizes are "large."

Using the guess estimator t_3, we were able to make $E_\mu[(\mathbf{T}_3-\mu)^2]$ small at $\mu = 30$ at the cost of poor behavior when μ was far from 30. By insisting on consistent estimators, we prohibit ideal behavior for some μ at the expense of extremely poor behavior elsewhere.

[1] In more advanced work a weaker criterion is used. This requires that $\lim_{n\to\infty} P\{\theta - a < \mathbf{T} < \theta + a \mid \theta\} = 1$ for each $a > 0$ and every θ. According to either definition, $\mathbf{T}$ is almost sure to be close to θ when n is large.

Can we compromise to this extent? Can we find a *consistent* estimator which is substantially better than $\bar{X}$ for some values of μ at the cost of being worse for other values of μ? The answer to this question is essentially "no." To be more precise, we introduce the following definition.

DEFINITION 10.2. An estimator t is said to be (asymptotically) *efficient* if there is no other consistent estimator t^* such that

$$\overline{\lim_{n\to\infty}} \frac{E_\theta[(\mathbf{T} - \theta)^2]}{E_\theta[(\mathbf{T}^* - \theta)^2]} > 1 \qquad \text{for } \theta \text{ in some interval} \tag{10.7}$$

where $\mathbf{T} = t(\mathbf{Z})$ and $\mathbf{T}^* = t^*(\mathbf{Z})$.

Roughly speaking this means that, for large samples, no consistent estimator can improve substantially on an efficient estimator t for any interval of θ.

Efficiency is a very strong property, but, as we shall soon see, for most practical estimation problems efficient estimators do exist. In particular $\bar{X}$ is efficient for estimating μ in Example 7.3.

The three methods of obtaining estimators described in Section 3 can now be partially evaluated. First, all three methods yield consistent estimators. Second, the analogue method does not always give efficient estimators. On the other hand, both the maximum-likelihood method and the Bayes method give efficient estimators. In fact, if we start with any *a priori* probability distribution $\bar{w}$, which has positive density for all θ, the Bayes estimator and the maximum-likelihood estimator will be extremely close to one another for large samples. For this reason, the maximum-likelihood method is often preferred since it is more convenient to apply.

At this point we may remark that in most (but not all) applications all three methods produce estimates which are *approximately normally distributed with mean* θ *and standard deviation of the order of magnitude of* $1/\sqrt{n}$.[1] In Section 2 we remarked that, in comparing estimators which are normally distributed with mean θ, the variance of the estimator is the relevant measure, whether or not the regret is given by $c(\theta)(T - \theta)^2$. Thus efficiency is an

[1] We say that $\{a_n\}$ is of the order of magnitude of $1/\sqrt{n}$ if $\{a_n/(1/\sqrt{n})\}$ is bounded.

important estimator property for estimation problems with large samples and any reasonable regret function.

It is interesting that, in many of the exceptional cases where the maximum-likelihood estimates are not approximately normally distributed with standard deviation of the order of magnitude of $1/\sqrt{n}$, even better results are obtained. For example, in the case of the rectangular distribution (Example 10.4), the distribution of the maximum-likelihood estimate is not approximately normal but the standard deviation is of the order of magnitude of $1/n$, which is considerably smaller than $1/\sqrt{n}$.

Exercise 10.8. For Example 7.3, find

$$\overline{\lim_{n \to \infty}} \frac{E_\mu[(\mathbf{T} - \mu)^2]}{E_\mu[(\mathbf{T}^* - \mu)^2]}$$

where (a) $t = t_1$, $t^* = t_2$; (b) $t = t_2$, $t^* = t_4$; and (c) $t = t_1$, $t^* = t_4$.

Exercise 10.9. Modify the cord problem (Example 7.3) so that $c(\theta) = 1$. Then $R(\mu, t) = E_\mu[(\mathbf{T} - \mu)^2]$. Use $\bar{\mathbf{X}}$ to estimate μ. What sample size should you use if $\sigma = 8$ and the cost of sampling is given by $C(n) = 0.4 + (0.01)n$?

Exercise 10.10. The parameter p is estimated by $\hat{\mathbf{p}}$. If the regret function is $(T - p)^2$, how large a sample size is required to ensure that the maximum value of the risk is no larger than 0.01?

5. SMALL-SAMPLE PROPERTIES OF ESTIMATORS

There are occasions when the large-sample theory is not appropriate, mainly because the sample size for a particular problem is small. Then one may prefer to regard consistency as being irrelevant. What other properties not involving large samples are of interest? In the following subsections we discuss several such properties.

5.1. Invariance Under Transformation of Parameter

In Example 4.6. we dealt with a problem where Mr. Heath was interested in the mode of a distribution of foot sizes so that he could adjust his machine to produce shoes of the corresponding size. Suppose that he estimates the mode to be 10.5. He tells the engineer to produce shoes of size 10.5. The engineer replies that his machine is calibrated in inches and that he must know the length of the shoes. Mr. Heath claims that a size 12 corresponds to a foot

12 inches long and a unit change in size represents 1/3 inch. Thus length and size are related by $L = (1/3)S + 8$. Therefore he tells his engineer to produce shoes of length $(1/3)(10.5) + 8 = 11.5$ inches.

If the modal size is called θ, Mr. Heath is interested in $\theta^* = (1/3)\theta + 8$. Since $T = 10.5$ is his estimate of θ (i.e., he wishes to act as though $\theta = 10.5$), he will have to estimate θ^* by $\mathbf{T}^* = (1/3)\mathbf{T} + 8$ (i.e., act as though $\theta^* = 11.5$ inches) to be consistent. In general it seems reasonable to require that, if we estimate θ by $\mathbf{T}$, we should estimate $\theta^* = g(\theta)$ by $\mathbf{T}^* = g(\mathbf{T})$.

DEFINITION 10.3. A method of producing estimators is said to be *invariant under transformation of parameters* if it leads to estimating $\theta^* = g(\theta)$ by $\mathbf{T}^* = g(\mathbf{T})$ whenever it leads to estimating θ by $\mathbf{T}$.

Further extensions of this concept of invariance would apply to the case where the measurements of the data are changed from one scale to another. For example, the mere fact that a French scientist uses centimeters instead of inches to measure certain random variables should not lead the French scientist to different conclusions or actions than an English scientist having the same information in inches.

The Bayes and maximum-likelihood estimators are invariant under transformation of parameter. Because of ambiguities inherent in the analogue method, this method may, on occasion, fail to be invariant.

5.2. Unbiased Estimators

In the Sheppard rifle problem (Example 4.3), we found that the value of a which minimizes $E[(\mathbf{X} - a)^2]$ is $a = E(\mathbf{X})$. Since the regret function

$$E_\theta[(\mathbf{T} - \theta)^2] = \sigma_\mathbf{T}^2 + [E_\theta(\mathbf{T}) - \theta]^2$$

it would seem desirable to have $E_\theta(\mathbf{T}) = \theta$. If we could modify an estimator so as to obtain $E_\theta(\mathbf{T}) = \theta$ without increasing $\sigma_\mathbf{T}^2$, it would be desirable to do so. However, this is not possible in general.

DEFINITION 10.4. An estimator t is said to be an *unbiased estimator* of θ if

$$E_\theta(\mathbf{T}) = \theta \qquad \text{for every } \theta. \tag{10.8}$$

We call $E_\theta(\mathbf{T}) - \theta$ the *bias* of t.

For large samples, consistent estimators are approximately unbiased. Unbiased estimators are desirable when they can be obtained without increasing $\sigma_{\mathbf{T}}^2$. Note that in Example 7.3, t_1, t_2, t_4, and t_5 are all unbiased estimators. Also $\hat{\mathbf{p}}$ is an unbiased estimate of p since $E_p(\hat{\mathbf{p}}) = p$. Connected with the reason that most statisticians find it more convenient to use

$$\mathbf{s}_{\mathbf{X}}^2 = \frac{1}{n-1} \sum_{i=1}^{n} (\mathbf{X}_i - \overline{\mathbf{X}})^2$$

instead of $\mathbf{d}_{\mathbf{X}}^2$ to estimate $\sigma_{\mathbf{X}}^2$ is the fact that $\mathbf{s}_{\mathbf{X}}^2$ is an unbiased estimate of $\sigma_{\mathbf{X}}^2$.

Note that, whereas $E_\mu(\overline{\mathbf{X}}) = \mu$,

$$E_\mu(\overline{\mathbf{X}}^2) = \sigma_{\overline{\mathbf{X}}}^2 + \mu^2 = \mu^2 + \sigma^2/n > \mu^2.$$

Thus if $\overline{\mathbf{X}}$ is an unbiased estimate of μ, $\overline{\mathbf{X}}^2$ will be a slightly biased estimate of μ^2. The criteria of unbiasedness and invariance cannot both be simultaneously satisfied. Of the two, the invariance criterion is definitely more fundamental.

Exercise 10.11. The Bay Cable Company was required to lay a telephone cable across the San Francisco Bay. Never having studied decision making in the face of uncertainty, the cable layers estimated the amount of cable necessary to cross the bay from one given point to another. As they were finishing their job, it turned out that they had underestimated, for the cable slipped off the reel and sank to the bottom of the bay. Divers were required to raise the end of the cable, splices were made, and the job was finished at great expense to the company.

Graph a reasonable regret function for the estimate T (using a reel with T feet of cable) when θ is the required length. Would a reasonable estimator be biased or unbiased? If biased, in which direction?

Exercise 10.12. Modify $\hat{\mathbf{p}}^2$ so as to yield an unbiased estimate of p^2.

Exercise 10.13. Prove that $\mathbf{s}^2$ is an unbiased estimate of σ^2 if the sample consists of two independent observations on a random variable with variance σ^2.

†5.3. Sufficiency

Example 10.6. Mr. Jones desires to estimate p the probability that a coin will fall heads. The coin is tossed n times and falls heads $\mathbf{m}$ times. Mr. Jones feels that $\hat{\mathbf{p}} = \mathbf{m}/n$ is a good estimate of p. It is unbiased, consistent, and asymptotically efficient. However, his friend Mr. Martin has noticed that this method completely ignores the order in which heads and tails appear. He asks whether Mr. Jones is really making good use of all the data. Mr. Jones replies that this method is efficient. But, argues Mr. Martin, there are only $n = 3$ observations, and this is hardly a large sample. Mr. Jones replies that $\hat{\mathbf{p}}$ is unbiased. Mr. Martin crossly reminds him that, had he done his homework in statistics, the concept of unbiasedness would not mean terribly much, especially for $n = 3$ and a highly unspecified loss function.

After some introspection, Mr. Jones claims that everyone knows that the order in which heads and tails fall is irrelevant. "Oh," replies Mr. Martin, "and how do you know this?" While Mr. Jones was thinking this over, along came his friend, Mr. Kurt. When Mr. Martin's question was explained to him he said: "I do not see much point in using $\hat{\mathbf{p}}$ as the estimate of p with a sample of size 3. First of all, I am not convinced that you have a real estimation problem and, if you did, I would feel unhappy about giving out an estimate of zero just because the coin fell tails 3 times. However, in any case, whatever I did would depend on $\hat{\mathbf{p}}$ and not on the order in which heads and tails appeared. The reason is essentially this: Suppose I knew that heads had appeared exactly once in the three tosses, i.e., $\hat{\mathbf{p}} = 1/3$. Then the three possible arrangements *HTT*, *THT*, and *TTH* are all equally likely (i.e., each have conditional probability 1/3 no matter what the value of p). In other words, any additional information besides the fact that heads had appeared once has a conditional probability distribution which does not depend on the state of nature. Now, clearly, information whose probability distribution does not depend on p cannot help to estimate p. Hence, the only useful information is the number of heads, or equivalently, $\hat{\mathbf{p}}$."

Suppose that $\mathbf{T}$ is the value of a statistic[1] given by $\mathbf{T} = t(\mathbf{Z})$. If

[1] A statistic is any random variable which can be computed from the observed data.

$\mathbf{T}'$ is the value of another statistic, and $\mathbf{T}$ and $\mathbf{T}'$ have joint density $f(x, x' \mid \theta)$, then

$$f(x, x' \mid \theta) = g(x \mid \theta)\, h(x' \mid x, \theta) \tag{10.9}$$

where $g(x \mid \theta)$ is the density of $\mathbf{T}$ and $h(x' \mid x, \theta)$ represents the conditional density of $\mathbf{T}'$ given $\mathbf{T} = x$.[1] Suppose that $h(x' \mid x, \theta)$ does not involve θ. Then, once $\mathbf{T}$ is known, the conditional distribution of $\mathbf{T}'$ does not depend on θ, and $\mathbf{T}'$ is irrelevant in the decision making problem. It can be shown that this will be the case for all $\mathbf{T}'$ if the density f of all the data $\mathbf{Z}$ can be represented as a product as follows:

$$f(\mathbf{Z} \mid \theta) = g(\mathbf{T}, \theta)\, h(\mathbf{Z}). \tag{10.10}$$

Then it makes sense to say that $\mathbf{T}$ summarizes all the data of the experiment relevant to θ, and we call $\mathbf{T}$ a *sufficient statistic*.

It can be shown that, if there is a sufficient statistic for θ, the maximum-likelihood estimate will be sufficient. Furthermore, if there is a sufficient statistic and an unbiased estimator of θ, these two can be combined under very general circumstances to give an estimate which is (1) sufficient, (2) unbiased, and (3) has the least variance among unbiased estimators.

6. SEVERAL PARAMETERS

Up to now we have considered only cases where the action and the state of nature can each be represented by a single number. In many problems, the action and the state of nature can be represented by several numbers.

Example 10.7. Estimate the mean μ and the standard deviation σ of a normal distribution on the basis of n independent observations $\mathbf{X}_1, \mathbf{X}_2, \cdots, \mathbf{X}_n$.

For this problem, both the analogue method and the maximum-likelihood method yield the same estimates of μ and σ. These are $\overline{\mathbf{X}}$ and $\mathbf{d}_{\mathbf{X}}$ respectively. However, for technical reasons, statisticians prefer to use $\mathbf{s}_{\mathbf{X}}$ instead of $\mathbf{d}_{\mathbf{X}}$ (for substantial n, $\mathbf{d}_{\mathbf{X}}$ and $\mathbf{s}_{\mathbf{X}}$ are almost equal).

In this problem, the state of nature θ can be represented by the pair of parameters (μ, σ) where $\sigma > 0$. The maximum-likelihood

[1] This is related to the fact that $P\{A \text{ and } B\} = P\{A\}\, P\{B \mid A\}$.

estimate of θ is $\hat{\theta} = (\bar{\mathbf{X}}, \mathbf{d}_{\mathrm{X}})$ but the usual estimator is $\mathbf{T} = (\bar{\mathbf{X}}, \mathbf{s}_{\mathrm{X}})$.

Example 10.8. The six faces of a die have probability $p_1, p_2, \cdots, p_6$. Estimate these parameters on the basis of n observations.

If we let $\hat{\mathbf{p}}_i$ be the proportion of observations with face i showing, the analogue method suggests $\hat{\mathbf{p}}_i$ as the estimate of p_i. It is possible to show (using calculus) that this is also the maximum-likelihood estimate. In this example the state of nature is represented by $\theta = (p_1, p_2, \cdots, p_6)$.

The general decision making model is the following. The set $\mathscr{N}$ of states of nature θ can be represented by k coordinates, i.e., $\theta = (\theta_1, \theta_2, \cdots, \theta_k)$. The best action to take if θ were known would depend on only h of these coordinates, and thus we label this best action by $T(\theta) = (\theta_1, \theta_2, \cdots, \theta_h)$. For example, we might be interested in only the mean of a normal distribution with unknown mean and standard deviation. The regret function $r(\theta, T)$ increases as $T = (T_1, T_2, \cdots, T_h)$ gets farther away from $T(\theta)$. In Example 10.7, a reasonable regret function might be approximated by

$$r(\theta, T) = c_1(T_1 - \mu)^2 + c_2(T_2 - \sigma)^2 + c_3(T_1 - \mu)(T_2 - \sigma)$$

when we act as though μ were T_1 and σ were T_2. The ideas of consistency, efficiency, invariance, unbiasedness, and sufficiency can be extended to this case.

7. CONFIDENCE INTERVALS: LARGE SAMPLES

To permit the possibility of hedging or to determine the extent of hedging that would be appropriate, we require some idea of how good an estimate is. When the sample size is large, we frequently have the following situation. The estimate $\mathbf{T}$ is approximately normally distributed with mean θ and variance σ_{T}^2 which can be estimated by some estimate, say $\hat{\boldsymbol{\sigma}}_{\mathrm{T}}^2$. For large sample size, $\hat{\boldsymbol{\sigma}}_{\mathrm{T}}^2$ is close to σ_{T}^2. Then $\hat{\boldsymbol{\sigma}}_{\mathrm{T}}$ is a measure of how good the estimate is. Furthermore, the following equations, among others, are approximately true.

$$P\{|\mathbf{T} - \theta| \leq 1.96\hat{\boldsymbol{\sigma}}_{\mathrm{T}} \mid \theta\} = 0.95$$

$$P\{|\mathbf{T} - \theta| \leq 2.58\hat{\boldsymbol{\sigma}}_{\mathrm{T}} \mid \theta\} = 0.99.$$

Thus with probability 0.95, $\mathbf{T}$ will be within $1.96\hat{\boldsymbol{\sigma}}_{\mathrm{T}}$ of θ and the interval $(\mathbf{T} - 1.96\hat{\boldsymbol{\sigma}}_{\mathrm{T}}, \mathbf{T} + 1.96\hat{\boldsymbol{\sigma}}_{\mathrm{T}})$ will contain θ.

Example 10.9. Binomial Case. To estimate p, we use $\hat{\mathbf{p}}$. Then $\sigma_{\hat{\mathbf{p}}} = \sqrt{p(1-p)/n}$ which can be estimated by $\sqrt{\hat{\mathbf{p}}(1-\hat{\mathbf{p}})/n}$. If out of 400 tosses, we obtain 160 heads, our estimates $\hat{\mathbf{p}}$ and $\hat{\sigma}_{\hat{\mathbf{p}}}$ are 0.4 and $\sqrt{(0.4)(0.6)/400} = 0.025$. The interval $(\hat{\mathbf{p}} - 1.96\hat{\sigma}_{\hat{\mathbf{p}}}, \hat{\mathbf{p}} + 1.96\hat{\sigma}_{\hat{\mathbf{p}}})$ is (0.351, 0.449).

DEFINITION 10.5. If $\mathbf{\Gamma}$ is a random interval depending on the data $\mathbf{Z}$ such that

$$P\{\mathbf{\Gamma} \text{ contains } \theta \mid \theta\} = \gamma \qquad \text{for all } \theta$$

then $\mathbf{\Gamma}$ is called a γ confidence interval for θ.

For large samples in the situation described above, the interval $(\mathbf{T} - 1.96\hat{\sigma}_{\mathbf{T}}, \mathbf{T} + 1.96\hat{\sigma}_{\mathbf{T}})$ *is an approximate 95% confidence interval.*

In the Example 10.9, the approximate 95% confidence interval corresponding to the data turned out to be (0.351, 0.449). In this case, we may say "(0.351, 0.449) covers θ" with confidence 0.95. In the long run, 95% of those statements which are made with confidence 0.95 will be correct.

Example 10.10. If $\mathbf{X}_1, \mathbf{X}_2, \cdots, \mathbf{X}_n$ are a large sample of independent observations on a random variable with unknown mean μ and unknown variance σ^2, $\overline{\mathbf{X}}$ is an estimate of μ with standard deviation $\sigma_{\overline{\mathbf{X}}} = \sigma/\sqrt{n}$. Since $\mathbf{s}_{\mathbf{X}}$ is an estimate of σ, the corresponding estimate of $\sigma_{\overline{\mathbf{X}}}$ is $\mathbf{s}_{\mathbf{X}}/\sqrt{n}$ and an approximate 95% confidence interval for μ is $(\overline{\mathbf{X}} - 1.96\mathbf{s}_{\mathbf{X}}/\sqrt{n}, \overline{\mathbf{X}} + 1.96\mathbf{s}_{\mathbf{X}}/\sqrt{n})$.

Exercise 10.14. Find the approximate 95% and 98% confidence intervals for p if: (a) $\hat{\mathbf{p}} = 0.4$, $n = 400$; (b) $\hat{\mathbf{p}} = 0.1$, $n = 400$; (c) $\hat{\mathbf{p}} = 0.5$, $n = 100$.

Exercise 10.15. Find the 90% confidence interval for $\mu = E(\mathbf{X})$ $(\overline{\mathbf{X}} = 20, \mathbf{s}_{\mathbf{X}} = 5)$ if: (a) $n = 100$; (b) $n = 400$; (c) $n = 900$.

8. CONFIDENCE INTERVALS: SMALL SAMPLES

In Section 7 we defined confidence intervals and indicated a method of obtaining approximate confidence intervals for θ when the sample size is large. There, all that is required is an estimate $\mathbf{T}$ which is approximately normally distributed with mean θ and variance $\sigma_{\mathbf{T}}^2$ approximated by an estimate $\hat{\sigma}_{\mathbf{T}}^2$. For small sample sizes, this approach does not work. In Section 4, Chapter 7, we discussed how a confidence interval for the mean of a normal distribution with unknown mean and unknown variance could be obtained for small samples using the t distribution.

We propose to indicate a general method of obtaining confidence intervals for specified samples. The idea is easily represented graphically if we assume that there is only one datum, although the method as described applies in general. For each possible value θ_0 of θ, construct a test of significance level $1-\gamma$ to test H_{θ_0}: $\theta=\theta_0$ versus the alternative that $\theta \neq \theta_0$. Suppose that the acceptance set (the set of observations leading to accepting H_{θ_0}) is A_{θ_0}. Then,

(10.11) $$P\{\mathbf{Z} \in A_{\theta_0} \mid \theta_0\} = \gamma \qquad \text{for all } \theta_0.$$

Now let

(10.12) $$\boldsymbol{\Gamma} = \{\theta_0 \colon \mathbf{Z} \in A_{\theta_0}\}.$$

That is to say, let $\boldsymbol{\Gamma}$ be the set of parameter values θ_0 for which the hypothesis H_{θ_0}: $\theta = \theta_0$ would not be rejected by the data.

Then

(10.13) $$P\{\boldsymbol{\Gamma} \text{ contains } \theta \mid \theta\} = \gamma \qquad \text{for all } \theta$$

and $\boldsymbol{\Gamma}$ is a γ confidence interval for θ. This statement follows from Equation (10.11) because

$$\{\mathbf{Z}\colon \boldsymbol{\Gamma} \text{ contains } \theta_0\} = \{\mathbf{Z}\colon \mathbf{Z} \in A_{\theta_0}\}.$$

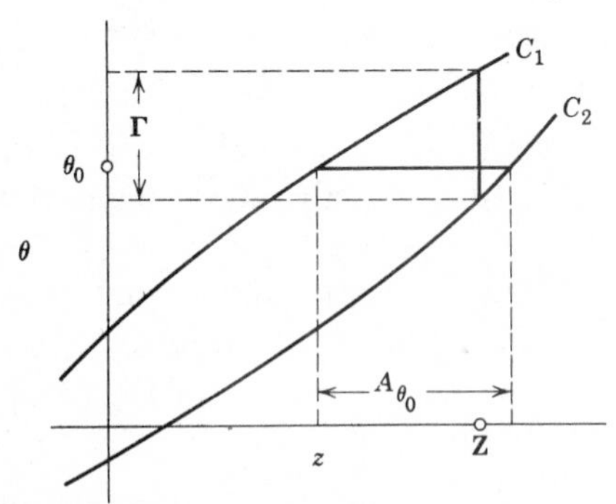

Figure 10.2. Schematic representation of method of constructing confidence intervals.

1. For each θ_0, construct acceptance set A_{θ_0}.
2. The boundaries are curves labeled C_1 and C_2.
3. If $\mathbf{Z}$ is observed, the corresponding ordinates between C_1 and C_2 form $\boldsymbol{\Gamma}$.

This definition is illustrated by the schematic representation in Figure 10.2. In Figure 10.3, we apply the schematic representation to indicate 95% confidence intervals for the binomial problem.

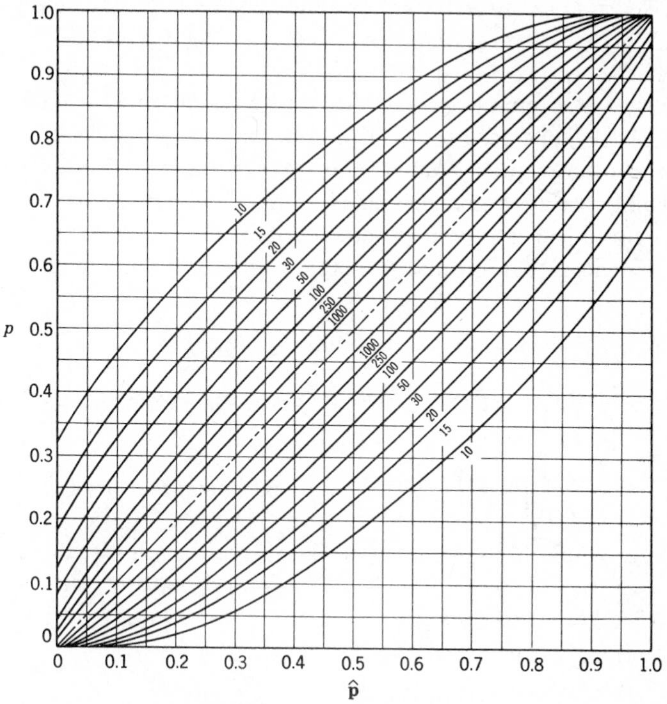

Figure 10.3. Ninety-five percent confidence interval for the proportion p based on various sample sizes. (Reproduced by permission of the authors and the editor from Clopper, C. and E. S. Pearson, "The Use of Confidence or Fiducial Limits Illustrated in the Case of the Binomial," *Biometrika*, Vol. 26, 1934, pp. 404–413.)

9. SUMMARY

We review briefly the one-parameter case. Here the state of nature can be represented by a number θ. The best action to take if θ were known can also be represented by the number θ, and there are no hedging actions. The regret function r has values given by $r(\theta, T)$, which presumably increases as T gets further away from θ. In many problems $r(\theta, T)$ not only is minimized when $T = \theta$ but also, for fixed θ, has a graph which is approximately parabolic near $T = \theta$. That is, for T close to θ, $r(\theta, T) = c(\theta)(T-\theta)^2$ approximately. Hence, for such examples, the risk is approximately $R(\theta, t) = c(\theta)\, E_\theta[(T - \theta)^2]$, where the strategy or estimator t gives $\mathbf{T} = t(\mathbf{Z})$ as the estimate or action when data $\mathbf{Z}$ are observed. Thus, a good estimator is one for which $E[(\mathbf{T}-\theta)^2] = \sigma_{\mathbf{T}}^2 + [E(\mathbf{T}-\theta)]^2$ is small.

For the large-sample case, we regard an estimator as a sequence of estimators, one for each sample size, and study the limiting behavior of the sequence as $n \to \infty$. We do this to facilitate the mathematician's job since he finds limiting behavior relatively easy to study and so that we may obtain insight into what are reasonable estimators for ordinary sample sizes. An estimator t is *consistent* for estimating θ if

$$\lim_{n\to\infty} E_\theta[(\mathbf{T} - \theta)^2] = 0 \qquad \text{for all } \theta.$$

An estimator t is *efficient* if there is no other consistent estimator t^* such that

$$\overline{\lim_{n\to\infty}} \frac{E_\theta[(\mathbf{T} - \theta)^2]}{E_\theta[(\mathbf{T}^* - \theta)^2]} > 1 \qquad \text{for } \theta \text{ in some interval.}$$

A method of obtaining estimators is *invariant under transformation of parameter* if it yields $g(\mathbf{T})$ as an estimate of $g(\theta)$ whenever it leads to $\mathbf{T}$ as an estimate of θ.

An estimator t is an *unbiased* estimator of θ if

$$E_\theta(\mathbf{T}) = \theta \qquad \text{for all } \theta.$$

An insistence on unbiasedness would conflict with the more fundamental invariance property.

$\mathbf{T}$ is a *sufficient statistic* for θ if it exhausts all the relevant information in the data. If the likelihood of the data can be decomposed as

$$f(\mathbf{Z} \mid \theta) = g(\mathbf{T}, \theta)\, h(\mathbf{Z})$$

then $\mathbf{T}$ is sufficient for θ.

The analogue method consists of estimating a parameter (property of the probability distribution) by the analogous property of the sample. Thus the mean of a sample is used to estimate the mean of the population.

The maximum-likelihood method consists of estimating θ by $\hat{\boldsymbol{\theta}}$

which is that possible value of the parameter maximizing the likelihood $f(\mathbf{Z} \mid \theta)$.

All three methods considered in Section 3 usually (but not always) yield estimates which are consistent and approximately normally distributed with mean θ and standard deviation of the order of magnitude of $1/\sqrt{n}$. Of these methods, the analogue method is ordinarily the easiest to apply, but does not generally yield efficient estimates. The maximum-likelihood method is ordinarily more difficult to apply but, in most problems, yields efficient estimates. Furthermore, the maximum-likelihood method is invariant under transformation of parameter and will yield a sufficient statistic if there is one.

It is frequently important to supplement an estimate with an evaluation of its reliability. A technique for so doing is that of confidence intervals. A $100\gamma\%$ confidence interval for θ is a random interval $\mathbf{\Gamma}$ depending on the data $\mathbf{Z}$ such that

$$P\{\mathbf{\Gamma} \text{ contains } \theta \mid \theta\} = \gamma \qquad \text{for all } \theta.$$

For large samples

$$(\mathbf{T} - k_\gamma \hat{\sigma}_\mathbf{T}, \mathbf{T} + k_\gamma \hat{\sigma}_\mathbf{T})$$

is approximately a γ confidence interval for θ if (1) $\mathbf{T}$ is approximately normally distributed with mean θ and standard deviation $\hat{\sigma}_\mathbf{T}$, and (2) k_γ is obtained from the table of the normal distribution according to the equation

$$P\{|\mathbf{X} - \mu| < k_\gamma \sigma_\mathbf{X}\} = \gamma.$$

For small-sample sizes, exact confidence intervals can be obtained by letting

$$\mathbf{\Gamma} = \{\theta_0 : \mathbf{Z} \in A_{\theta_0}\}$$

where A_{θ_0} is selected for each θ_0 so that

$$P\{\mathbf{Z} \in A_{\theta_0} \mid \theta_0\} = \gamma.$$

SUGGESTED READINGS

See the readings suggested for Chapter 9.

APPENDIX A

Notation

RANDOM VARIABLES AND STATISTICS

A number which depends on the outcome of an experiment is a *random variable*. If we can compute it from observable data, it is called a *statistic*. All random variables and statistics are always indicated by boldfaced symbols. Since boldfaced symbols cannot easily be written we recommend for use on paper or at the blackboard that they be replaced by a corresponding symbol with an extra line through it. For example, we can use $\not{X}, \not{Y}, \not{s}, \not{\hat{p}}$ for $\mathbf{X}$, $\mathbf{Y}$, $\mathbf{s}$, $\hat{\mathbf{p}}$. We usually use capital roman letters for such symbols but there are occasional exceptions.

PARAMETERS

A property of the relevant probability distributions or of the state of nature is called a parameter. *Parameters* are usually denoted by Greek symbols. For example, we use μ, σ, ν, θ. An occasional exception such as p is made for the sake of tradition.

MODIFYING SYMBOLS

A dagger (†) next to a section indicates that this section is to be included in the course at the instructor's option.

An asterisk (*) next to an exercise indicates that this exercise is important and should definitely be assigned to the class.

A circle (°) next to an exercise indicates that a solution of the exercise may require mathematical knowledge beyond the material usually covered in two years of high school mathematics.

A bar above a letter usually stands for the sample mean such as $\bar{\mathbf{X}}$. We sometimes use it as an abbreviation for a point such as $\bar{u} = (x, y, z)$.

A breve above a letter stands for the sample median such as $\breve{\mathbf{X}}$.

A circumflex is used for an estimate, usually for a maximum-likelihood estimate, e. g., $\hat{\mathbf{p}}$, $\hat{\theta}$.

A tilde is used to denote the complement of a set. Thus $\tilde{E}$ is the set of elements not in E.

The symbols following a vertical bar represent an assumed condition or state of nature. For example, $P\{A \mid B\}$ is the conditional probability of A given B, and $f(z \mid \theta)$ represents the probability density function of $\mathbf{Z}$ when θ is the state of nature.

GLOSSARY OF LETTERS

The following is a highly detailed list of almost every letter which occurs with a special symbolic significance in this text. In studying the text, this glossary should not be often necessary. Few of these letters occur frequently. In no section of the book do many distinct symbols occur simultaneously. In context, the meaning of the symbols should ordinarily be clear. However, students who use this book for reference or for a brief review may find the following list occasionally helpful.

Lower Case Letters

- a action
- $\mathbf{d}$ sample standard deviation
- $\mathbf{d}^2$ sample variance
- f probability density function[1], $f(x)$, $f(z \mid \theta)$
- $\mathbf{f}$ observed frequency
- l loss function, $l(\theta, a)$
- m slope
- $\mathbf{m}$ binomial random variable (e.g., numbers of heads in n tosses of a coin)
- n sample size
- p probability
- $\hat{\mathbf{p}}$ observed proportion
- r regret, $r(\theta, a)$
- s strategy, $s(\mathbf{Z})$
- $\mathbf{s}$ sample standard deviation[2]

[1] If a symbol represents a function, the value of this function will be indicated to show the set on which the function is defined. Thus the value $f(x)$ of the probability density function is indicated.

[2] $\mathbf{s}$ is related to $\mathbf{d}$ by $\mathbf{s} = \mathbf{d}\sqrt{n/(n-1)}$. Statisticians generally prefer to use $\mathbf{s}$.

$\mathbf{s}^2$	sample variance
t	estimator
t_n	the t distribution with n degrees of freedom
u	utility function, $u(P)$
$\bar{u}$	point
$\bar{w}$	set of weights or of *a priori* probabilities
$\bar{\mathbf{w}}$	set of *a posteriori* probabilities
x	possible value of a random variable $\mathbf{X}$
z	possible value for the entire collection of data $\mathbf{Z}$

Capital Letters

A_1	acceptance set for H_1
$\mathbf{A}$	action taken, $\mathbf{A} = s(\mathbf{Z})$
B	risk corresponding to the *a priori probability*; $B(\bar{w}, a)$
E	set or expectation (E_θ represents expectation when θ is the state of nature)
F	cumulative probability distribution function (*cdf*), $F(x)$
$\mathbf{F}$	sample cumulative frequency function, $\mathbf{F}(x)$
H	history or hypothesis
L	expected loss, $L(\theta, s)$
$\mathbf{N}$	random size of sample
P	probability or point or prospect
R	risk, $R(\theta, s)$
T	estimate
$\mathbf{T}$	value of estimator t
$\mathbf{X}$	random variable
$\mathbf{Z}$	collection of data

Capital Script Letters

$\mathscr{B}$	Bayes risk corresponding to the *best* action in a no-data problem, $\mathscr{B}(\bar{w})$
$\boldsymbol{\mathscr{B}}$	Bayes risk corresponding to the *a posteriori* probability; $\boldsymbol{\mathscr{B}} = \mathscr{B}(\bar{\mathbf{w}})$
$\mathscr{L}$	weighted average of expected losses, $\mathscr{L}(s)$
$\mathscr{N}$	set of states of nature
$\mathscr{R}$	weighted average of risks, $\mathscr{R}(s)$
$\mathscr{T}$	a set of tests
$\mathscr{X}$	set of possible outcomes of an experiment

Greek Letters

α (alpha)	action probability, $\alpha_s(\theta, a)$
γ (gamma)	confidence coefficient
ε (epsilon)	error probabilities, $\varepsilon(\theta, s)$
λ (lambda)	likelihood ratio, $\lambda(\mathbf{Z})$
θ (theta)	state of nature
$\boldsymbol{\theta}$	θ considered as random for computing *a posteriori* probabilities
$\hat{\boldsymbol{\theta}}$	maximum-likelihood estimate of θ
μ (mu)	population mean
σ (sigma)	population standard deviation
σ^2	population variance
$\hat{\boldsymbol{\sigma}}^2$	estimated value of variance
ν (nu)	median
Γ (gamma)	confidence set
Σ (sigma)	summation

SYMBOLS USED IN SET NOTATION

$\cup$	union (or)
$\cap$	intersection (and)
ϕ	null set
$\in$	"is an element of"
$\{x : x \text{ has a given property}\}$	the set of elements with a given property
(a, b)	the interval between a and b (not including the end points)

MISCELLANEOUS

$max\,(x, y, z, \cdots)$	represents the largest of the numbers $x, y, z, \cdots$
$min\,(x, y, z, \cdots)$	represents the smallest of the numbers $x, y, z, \cdots$

APPENDIX B_1

Short Table of Squares and Square Roots

n	n^2	$\sqrt{n}$	$\sqrt{10n}$	n	n^2	$\sqrt{n}$	$\sqrt{10n}$
1	1	1.000 000	3.16228	51	2601	7.141 428	22.58318
2	4	1.414 214	4.47214	52	2704	7.211 103	22.80351
3	9	1.732 051	5.47723	53	2809	7.280 110	23.02173
4	16	2.000 000	6.32456	54	2916	7.348 469	23.23790
5	25	2.236 068	7.07107	55	3025	7.416 198	23.45208
6	36	2.449 490	7.74597	56	3136	7.483 315	23.66432
7	49	2.645 751	8.36660	57	3249	7.549 834	23.87467
8	64	2.828 427	8.94427	58	3364	7.615 773	24.08319
9	81	3.000 000	9.48683	59	3481	7.681 146	24.28992
10	100	3.162 278	10.00000	60	3600	7.745 967	24.49490
11	121	3.316 625	10.48809	61	3721	7.810 250	24.69818
12	144	3.464 102	10.95445	62	3844	7.874 008	24.89980
13	169	3.605 551	11.40175	63	3969	7.937 254	25.09980
14	196	3.741 657	11.83216	64	4096	8.000 000	25.29822
15	225	3.872 983	12.24745	65	4225	8.062 258	25.49510
16	256	4.000 000	12.64911	66	4356	8.124 038	25.69047
17	289	4.123 106	13.03840	67	4489	8.185 353	25.88436
18	324	4.242 641	13.41641	68	4624	8.246 211	26.07681
19	361	4.358 899	13.78405	69	4761	8.306 624	26.26785
20	400	4.472 136	14.14214	70	4900	8.366 600	26.45751
21	441	4.582 576	14.49138	71	5041	8.426 150	26.64583
22	484	4.690 416	14.83240	72	5184	8.485 281	26.83282
23	529	4.795 832	15.16575	73	5329	8.544 004	27.01851
24	576	4.898 979	15.49193	74	5476	8.602 325	27.20294
25	625	5.000 000	15.81139	75	5625	8.660 254	27.38613
26	676	5.099 020	16.12452	76	5776	8.717 798	27.56810
27	729	5.196 152	16.43168	77	5929	8.774 964	27.74887
28	784	5.291 503	16.73320	78	6084	8.831 761	27.92848
29	841	5.385 165	17.02939	79	6241	8.888 194	28.10694
30	900	5.477 226	17.32051	80	6400	8.944 272	28.28427
31	961	5.567 764	17.60682	81	6561	9.000 000	28.46050
32	1024	5.656 854	17.88854	82	6724	9.055 385	28.63564
33	1089	5.744 563	18.16590	83	6889	9.110 434	28.80972
34	1156	5.830 952	18.43909	84	7056	9.165 151	28.98275
35	1225	5.916 080	18.70829	85	7225	9.219 544	29.15476
36	1296	6.000 000	18.97367	86	7396	9.273 618	29.32576
37	1369	6.082 763	19.23538	87	7569	9.327 379	29.49576
38	1444	6.164 414	19.49359	88	7744	9.380 832	29.66479
39	1521	6.244 998	19.74842	89	7921	9.433 981	29.83287
40	1600	6.324 555	20.00000	90	8100	9.486 833	30.00000
41	1681	6.403 124	20.24846	91	8281	9.539 392	30.16621
42	1764	6.480 741	20.49390	92	8464	9.591 663	30.33150
43	1849	6.557 439	20.73644	93	8649	9.643 651	30.49590
44	1936	6.633 250	20.97618	94	8836	9.695 360	30.65942
45	2025	6.708 204	21.21320	95	9025	9.746 794	30.82207
46	2116	6.782 330	21.44761	96	9216	9.797 959	30.98387
47	2209	6.855 655	21.67948	97	9409	9.848 858	31.14482
48	2304	6.928 203	21.90890	98	9604	9.899 495	31.30495
49	2401	7.000 000	22.13594	99	9801	9.949 874	31.46427
50	2500	7.071 068	22.36068	100	10000	10.000 000	31.62278

APPENDIX B_2

Logarithms

N	0	1	2	3	4	5	6	7	8	9
10	0000	0043	0086	0128	0170	0212	0253	0294	0334	0374
11	0414	0453	0492	0531	0569	0607	0645	0682	0719	0755
12	0792	0828	0864	0899	0934	0969	1004	1038	1072	1106
13	1139	1173	1206	1239	1271	1303	1335	1367	1399	1430
14	1461	1492	1523	1553	1584	1614	1644	1673	1703	1732
15	1761	1790	1818	1847	1875	1903	1931	1959	1987	2014
16	2041	2068	2095	2122	2148	2175	2201	2227	2253	2279
17	2304	2330	2355	2380	2405	2430	2455	2480	2504	2529
18	2553	2577	2601	2625	2648	2672	2695	2718	2742	2765
19	2788	2810	2833	2856	2878	2900	2923	2945	2967	2989
20	3010	3032	3054	3075	3096	3118	3139	3160	3181	3201
21	3222	3243	3263	3284	3304	3324	3345	3365	3385	3404
22	3424	3444	3464	3483	3502	3522	3541	3560	3579	3598
23	3617	3636	3655	3674	3692	3711	3729	3747	3766	3784
24	3802	3820	3838	3856	3874	3892	3909	3927	3945	3962
25	3979	3997	4014	4031	4048	4065	4082	4099	4116	4133
26	4150	4166	4183	4200	4216	4232	4249	4265	4281	4298
27	4314	4330	4346	4362	4378	4393	4409	4425	4440	4456
28	4472	4487	4502	4518	4533	4548	4564	4579	4594	4609
29	4624	4639	4654	4669	4683	4698	4713	4728	4742	4757
30	4771	4786	4800	4814	4829	4843	4857	4871	4886	4900
31	4914	4928	4942	4955	4969	4983	4997	5011	5024	5038
32	5051	5065	5079	5092	5105	5119	5132	5145	5159	5172
33	5185	5198	5211	5224	5237	5250	5263	5276	5289	5302
34	5315	5328	5340	5353	5366	5378	5391	5403	5416	5428
35	5441	5453	5465	5478	5490	5502	5514	5527	5539	5551
36	5563	5575	5587	5599	5611	5623	5635	5647	5658	5670
37	5682	5694	5705	5717	5729	5740	5752	5763	5775	5786
38	5798	5809	5821	5832	5843	5855	5866	5877	5888	5899
39	5911	5922	5933	5944	5955	5966	5977	5988	5999	6010
40	6021	6031	6042	6053	6064	6075	6085	6096	6107	6117
41	6128	6138	6149	6160	6170	6180	6191	6201	6212	6222
42	6232	6243	6253	6263	6274	6284	6294	6304	6314	6325
43	6335	6345	6355	6365	6375	6385	6395	6405	6415	6425
44	6435	6444	6454	6464	6474	6484	6493	6503	6513	6522
45	6532	6542	6551	6561	6571	6580	6590	6599	6609	6618
46	6628	6637	6646	6656	6665	6675	6684	6693	6702	6712
47	6721	6730	6739	6749	6758	6767	6776	6785	6794	6803
48	6812	6821	6830	6839	6848	6857	6866	6875	6884	6893
49	6902	6911	6920	6928	6937	6946	6955	6964	6972	6981
50	6990	6998	7007	7016	7024	7033	7042	7050	7059	7067
51	7076	7084	7093	7101	7110	7118	7126	7135	7143	7152
52	7160	7168	7177	7185	7193	7202	7210	7218	7226	7235
53	7243	7251	7259	7267	7275	7284	7292	7300	7308	7316
54	7324	7332	7340	7348	7356	7364	7372	7380	7388	7396

N	0	1	2	3	4	5	6	7	8	9
55	7404	7412	7419	7427	7435	7443	7451	7459	7466	7474
56	7482	7490	7497	7505	7513	7520	7528	7536	7543	7551
57	7559	7566	7574	7582	7589	7597	7604	7612	7619	7627
58	7634	7642	7649	7657	7664	7672	7679	7686	7694	7701
59	7709	7716	7723	7731	7738	7745	7752	7760	7767	7774
60	7782	7789	7796	7803	7810	7818	7825	7832	7839	7846
61	7853	7860	7868	7875	7882	7889	7896	7903	7910	7917
62	7924	7931	7938	7945	7952	7959	7966	7973	7980	7987
63	7993	8000	8007	8014	8021	8028	8035	8041	8048	8055
64	8062	8069	8075	8082	8089	8096	8102	8109	8116	8122
65	8129	8136	8142	8149	8156	8162	8169	8176	8182	8189
66	8195	8202	8209	8215	8222	8228	8235	8241	8248	8254
67	8261	8267	8274	8280	8287	8293	8299	8306	8312	8319
68	8325	8331	8338	8344	8351	8357	8363	8370	8376	8382
69	8388	8395	8401	8407	8414	8420	8426	8432	8439	8445
70	8451	8457	8463	8470	8476	8482	8488	8494	8500	8506
71	8513	8519	8525	8531	8537	8543	8549	8555	8561	8567
72	8573	8579	8585	8591	8597	8603	8609	8615	8621	8627
73	8633	8639	8645	8651	5657	8663	8669	8675	8681	8686
74	8692	8698	8704	8710	8716	8722	8727	8733	8739	8745
75	8751	8756	8762	8768	8774	8779	8785	8791	8797	8802
76	8808	8814	8820	8825	8831	8837	8842	8848	8854	8859
77	8865	8871	8876	8882	8887	8893	8899	8904	8910	8915
78	8921	8927	8932	8988	8943	8949	8954	8960	8965	8971
79	8976	8982	8987	8993	8998	9004	9009	9015	9020	9025
80	9031	9036	9042	9047	9053	9058	9063	9069	9074	9079
81	9085	9090	9096	9101	9106	9112	9117	9122	9128	9133
82	9138	9143	9149	9154	9159	9165	9170	9175	9180	9186
83	9191	9196	9201	9206	9212	9217	9222	9227	9232	9238
84	9243	9248	9253	9258	9263	9269	9274	9279	9284	9289
85	9294	9299	9304	9309	9315	9320	9325	9330	9335	9340
86	9345	9350	9355	9360	9365	9370	9375	9380	9385	9390
87	9395	9400	9405	9410	9415	9420	9425	9430	9435	9440
88	9445	9450	9455	9460	9465	9469	9474	9479	9484	9489
89	9494	9499	9504	9509	9513	9518	9523	9528	9533	9538
90	9542	9547	9552	9557	9562	9566	9571	9576	9581	9586
91	9590	9595	9600	9605	9609	9614	9619	9624	9628	9633
92	9638	9643	9647	9652	9657	9661	9666	9671	9675	9680
93	9685	9689	9694	9699	9703	9708	9713	9717	9722	9727
94	9731	9736	9741	9745	9750	9754	9759	9763	9768	9773
95	9777	9782	9786	9791	9795	9800	9805	9809	9814	9818
96	9823	9827	9832	9836	9841	9845	9850	9854	9859	9863
97	9868	9872	9877	9881	9886	9890	9894	9899	9903	9908
98	9912	9917	9921	9926	9930	9934	9939	9943	9948	9952
99	9956	9961	9965	9969	9974	9978	9983	9987	9991	9996

APPENDIX C_1

Table of Random Digits

03 47 43 73 86	36 96 47 36 61	46 98 63 71 62	33 26 16 80 45	60 11 14 10 95
97 74 24 67 62	42 81 14 57 20	42 53 32 37 32	27 07 36 07 51	24 51 79 89 73
16 76 62 27 66	56 50 26 71 07	32 90 79 78 53	13 55 38 58 59	88 97 54 14 10
12 56 85 99 26	96 96 68 27 31	05 03 72 93 15	57 12 10 14 21	88 26 49 81 76
55 59 56 35 64	38 54 82 46 22	31 62 43 09 90	06 18 44 32 53	23 83 01 30 30
16 22 77 94 39	49 54 43 54 82	17 37 93 23 78	87 35 20 96 43	84 26 34 91 64
84 42 17 53 31	57 24 55 06 88	77 04 74 47 67	21 76 33 50 25	83 92 12 06 76
63 01 63 78 59	16 95 55 67 19	98 10 50 71 75	12 86 73 58 07	44 39 52 38 79
33 21 12 34 29	78 64 56 07 82	52 42 07 44 38	15 51 00 13 42	99 66 02 79 54
57 60 86 32 44	09 47 27 96 54	49 17 46 09 62	90 52 84 77 27	08 02 73 43 28
18 18 07 92 46	44 17 16 58 09	79 83 86 19 62	06 76 50 03 10	55 23 64 05 05
26 62 38 97 75	84 16 07 44 99	83 11 46 32 24	20 14 85 88 45	10 93 72 88 71
23 42 40 64 74	82 97 77 77 81	07 45 32 14 08	32 98 94 07 72	93 85 79 10 75
52 36 28 19 95	50 92 26 11 97	00 56 76 31 38	80 22 02 53 53	86 60 42 04 53
37 85 94 35 12	83 39 50 08 30	42 34 07 96 88	54 42 06 87 98	35 85 29 48 39
70 29 17 12 13	40 33 20 38 26	13 89 51 03 74	17 76 37 13 04	07 74 21 19 30
56 62 18 37 35	96 83 50 87 75	97 12 25 93 47	70 33 24 03 54	97 77 46 44 80
99 49 57 22 77	88 42 95 45 72	16 64 36 16 00	04 43 18 66 79	94 77 24 21 90
16 08 15 04 72	33 27 14 34 09	45 59 34 68 49	12 72 07 34 45	99 27 72 95 14
31 16 93 32 43	50 27 89 87 19	20 15 37 00 49	52 85 66 60 44	38 68 88 11 80
68 34 30 13 70	55 74 30 77 40	44 22 78 84 26	04 33 46 09 52	68 07 97 06 57
74 57 25 65 76	59 29 97 68 60	71 91 38 67 54	13 58 18 24 76	15 54 55 95 52
27 42 37 86 53	48 55 90 65 72	96 57 69 36 10	96 46 92 42 45	97 60 49 04 91
00 39 68 29 61	66 37 32 20 30	77 84 57 03 29	10 45 65 04 26	11 04 96 67 24
29 94 98 94 24	68 49 69 10 82	53 75 91 93 30	34 25 20 57 27	40 48 73 51 92
16 90 82 66 59	83 62 64 11 12	67 19 00 71 74	60 47 21 29 68	02 02 37 03 31
11 27 94 75 06	06 09 19 74 66	02 94 37 34 02	76 70 90 30 86	38 45 94 30 38
35 24 10 16 20	33 32 51 26 38	79 78 45 04 91	16 92 53 56 16	02 75 50 95 98
38 23 16 86 38	42 38 97 01 50	87 75 66 81 41	40 01 74 91 62	48 51 84 08 32
31 96 25 91 47	96 44 33 49 13	34 86 82 53 91	00 52 43 48 85	27 55 26 89 62
66 67 40 67 14	64 05 71 95 86	11 05 65 09 68	76 83 20 37 90	57 16 00 11 66
14 90 84 45 11	75 73 88 05 90	52 27 41 14 86	22 98 12 22 08	07 52 74 95 80
68 05 51 18 00	33 96 02 75 19	07 60 62 93 55	59 33 82 43 90	49 37 38 44 59
20 46 78 73 90	97 51 40 14 02	04 02 33 31 08	39 54 16 49 36	47 95 93 13 30
64 19 58 97 79	15 06 15 93 20	01 90 10 75 06	40 78 78 89 62	02 67 74 17 33
05 26 93 70 60	22 35 85 15 13	92 03 51 59 77	59 56 78 06 83	52 91 05 70 74
07 97 10 88 23	09 98 42 99 64	61 71 62 99 15	06 51 29 16 93	58 05 77 09 51
68 71 86 85 85	54 87 66 47 54	73 32 08 11 12	44 95 92 63 16	29 56 24 29 48
26 99 61 65 53	58 37 78 80 70	42 10 50 67 42	32 17 55 85 74	94 44 67 16 94
14 65 52 68 75	87 59 36 22 41	26 78 63 06 55	13 08 27 01 50	15 29 39 39 43

Source: Abridged from Table XXXIII of Fisher and Yates, *Statistical Tables for Biological, Agricultural, and Medical Research*, published by Oliver and Boyd Ltd., Edinburgh, by permission of the authors and publishers.

APPENDIX C_2

Table of Random Normal Deviates

1	2	3	4	5	6	7	8
−0.538	+0.424	−0.527	+2.040	+0.835	+0.230	−0.476	+0.157
−0.211	+1.670	−1.018	−1.082	+0.584	+0.089	−1.126	−0.615
−0.604	+0.669	−0.410	−0.594	−0.366	−0.880	+0.499	+0.736
−1.590	+0.062	−1.220	+0.281	+0.255	+0.888	−0.194	+1.163
−0.597	+1.368	−1.078	+1.296	+0.525	+0.282	+1.183	+0.371
+0.220	−0.290	−0.003	−0.971	−0.547	+0.297	+0.261	−0.316
+1.449	−0.395	−0.413	+0.111	+1.145	+0.261	+0.012	−0.336
−0.841	+0.786	−0.954	+0.676	−1.726	+0.107	+1.155	+0.556
+1.714	−0.573	+1.578	−0.340	−0.645	+1.185	−0.858	+0.399
+0.932	−1.003	−0.867	+0.007	−0.406	+0.550	−0.256	+0.568
+0.168	−0.382	+1.454	+0.331	+0.357	+0.615	+0.806	+0.787
−1.271	+0.493	−1.169	+0.402	−0.762	+0.003	−1.454	−0.705
+0.479	+0.028	+0.173	−0.312	+0.629	−0.403	+0.964	+0.367
−0.827	−0.770	−0.030	+0.627	−0.288	−1.015	+0.243	+0.120
+0.142	−0.059	−0.639	+0.071	−0.888	+0.385	+0.188	+1.723
−0.124	−0.912	−0.443	−0.255	+1.631	−0.192	−0.573	+2.616
−0.658	+0.378	+0.174	+1.480	+0.726	−0.967	+0.108	+0.725
+0.501	−0.224	+0.625	+0.483	−1.189	+0.592	−1.137	+0.021
+0.931	−0.221	−0.219	+1.645	+0.168	−0.271	+0.238	−0.435
+0.811	−0.329	+0.239	+1.983	−0.317	−0.199	−0.001	+1.105
+1.543	+0.561	−0.461	−1.449	−0.537	−1.274	+0.350	−0.127
−0.403	+0.330	+0.260	+1.542	−0.428	−1.242	−1.050	−0.050
+1.238	−0.981	+0.018	−1.504	+0.388	−1.330	−0.100	+0.178
+1.423	+1.473	−0.584	+0.553	−0.239	−0.816	+0.331	−0.648
+0.766	−0.316	−0.555	+0.724	−2.360	+0.528	−1.123	−0.861
+1.947	+1.873	+0.625	−2.930	+1.720	−0.897	−2.270	−0.879
+1.573	+1.412	+1.169	+1.535	−0.085	−1.756	+0.445	+0.142
−1.186	−0.366	+0.251	−0.508	+1.290	+0.153	−0.723	+0.894
−1.470	−0.251	−0.239	−1.015	−0.965	−1.091	+0.061	−0.144
+0.345	−0.254	−0.307	−0.780	+0.909	−0.122	+0.345	−0.390
−1.074	+0.569	−0.343	−0.980	−1.254	−0.401	−0.141	−0.500
+0.537	+1.273	+0.528	+0.170	+0.697	+0.436	−0.925	−0.481
−0.456	−0.310	−1.379	+1.312	+1.207	+0.043	+1.008	+1.351
+0.814	−0.017	+0.190	+0.295	+0.403	+1.081	−0.406	+1.325
+1.160	−0.382	+1.268	−1.419	+0.354	+2.760	−0.638	+0.249
−0.383	+0.605	+1.147	−0.390	−0.897	−0.704	−0.358	+0.045
−2.007	−0.411	+2.080	+0.423	+1.930	−0.969	+1.377	−1.079
−0.522	+1.043	+0.596	+1.563	−0.294	−1.463	−2.448	−0.485
+0.671	−2.022	+0.814	−0.722	−0.333	−0.024	−0.680	−0.288
+0.047	−0.906	−0.766	+1.540	−1.615	−0.873	−0.919	−0.813

Source: Abridged from P. C. Mahalanobis, "Tables of Random Samples From a Normal Population," *Sankhyā*, Vol. 1, pp. 289-328, by permission of the author and publishers.

The numbers in this table are independent observations $\mathbf{X}_1, \mathbf{X}_2, \cdots$ from a normal population with mean $\mu = 0$ and standard deviation $\sigma = 1$. To obtain random variables $\mathbf{Y}_1, \mathbf{Y}_2, \cdots$ from a normal population with arbitrary mean μ and standard deviation σ let $\mathbf{Y}_i = \mu + \sigma \mathbf{X}_i$.

APPENDIX D_1

Areas Under the Normal Curve From $\mu + a\sigma$ to ∞

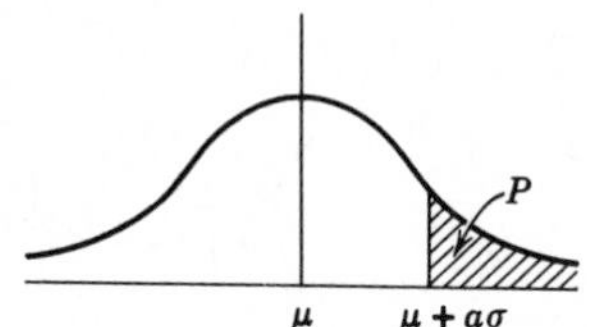

a	0.00	0.01	0.02	0.03	0.04	0.05	0.06	0.07	0.08	0.09
0.0	0.5000	0.4960	0.4920	0.4880	0.4840	0.4801	0.4761	0.4721	0.4681	0.4641
0.1	0.4602	0.4562	0.4522	0.4483	0.4443	0.4404	0.4364	0.4325	0.4286	0.4247
0.2	0.4207	0.4168	0.4129	0.4090	0.4052	0.4013	0.3974	0.3936	0.3897	0.3859
0.3	0.3821	0.3783	0.3745	0.3707	0.3669	0.3632	0.3594	0.3557	0.3520	0.3483
0.4	0.3446	0.3409	0.3372	0.3336	0.3300	0.3264	0.3228	0.3192	0.3156	0.3121
0.5	0.3085	0.3050	0.3015	0.2981	0.2946	0.2912	0.2877	0.2843	0.2810	0.2776
0.6	0.2743	0.2709	0.2676	0.2643	0.2611	0.2578	0.2546	0.2514	0.2483	0.2451
0.7	0.2420	0.2389	0.2358	0.2327	0.2296	0.2266	0.2236	0.2206	0.2177	0.2148
0.8	0.2119	0.2090	0.2061	0.2033	0.2005	0.1977	0.1949	0.1922	0.1894	0.1867
0.9	0.1841	0.1814	0.1788	0.1762	0.1736	0.1711	0.1685	0.1660	0.1635	0.1611
1.0	0.1587	0.1562	0.1539	0.1515	0.1492	0.1469	0.1446	0.1423	0.1401	0.1379
1.1	0.1357	0.1335	0.1314	0.1292	0.1271	0.1251	0.1230	0.1210	0.1190	0.1170
1.2	0.1151	0.1131	0.1112	0.1093	0.1075	0.1056	0.1038	0.1020	0.1003	0.0985
1.3	0.0968	0.0951	0.0934	0.0918	0.0901	0.0885	0.0869	0.0853	0.0838	0.0823
1.4	0.0808	0.0793	0.0778	0.0764	0.0749	0.0735	0.0721	0.0708	0.0694	0.0681
1.5	0.0668	0.0655	0.0643	0.0630	0.0618	0.0606	0.0594	0.0582	0.0571	0.0559
1.6	0.0548	0.0537	0.0526	0.0516	0.0505	0.0495	0.0485	0.0475	0.0465	0.0455
1.7	0.0446	0.0436	0.0427	0.0418	0.0409	0.0401	0.0392	0.0384	0.0375	0.0367
1.8	0.0359	0.0351	0.0344	0.0336	0.0329	0.0322	0.0314	0.0307	0.0301	0.0294
1.9	0.0287	0.0281	0.0274	0.0268	0.0262	0.0256	0.0250	0.0244	0.0239	0.0233
2.0	0.0228	0.0222	0.0217	0.0212	0.0207	0.0202	0.0197	0.0192	0.0188	0.0183
2.1	0.0179	0.0174	0.0170	0.0166	0.0162	0.0158	0.0154	0.0150	0.0146	0.0143
2.2	0.0139	0.0136	0.0132	0.0129	0.0125	0.0122	0.0119	0.0116	0.0113	0.0110
2.3	0.0107	0.0104	0.0102	0.00990	0.00964	0.00939	0.00914	0.00889	0.00866	0.00842
2.4	0.00820	0.00798	0.00776	0.00755	0.00734	0.00714	0.00695	0.00676	0.00657	0.00639
2.5	0.00621	0.00604	0.00587	0.00570	0.00554	0.00539	0.00523	0.00508	0.00494	0.00480
2.6	0.00466	0.00453	0.00440	0.00427	0.00415	0.00402	0.00391	0.00379	0.00368	0.00357
2.7	0.00347	0.00336	0.00326	0.00317	0.00307	0.00298	0.00289	0.00280	0.00272	0.00264
2.8	0.00256	0.00248	0.00240	0.00233	0.00226	0.00219	0.00212	0.00205	0.00199	0.00193
2.9	0.00187	0.00181	0.00175	0.00169	0.00164	0.00159	0.00154	0.00149	0.00144	0.00139

a	0.0	0.1	0.2	0.3	0.4	0.5	0.6	0.7	0.8	0.9
3	0.00135	$0.0^{3}968$	$0.0^{3}687$	$0.0^{3}483$	$0.0^{3}337$	$0.0^{3}233$	$0.0^{3}159$	$0.0^{3}108$	$0.0^{4}723$	$0.0^{4}481$
4	$0.0^{4}317$	$0.0^{4}207$	$0.0^{4}133$	$0.0^{5}854$	$0.0^{5}541$	$0.0^{5}340$	$0.0^{5}211$	$0.0^{5}130$	$0.0^{6}793$	$0.0^{6}479$
5	$0.0^{6}287$	$0.0^{6}170$	$0.0^{7}996$	$0.0^{7}579$	$0.0^{7}333$	$0.0^{7}190$	$0.0^{7}107$	$0.0^{8}599$	$0.0^{8}332$	$0.0^{3}182$
6	$0.0^{9}987$	$0.0^{9}530$	$0.0^{9}282$	$0.0^{9}149$	$0.0^{10}777$	$0.0^{10}402$	$0.0^{10}206$	$0.0^{10}104$	$0.0^{11}523$	$0.0^{11}260$

Source: Reproduced by permission from *Tables of Areas in Two Tails and in One Tail of the Normal Curve*, by Frederick E. Croxton. Copyright, 1949, by Prentice-Hall, Inc., Englewood Cliffs, N. J.

APPENDIX D_2

Chi Square Distribution

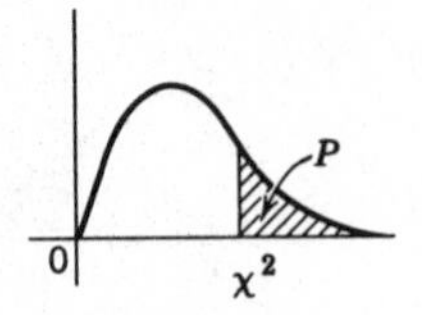

VALUES OF χ^2 FOR VARIOUS VALUES OF P AND DEGREES OF FREEDOM n

P

Degrees of Freedom n	0.99	0.95	0.50	0.20	0.10	0.05	0.02	0.01
1	0.000157	0.00393	0.455	1.642	2.706	3.841	5.412	6.635
2	0.0201	0.103	1.386	3.219	4.605	5.991	7.824	9.210
3	0.115	0.352	2.366	4.642	6.251	7.815	9.837	11.341
4	0.297	0.711	3.357	5.989	7.779	9.488	11.668	13.277
5	0.554	1.145	4.351	7.289	9.236	11.070	13.388	15.086
6	0.872	1.635	5.348	8.558	10.645	12.592	15.033	16.812
7	1.239	2.167	6.346	9.803	12.017	14.067	16.622	18.475
8	1.646	2.733	7.344	11.030	13.362	15.507	18.168	20.090
9	2.088	3.325	8.343	12.242	14.684	16.919	19.679	21.666
10	2.558	3.940	9.343	13.442	15.987	18.307	21.161	23.209
11	3.053	4.575	10.341	14.631	17.275	19.675	22.618	24.725
12	3.571	5.226	11.340	15.812	18.549	21.026	24.054	26.217
13	4.107	5.892	12.340	16.985	19.812	22.362	25.472	27.688
14	4.660	6.571	13.339	18.151	21.064	23.685	26.873	29.141
15	5.229	7.261	14.339	19.311	22.307	24.996	28.259	30.578
16	5.812	7.962	15.338	20.465	23.542	26.296	29.633	32.000
17	6.408	8.672	16.338	21.615	24.769	27.587	30.995	33.409
18	7.015	9.390	17.338	22.760	25.989	28.869	32.346	34.805
19	7.633	10.117	18.338	23.900	27.204	30.144	33.687	36.191
20	8.260	10.851	19.337	25.038	28.412	31.410	35.020	37.566
21	8.897	11.591	20.337	26.171	29.615	32.671	36.343	38.932
22	9.542	12.338	21.337	27.301	30.813	33.924	37.659	40.289
23	10.196	13.091	22.337	28.429	32.007	35.172	38.968	41.638
24	10.856	13.848	23.337	29.553	33.196	36.415	40.270	42.980
25	11.524	14.611	24.337	30.675	34.382	37.652	41.566	44.314
26	12.198	15.379	25.336	31.795	35.563	38.885	42.856	45.642
27	12.879	16.151	26.336	32.912	36.741	40.113	44.140	46.963
28	13.565	16.928	27.336	34.027	37.916	41.337	45.419	48.278
29	14.256	17.708	28.336	35.139	39.087	42.557	46.693	49.588
30	14.953	18.493	29.336	36.250	40.256	43.773	47.962	50.892

Source: Abridged from Table IV of Fisher and Yates, *Statistical Tables for Biological, Agricultural, and Medical Research,* published by Oliver and Boyd, Ltd., Edinburgh, and from Catherine M. Thompson, *Biometrika*, Vol. XXXII, Part II, October 1941, pp. 187-191, "Tables of Percentage Points of the χ^2 Distribution," by permission of the authors and publishers.

APPENDIX D_3

Exponential Distribution

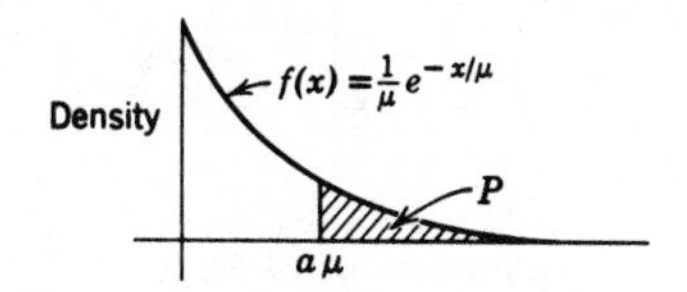

a	$P\{X > a\mu\}$	a	$P\{X > a\mu\}$	a	$P\{X > a\mu\}$
0.00	1.000000	1.60	0.201897	3.40	0.033373
0.01	0.990050	1.65	0.192050	3.45	0.031746
0.02	0.980199	1.70	0.182684	3.50	0.030197
0.03	0.970446	1.75	0.173774	3.55	0.028725
0.04	0.960789	1.80	0.165299	3.60	0.027324
0.05	0.951229	1.85	0.157237	3.65	0.025991
0.10	0.904837	1.90	0.149569	3.70	0.024724
0.15	0.860708	1.95	0.142274	3.75	0.023518
0.20	0.818731	2.00	0.135335	3.80	0.022371
0.25	0.778801	2.05	0.128735	3.85	0.021280
0.30	0.740818	2.10	0.122456	3.90	0.020242
0.35	0.704688	2.15	0.116484	3.95	0.019255
0.40	0.670320	2.20	0.110803	4.00	0.018316
0.45	0.637628	2.25	0.105399	4.05	0.017422
0.50	0.606531	2.30	0.100259	4.10	0.016573
0.55	0.576950	2.35	0.095369	4.15	0.015764
0.60	0.548812	2.40	0.090718	4.20	0.014996
0.65	0.522046	2.45	0.086294	4.25	0.014264
0.70	0.496585	2.50	0.082085	4.30	0.013569
0.75	0.472367	2.55	0.078085	4.35	0.012907
0.80	0.449329	2.60	0.074274	4.40	0.012277
0.85	0.427415	2.65	0.070651	4.45	0.011679
0.90	0.406570	2.70	0.067206	4.50	0.011109
0.95	0.386741	2.75	0.063928	4.55	0.010567
1.00	0.367879	2.80	0.060810	4.60	0.010052
1.05	0.349938	2.85	0.057844	4.65	0.009562
1.10	0.332871	2.90	0.055023	4.70	0.009095
1.15	0.316637	2.95	0.052340	4.75	0.008652
1.20	0.301194	3.00	0.049787	4.80	0.008230
1.25	0.286505	3.05	0.047359	4.85	0.007828
1.30	0.272532	3.10	0.045049	4.90	0.007447
1.35	0.259240	3.15	0.042852	4.95	0.007083
1.40	0.246597	3.20	0.040762	5.00	0.006738
1.45	0.234570	3.25	0.038774	5.50	0.004087
1.50	0.223130	3.30	0.036883	6.00	0.002479
1.55	0.212248	3.35	0.035084		

APPENDIX D_4

Student's *t* Distribution

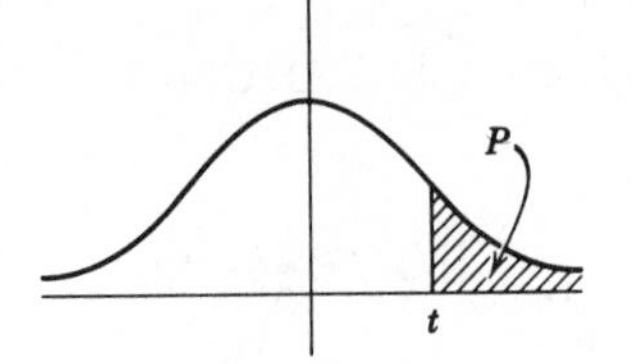

VALUES OF t CORRESPONDING TO GIVEN VALUES OF P AND n DEGREES OF FREEDOM

Degrees of Freedom n	P 0.005	0.01	0.025	0.05	0.10	0.25
1	63.657	31.821	12.706	6.314	3.078	1.000
2	9.925	6.965	4.303	2.920	1.886	0.816
3	5.841	4.541	3.182	2.353	1.638	0.765
4	4.604	3.747	2.776	2.132	1.533	0.741
5	4.032	3.365	2.571	2.015	1.476	0.727
6	3.707	3.143	2.447	1.943	1.440	0.718
7	3.499	2.998	2.365	1.895	1.415	0.711
8	3.355	2.896	2.306	1.860	1.397	0.706
9	3.250	2.821	2.262	1.833	1.383	0.703
10	3.169	2.764	2.228	1 812	1.372	0.700
11	3.106	2.718	2.201	1.796	1.363	0.697
12	3.055	2.681	2.179	1.782	1.356	0.695
13	3.012	2.650	2.160	1.771	1.350	0.694
14	2.977	2.624	2.145	1.761	1.345	0.692
15	2.947	2.602	2.131	1.753	1.341	0.691
16	2.921	2.583	2.120	1.746	1.337	0.690
17	2.898	2.567	2.110	1.740	1.333	0.689
18	2.878	2.552	2.101	1.734	1.330	0.688
19	2.861	2.539	2.093	1.729	1.328	0.688
20	2.845	2.528	2.086	1.725	1.325	0.687
21	2.831	2.518	2.080	1.721	1.323	0.686
22	2.819	2.508	2.074	1.717	1.321	0.686
23	2.807	2.500	2.069	1.714	1.319	0.685
24	2.797	2.492	2.064	1.711	1.318	0.685
25	2.787	2.485	2.060	1.708	1.316	0.684
26	2.779	2.479	2.056	1.706	1.315	0.684
27	2.771	2.473	2.052	1.703	1.314	0.684
28	2.763	2.467	2.048	1.701	1.313	0.683
29	2.756	2.462	2.045	1.699	1.311	0.683
30	2.750	2.457	2.042	1.697	1.310	0.683
∞	2.576	2.326	1.960	1.645	1.282	0.674

Source: Adapted from Table III of Fisher and Yates, *Statistical Tables for Biological, Agricultural, and Medical Research*, published by Oliver and Boyd, Ltd., Edinburgh, by permission of the authors and publishers.

APPENDIX E_1

Derivation of an Equation For the Sample Variance

$$\mathbf{d}_{\mathbf{X}}^2 = \frac{1}{n} \sum_{i=1}^{n} (\mathbf{X}_i - \overline{\mathbf{X}})^2$$

$$\begin{aligned}
\sum_{i=1}^{n} (\mathbf{X}_i - \overline{\mathbf{X}})^2 &= (\mathbf{X}_1 - \overline{\mathbf{X}})^2 + (\mathbf{X}_2 - \overline{\mathbf{X}})^2 + \cdots + (\mathbf{X}_n - \overline{\mathbf{X}})^2 \\
&= \mathbf{X}_1^2 - 2\mathbf{X}_1\overline{\mathbf{X}} + \overline{\mathbf{X}}^2 + \mathbf{X}_2^2 - 2\mathbf{X}_2\overline{\mathbf{X}} + \overline{\mathbf{X}}^2 + \cdots \\
&\qquad + \mathbf{X}_n^2 - 2\mathbf{X}_n\overline{\mathbf{X}} + \overline{\mathbf{X}}^2 \\
&= (\mathbf{X}_1^2 + \mathbf{X}_2^2 + \cdots + \mathbf{X}_n^2) - 2\overline{\mathbf{X}}(\mathbf{X}_1 + \mathbf{X}_2 + \cdots + \mathbf{X}_n) \\
&\qquad + \overline{\mathbf{X}}^2 + \overline{\mathbf{X}}^2 + \cdots + \overline{\mathbf{X}}^2 \\
&= \sum_{i=1}^{n} \mathbf{X}_i^2 - 2\overline{\mathbf{X}}\left(\sum_{i=1}^{n} \mathbf{X}_i\right) + n\overline{\mathbf{X}}^2.
\end{aligned}$$

But $\sum_{i=1}^{n} \mathbf{X}_i = n\overline{\mathbf{X}}$, and so

$$\sum_{i=1}^{n} (\mathbf{X}_i - \overline{\mathbf{X}})^2 = \sum_{i=1}^{n} \mathbf{X}_i^2 - 2n\overline{\mathbf{X}}^2 + n\overline{\mathbf{X}}^2 = \sum_{i=1}^{n} \mathbf{X}_i^2 - n\overline{\mathbf{X}}^2$$

and

$$\mathbf{d}_{\mathbf{X}}^2 = \frac{1}{n} \sum_{i=1}^{n} (\mathbf{X}_i - \overline{\mathbf{X}})^2 = \left(\frac{1}{n} \sum_{i=1}^{n} \mathbf{X}_i^2\right) - \overline{\mathbf{X}}^2.$$

APPENDIX E_2

Simplified Computation Scheme for the Sample Mean and Variance Using Grouped Data

We indicate here why the simplified scheme of Section 7, Chapter 2 works. Let **W** be defined by $\mathbf{X} = a + b\mathbf{W}$. Then when $\mathbf{W} = 0$, $\mathbf{X} = a$, and when **W** changes by one, **X** changes by b. Since a is a cell mid-point and b is the cell length for the **X**, the mid-point values w_i of **W** are $\cdots -2, -1, 0, 1, 2, \cdots$ and correspond to the x_i. First we compute $\overline{\mathbf{W}}$ and $\mathbf{d}_{\mathbf{W}}^2$. Assuming for simplicity that all $\mathbf{f}_i$ observations in the ith cell are at the mid-point (i.e., $\mathbf{W} = w_i$), there are $\mathbf{f}_1$ observations w_1, $\mathbf{f}_2$ observations w_2, etc. The sum of these observations

$$\sum_{i=1}^{n} \mathbf{W}_i = \underbrace{w_1 + w_1 + \cdots + w_1}_{\mathbf{f}_1 \text{ times}} + \underbrace{w_2 + \cdots + w_2}_{\mathbf{f}_2 \text{ times}} + \cdots + \underbrace{w_k + \cdots + w_k}_{\mathbf{f}_k \text{ times}}$$

$$\sum_{i=1}^{n} \mathbf{W}_i = \mathbf{f}_1 w_1 + \mathbf{f}_2 w_2 + \cdots + \mathbf{f}_k w_k = \sum_{i=1}^{k} \mathbf{f}_i w_i$$

and

$$\overline{\mathbf{W}} = \frac{1}{n} \sum_{i=1}^{k} \mathbf{f}_i w_i.$$

Similarly

$$\sum_{i=1}^{n} \mathbf{W}_i^2 = \mathbf{f}_1 w_1^2 + \mathbf{f}_2 w_2^2 + \cdots + \mathbf{f}_k w_k^2 = \sum_{i=1}^{k} \mathbf{f}_i w_i^2$$

and applying Equation (2.12),

$$\mathbf{d}_{\mathbf{W}}^2 = \frac{1}{n} \sum_{i=1}^{k} \mathbf{f}_i w_i^2 - \overline{\mathbf{W}}^2.$$

Since $\mathbf{X} = a + b\mathbf{W}$, Equation (2.15) yields $\overline{\mathbf{X}} = a + b\overline{\mathbf{W}}$ and $\mathbf{d}_{\mathbf{X}} = b\mathbf{d}_{\mathbf{W}}$.

APPENDIX E_3

Axioms of Probability

Let $\mathscr{X}$ be the set of possible outcomes for an experiment. The probability distribution for the experiment is a function P on the subsets of $\mathscr{X}$ to the numbers between 0 and 1, which satisfies the following axioms. These axioms seem relevant for our long-run frequency interpretation of probability.

Axiom 1. $0 \leq P\{A\}$

Axiom 2. $P\{\mathscr{X}\} = 1$

Axiom 3. $P\{A_1 \text{ or } A_2 \text{ or } \cdots\} = P\{A_1\} + P\{A_2\} + \cdots$ if $A_1, A_2, \cdots$ are nonoverlapping.

From these axioms the following consequences are derived:

CONSEQUENCE 1. $P\{A\} + P\{\tilde{A}\} = 1$.

Proof: A and $\tilde{A}$ are nonoverlapping sets.

$$\{A \quad \text{or} \quad \tilde{A}\} = \mathscr{X}$$

(i.e., all possible outcomes are in A or in $\tilde{A}$)

$$P\{A\} + P\{\tilde{A}\} = P\{\mathscr{X}\} \qquad \text{by Axiom 3.}$$

$$P\{A\} + P\{\tilde{A}\} = 1 \qquad \text{by Axiom 2.}$$

CONSEQUENCE 2. $P\{A\} \leq 1$.

Proof: By Axiom 1, $P\{\tilde{A}\} \geq 0$ and hence by Consequence 1, $P\{A\} \leq 1$.

CONSEQUENCE 3. $P\{\phi\} = 0$.

Proof: Since $\phi = \tilde{\mathscr{X}}$, $P\{\phi\} = 1 - P\{\mathscr{X}\} = 1 - 1 = 0$.

CONSEQUENCE 4. *If B is a subset of A*, $P\{B\} \leq P\{A\}$.

Proof: Suppose B is a subset of A. Then let C be the set of points in A which are not in B. Then $A = \{B \text{ or } C\}$ where B and C are nonoverlapping sets. Hence, by Axiom 3,

$$P\{B\} + P\{C\} = P\{A\}.$$

But by Axiom 1, $P\{C\} \geq 0$. Hence $P\{B\} \leq P\{A\}$.

CONSEQUENCE 5. $P\{A_1 \text{ or } A_2 \text{ or } \cdots\} \leq P\{A_1\} + P\{A_2\} + \cdots$.

Proof: Let $B_1 = A_1$, B_2 equal the set of possible outcomes in A_2 which are not in A_1, B_3 equal the set of possible outcomes in A_3 which are in neither A_1 nor A_2, etc. The set B_1 is a subset of A_1 (in a rather trivial way since $B_1 = A_1$), B_2 is a subset of A_2, B_3 is a subset of A_3, etc. The sets $B_1, B_2, B_3 \cdots$ are nonoverlapping. Any outcome which is in at least one of the A's is in one of the B's and vice versa. Thus,

$$\{A_1 \text{ or } A_2 \text{ or } \cdots\} = \{B_1 \text{ or } B_2 \text{ or } \cdots\}$$

$$P\{A_1 \text{ or } A_2 \text{ or } \cdots\} = P\{B_1 \text{ or } B_2 \text{ or } \cdots\}$$

$$= P\{B_1\} + P\{B_2\} + \cdots$$

by Axiom 3.

$$P\{A_1 \text{ or } A_2 \text{ or } \cdots\} \leq P\{A_1\} + P\{A_2\} + \cdots$$

by Consequence 4.

APPENDIX E_4

Properties of Expectation

Ordinarily, the expectation of a random variable $\mathbf{X}$ is defined in terms of the probability distribution for the experiment. For simple cases where the experiment has only a finite number of possible outcomes, this definition is quite simple and essentially the one implied in the discussion of Example 4.2. For more complicated examples, this definition has a relatively sophisticated form from a mathematical point of view.

Instead of presenting the definition of expectation, we shall list its basic properties as axioms and derive consequences of these.

Axiom 1. $E(\mathbf{X} + \mathbf{Y}) = E(\mathbf{X}) + E(\mathbf{Y})$.

Axiom 2. $E(c\mathbf{X}) = cE(\mathbf{X})$.

Axiom 3. $E(\mathbf{X}) > E(\mathbf{Y})$ if $\mathbf{X} > \mathbf{Y}$.

Axiom 4. $E(\mathbf{X}) = p$ *if* $\mathbf{X}$ *takes the values* 1 *and* 0 *with probabilities* p *and* $1 - p$ *respectively.*

CONSEQUENCE 1. $E(c) = c$.

Proof: Let $\mathbf{X} = 1$ for all possible outcomes. By Axiom 4, $E(\mathbf{X}) = 1$ and

$$E(c\mathbf{X}) = E(c) = c \qquad \text{by Axiom 2.}$$

CONSEQUENCE 2. $E(a\mathbf{X} + b\mathbf{Y}) = a\,E(\mathbf{X}) + b\,E(\mathbf{Y})$.

Proof:

$$\begin{aligned} E(a\mathbf{X} + b\mathbf{Y}) &= E(a\mathbf{X}) + E(b\mathbf{Y}) && \text{by Axiom 1.} \\ &= a\,E(\mathbf{X}) + b\,E(\mathbf{Y}) && \text{by Axiom 2.} \end{aligned}$$

CONSEQUENCE 3. $E(a_1\mathbf{X}_1 + a_2\mathbf{X}_2 + \cdots + a_n\mathbf{X}_n) = a_1\,E(\mathbf{X}_1) + a_2\,E(\mathbf{X}_2) + \cdots + a_n\,E(\mathbf{X}_n)$.

Proof: For $n = 3$, we have

$$\begin{aligned} E(a_1\mathbf{X}_1 + a_2\mathbf{X}_2 + a_3\mathbf{X}_3) &= E(a_1\mathbf{X}_1 + a_2\mathbf{X}_2) + E(a_3\mathbf{X}_3) && \text{by Axiom 1.} \\ &= a_1\,E(\mathbf{X}_1) + a_2\,E(\mathbf{X}_2) + a_3\,E(\mathbf{X}_3) && \text{by Axiom 2 and Consequence 2.} \end{aligned}$$

By a similar argument we can extend this result for arbitrary n.

In fact this result can, under mild conditions, be extended to the case where there are infinitely many $\mathbf{X}_i$. However, technical difficulties sometimes arise. It pays to point out now that $E(\mathbf{X})$ is not always defined. For example, it could be that $\mathbf{X} = \mathbf{X}_1 + \mathbf{X}_2 + \mathbf{X}_3 + \cdots$ where $E(\mathbf{X}_1) = E(\mathbf{X}_3) = E(\mathbf{X}_5) = \cdots = 1$ and $E(\mathbf{X}_2) = E(\mathbf{X}_4) = E(\mathbf{X}_6) = \cdots = -1$. In this case,

$$E(\mathbf{X}_1) + E(\mathbf{X}_2) + E(\mathbf{X}_3) + \cdots = 1 - 1 + 1 - 1 + 1 - 1 + \cdots$$

has no special meaning since the partial sums $1, 1 - 1, 1 - 1 + 1, 1 - 1 + 1 - 1, 1 - 1 + 1 - 1 + 1, \cdots$ fluctuate between 1 and 0, and do not approach some single number. The standard mathematical definition for expectation does not apply to this random variable $\mathbf{X}$. Consequently, in stating Axioms 1, 2, and 3, one ought to add the condition that $E(\mathbf{X})$ and $E(\mathbf{Y})$ have meaning.

CONSEQUENCE 4. *If* $\mathbf{X}$ *is the discrete ramdom variable which assumes the values* $x_1, x_2 \cdots$ *with probabilities* $p_1, p_2, \cdots$, *then*

$$E(\mathbf{X}) = p_1x_1 + p_2x_2 + p_3x_3 + \cdots$$

or

$$E(\mathbf{X}) = \sum x_i P\{\mathbf{X} = x_i\}.$$

Proof: Let $\mathbf{Y}_1 = 1$ if $\mathbf{X} = x_1$ and 0 otherwise;
$\mathbf{Y}_2 = 1$ if $\mathbf{X} = x_2$ and 0 otherwise;
$\mathbf{Y}_3 = 1$ if $\mathbf{X} = x_3$ and 0 otherwise, etc.

Then

$$\mathbf{X} = x_1\mathbf{Y}_1 + x_2\mathbf{Y}_2 + x_3\mathbf{Y}_3 + \cdots$$

and

$$E(\mathbf{X}) = x_1 E(\mathbf{Y}_1) + x_2 E(\mathbf{Y}_2) + x_3 (\mathbf{Y}_3) + \cdots.$$

But by Axiom 4, $E(\mathbf{Y}_1) = p_1$, $E(\mathbf{Y}_2) = p_2, \cdots$, and thus:

$$E(\mathbf{X}) = p_1x_1 + p_2x_2 + \cdots.$$

CONSEQUENCE 5. *Suppose* $\mathbf{Y} = h(\mathbf{X})$, *where* $\mathbf{X}$ *is a discrete random variable which assumes the values* $x_1, x_2, \cdots$ *with probabilities* $p_1, p_2, \cdots$. *Then*

$$E(\mathbf{Y}) = E[h(\mathbf{X})] = \sum_i p_i\, h(x_i).$$

Proof: $$\mathbf{Y} = \sum h(x_i)\mathbf{Y}_i$$

where $\mathbf{Y}_i$ is defined as in the proof of Consequence 4. Then by Consequence 3,

$$E(\mathbf{Y}) = \sum h(x_1)\, E(\mathbf{Y}_i) = \sum p_i\, h(x_i).$$

As an example, we have $E(\mathbf{X}^2) = p_1x_1^2 + p_2x_2^2 + \cdots$.

CONSEQUENCE 6. *If* $\mathbf{X} \geq \mathbf{Y}$, $E(\mathbf{X}) \geq E(\mathbf{Y})$.

Proof: If $\mathbf{X} \geq \mathbf{Y}$ then $\mathbf{X} > \mathbf{Y} - 0.001$, and

$$E(\mathbf{X}) > E(\mathbf{Y} - 0.001) = E(\mathbf{Y}) - 0.001$$

by Axiom 3 and Consequence 2.

Similarly, $E(\mathbf{X})$ exceeds each number smaller than $E(\mathbf{Y})$. But this is possible only if $E(\mathbf{X}) > E(\mathbf{Y})$ or $E(\mathbf{X}) = E(\mathbf{Y})$. In other words, $E(\mathbf{X}) \geq E(\mathbf{Y})$.

Although Consequence 4 gives us the rule for computing the expectation of a discrete random variable, it does not apply for a continuous one. We present here a brief outline of the principle applied to general random variables. As the reader may recall, in Chapter 3, we briefly indicated how to approximate a continuous random variable by a discrete one by using a rounding-off process. Here we shall let $\mathbf{X}^*$ be $\mathbf{X}$ rounded off to the second position after the decimal point, compute $E(\mathbf{X}^*)$, and show that it is close to $E(\mathbf{X})$. The rounded-off value of $\mathbf{X}$ is given by

$$\mathbf{X}^* = \frac{i}{10^2} \qquad \text{if } \frac{i - 1/2}{10^2} \leq \mathbf{X} < \frac{i + 1/2}{10^2}$$

$$i = \cdots -2, -1, 0, 1, 2, \cdots.$$

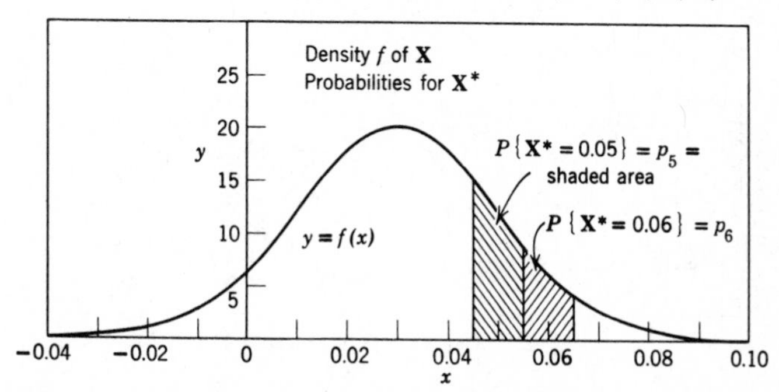

Figure E_4-1. Construction of an approximating discrete distribution for a continuous distribution with density f.

(Thus if $\mathbf{X} = 0.02435916\cdots$, $\mathbf{X}^* = 0.02$.) Then (see Figure E_4-1)

$$P\left\{\mathbf{X}^* = \frac{i}{10^2}\right\} = P\left\{\frac{i - 1/2}{10^2} \leq \mathbf{X} < \frac{i + 1/2}{10^2}\right\} = p_i$$

and

$$\mathbf{X}^* - 1/200 \le \mathbf{X} \le \mathbf{X}^* + 1/200.$$

By Consequence 4,

$$E(\mathbf{X}^*) = \sum_i \frac{i}{10^2} P\left\{\mathbf{X}^* = \frac{i}{10^2}\right\} = \sum_i \left(\frac{i}{10^2}\right) p_i.$$

By Consequence 6,

$$E(\mathbf{X}^* - 1/200) \le E(\mathbf{X}) \le E(\mathbf{X}^* + 1/200).$$

By Axiom 1,

$$E(\mathbf{X}^*) - 1/200 \le E(\mathbf{X}) \le E(\mathbf{X}^*) + 1/200.$$

Thus

$$E(\mathbf{X}) \text{ is within } 1/200 \text{ of } E(\mathbf{X}^*) = \sum_i \frac{i}{10^2} p_i.$$

In this way, $E(\mathbf{X})$ has been approximated within 0.005. By the same technique, $E(\mathbf{X})$ can be arbitrarily well approximated and evaluated as the limit of its approximations. Although this is the principle of evaluating $E(\mathbf{X})$, its application is sometimes difficult. However, it has a geometric interpretation if $\mathbf{X}$ has a continuous distribution. Draw the graph of g given by $g(x) = xf(x)$, where f is the density of $\mathbf{X}$, see Figure E$_4$–2. The area between this curve

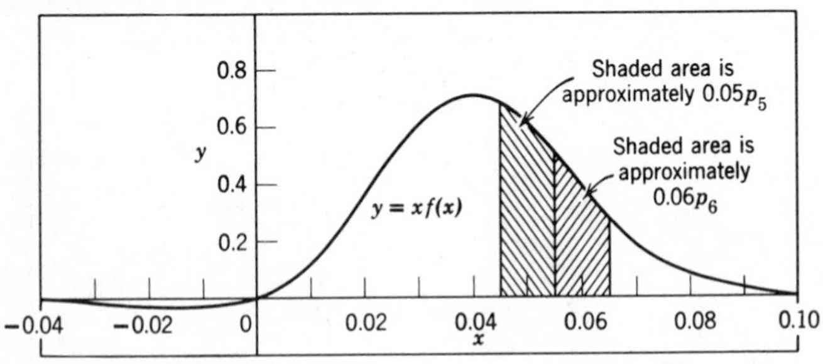

Figure E$_4$–2. Graphical representation of $E(\mathbf{X})$.

and the horizontal axis is made of little pieces, with areas approximately $(i/10^2)p_i$. But then the sum of these pieces is approximately $E(\mathbf{X}^*)$. *Thus $E(\mathbf{X})$ is the area between the curve given by $y = xf(x)$ and the horizontal axis.* (For negative x, $g(x)$ is negative and the corresponding area below the axis is subtracted from the part where x is positive.)

Similarly one can argue that $E(\mathbf{X}^2)$ *is the area between the curve given by* $y = x^2 f(x)$ *and the horizontal axis.*

Some other results of importance are the following. If we define $\mu_{\mathbf{X}} = E(\mathbf{X})$ and $\sigma_{\mathbf{X}}^2 = E[(\mathbf{X} - \mu_{\mathbf{X}})^2]$ and $\sigma_{\mathbf{X}}$ = positive square root of $\sigma_{\mathbf{X}}^2$, we have

CONSEQUENCE 7. $E(\mathbf{X} - \mu_{\mathbf{X}}) = 0.$

Proof: $E(\mathbf{X} - \mu_{\mathbf{X}}) = E(\mathbf{X}) - \mu_{\mathbf{X}} = \mu_{\mathbf{X}} - \mu_{\mathbf{X}} = 0.$

CONSEQUENCE 8. $\sigma_{\mathbf{X}}^2 = E(\mathbf{X}^2) - \mu_{\mathbf{X}}^2.$

Proof:

$$(\mathbf{X} - \mu_{\mathbf{X}})^2 = \mathbf{X}^2 - 2\mu_{\mathbf{X}}\mathbf{X} + \mu_{\mathbf{X}}^2$$
$$\sigma_{\mathbf{X}}^2 = E[(\mathbf{X} - \mu_{\mathbf{X}})^2] = E(\mathbf{X}^2) - 2\mu_{\mathbf{X}}\, E(\mathbf{X}) + \mu_{\mathbf{X}}^2$$
$$\sigma_{\mathbf{X}}^2 = E(\mathbf{X}^2) - 2\mu_{\mathbf{X}}^2 + \mu_{\mathbf{X}}^2 = E(\mathbf{X}^2) - \mu_{\mathbf{X}}^2.$$

CONSEQUENCE 9. *If* $\mathbf{Y} = a + b\mathbf{X}$, *and* b *is positive*

$$\mu_{\mathbf{Y}} = a + b\mu_{\mathbf{X}}, \qquad \sigma_{\mathbf{Y}}^2 = b^2\sigma_{\mathbf{X}}^2, \qquad \sigma_{\mathbf{Y}} = b\sigma_{\mathbf{X}}.$$

Proof:

$$\mu_{\mathbf{Y}} = E(a + b\mathbf{X}) = a + b\, E(\mathbf{X}) = a + b\mu_{\mathbf{X}}$$
$$\sigma_{\mathbf{Y}}^2 = E[(\mathbf{Y} - \mu_{\mathbf{Y}})^2] = E\{[(a + b\mathbf{X}) - (a + b\mu_{\mathbf{X}})]^2\}$$
$$= E\{[b(\mathbf{X} - \mu_{\mathbf{X}})]^2\} = b^2\, E[(\mathbf{X} - \mu_{\mathbf{X}})^2] = b^2\sigma_{\mathbf{X}}^2$$
$$\sigma_{\mathbf{Y}} = b\sigma_{\mathbf{X}}.$$

CONSEQUENCE 10. *If* $\mathbf{Y} = (\mathbf{X} - \mu_{\mathbf{X}})/\sigma_{\mathbf{X}}$, *then* $\mu_{\mathbf{Y}} = 0$ *and* $\sigma_{\mathbf{Y}} = 1.$

Proof:

$$\mu_{\mathbf{Y}} = E\left(\frac{\mathbf{X} - \mu_{\mathbf{X}}}{\sigma_{\mathbf{X}}}\right) = \frac{1}{\sigma_{\mathbf{X}}}\, E(\mathbf{X} - \mu_{\mathbf{X}}) = 0$$
$$\sigma_{\mathbf{Y}}^2 = E\left[\left(\frac{\mathbf{X} - \mu_{\mathbf{X}}}{\sigma_{\mathbf{X}}}\right)^2\right] = \frac{1}{\sigma_{\mathbf{X}}^2}\, E[(\mathbf{X} - \mu_{\mathbf{X}})^2]$$
$$= \frac{\sigma_{\mathbf{X}}^2}{\sigma_{\mathbf{X}}^2} = 1.$$

CONSEQUENCE 11. *The value of* a *which minimizes* $E[(\mathbf{X} - a)^2]$ *is* $a = \mu_{\mathbf{X}}$.

Proof:

$$\mathbf{X} - a = (\mathbf{X} - \mu_{\mathbf{X}}) + (\mu_{\mathbf{X}} - a)$$
$$(\mathbf{X} - a)^2 = (\mathbf{X} - \mu_{\mathbf{X}})^2 + 2(\mu_{\mathbf{X}} - a)(\mathbf{X} - \mu_{\mathbf{X}}) + (\mu_{\mathbf{X}} - a)^2$$
$$E[(\mathbf{X} - a)^2] = E[(\mathbf{X} - \mu_{\mathbf{X}})^2] + 2(\mu_{\mathbf{X}} - a)\, E(\mathbf{X} - \mu_{\mathbf{X}}) + (\mu_{\mathbf{X}} - a)^2$$
$$E[(\mathbf{X} - a)^2] = \sigma_{\mathbf{X}}{}^2 + 0 + (\mu_{\mathbf{X}} - a)^2.$$

Thus $E[(\mathbf{X} - a)^2]$ is at least equal to $\sigma_{\mathbf{X}}^2$, and is equal to $\sigma_{\mathbf{X}}^2$ only when $\mu_{\mathbf{X}} - a = 0$. Thus $a = \mu_{\mathbf{X}}$ minimizes this expression.

CONSEQUENCE 12. *The value of* a *which minimizes* $E(|\,\mathbf{X} - a\,|)$ *is* $a = \nu_{\mathbf{X}}$ = median of $\mathbf{X}$.[1]

[1] The median of $\mathbf{X}$ is a number $\nu_{\mathbf{X}}$ such that $P\{\mathbf{X} < \nu_{\mathbf{X}}\} \leq 0.5 \leq P\{\mathbf{X} \leq \nu_{\mathbf{X}}\}$. Sometimes $\nu_{\mathbf{X}}$ is not uniquely determined.

Proof: Suppose $a = \nu_{\mathbf{X}} + c > \nu_{\mathbf{X}}$. Then

$$\begin{aligned} |\mathbf{X} - a| - |\mathbf{X} - \nu_{\mathbf{X}}| &= c && \text{if } \mathbf{X} \leq \nu_{\mathbf{X}} \\ -c \leq |\mathbf{X} - a| - |\mathbf{X} - \nu_{\mathbf{X}}| &\leq c && \text{if } \nu_{\mathbf{X}} < \mathbf{X} < \nu_{\mathbf{X}} + c \\ |\mathbf{X} - a| - |\mathbf{X} - \nu_{\mathbf{X}}| &= -c && \text{if } \mathbf{X} \geq \nu_{\mathbf{X}} + c. \end{aligned}$$

Let

$$\begin{aligned} \mathbf{Y} &= c && \text{if } \mathbf{X} \leq \nu_{\mathbf{X}} \\ \mathbf{Y} &= -c && \text{if } \mathbf{X} > \nu_{\mathbf{X}}. \end{aligned}$$

Then

$$|\mathbf{X} - a| - |\mathbf{X} - \nu_{\mathbf{X}}| \geq \mathbf{Y}$$

and

$$E(\mathbf{Y}) = cP\{\mathbf{Y} = c\} - cP\{\mathbf{Y} = -c\} = cP\{\mathbf{X} \leq \nu_{\mathbf{X}}\} - cP\{\mathbf{X} > \nu_{\mathbf{X}}\}$$

and

$$E(\mathbf{Y}) \geq 0$$

since

$$P\{\mathbf{X} \leq \nu_{\mathbf{X}}\} \geq 0.5.$$

But then

$$E(|\mathbf{X} - a|) - E(|\mathbf{X} - \nu_{\mathbf{X}}|) \geq E(\mathbf{Y}) \geq 0$$

and thus $E(|\mathbf{X} - a|)$ is at least equal to $E(|\mathbf{X} - \nu_{\mathbf{X}}|)$. A similar argument applies for $a < \nu_{\mathbf{X}}$. With a slight refinement, the above argument yields the fact that, if a is not a candidate for the median, $E(|\mathbf{X} - a|)$ actually exceeds $E(|\mathbf{X} - \nu_{\mathbf{X}}|)$.

APPENDIX E_5

The Convex Set Generated by A is the Set of Weighted Averages of Elements of A

1. First we prove that the set S of weighted averages of elements of A is convex. Suppose $\bar{u}$ and $\bar{u}^*$ are weighted averages of some points of A. Then we can write

$$\bar{u} = w_1\bar{v}_1 + w_2\bar{v}_2 + \cdots + w_n\bar{v}_n$$
$$\bar{u}^* = w_1^*\bar{v}_1 + w_2^*\bar{v}_2 + \cdots + w_n^*\bar{v}_n$$

where the $\bar{v}_i$ are points of A and $w_i \geq 0$, $\sum_{i=1}^{n} w_i = 1$, $w_i^* \geq 0$, and $\sum_{i=1}^{n} w_i^* = 1$. Now we must show that

$$\bar{u}^{**} = (1 - w)\bar{u} + w\bar{u}^*$$

is a point of S (i.e., a weighted average of points of A) if $0 \leq w \leq 1$. But

$$\bar{u}^{**} = [(1 - w)w_1 + ww_1^*]\bar{v}_1 + [(1 - w)w_2 + ww_2^*]\bar{v}_2 + \cdots + [(1 - w)w_n + ww_n^*]\bar{v}_n.$$
$$\bar{u}^{**} = w_1^{**}\bar{v}_1 + w_2^{**}\bar{v}_2 + \cdots + w_n^{**}\bar{v}_n.$$

Now $\bar{u}^{**}$ is a weighted average of $\bar{v}_1, \bar{v}_2, \cdots$ provided only that $w_i^{**} \geq 0$ and $\sum_{i=1}^{n} w_i^{**} = 1$. But since w_i^{**} is a weighted average of the two non-negative numbers w_i and w_i^*, w_i^{**} is non-negative. Furthermore

$$\sum_{i=1}^{n} w_i^{**} = \sum_{i=1}^{n} [(1 - w)w_i + ww_i^*] = (1 - w)\sum_{i=1}^{n} w_i + w\sum_{i=1}^{n} w_i^*$$
$$= (1 - w) \cdot 1 + w \cdot 1 = 1.$$

We have shown that any weighted average of two points, $\bar{u}$ and $\bar{u}^*$, of S is itself a point of S and that therefore S is convex.

2. Next we prove that, if T is a convex set such that A is a subset of T, then S is a subset of T (i.e., any convex set containing A contains S). To do so, it suffices to show that an arbitrary weighted average of points of A,

$$\bar{u} = w_1\bar{v}_1 + w_2\bar{v}_2 + \cdots + w_n\bar{v}_n$$

is in T. There is no loss of generality in assuming $w_1 > 0$. First we note that $\bar{v}_1, \bar{v}_2, \cdots, \bar{v}_n$ are in A and hence in T. Next,

$$\bar{u}_2 = \frac{w_1}{w_1 + w_2}\bar{v}_1 + \frac{w_2}{w_1 + w_2}\bar{v}_2$$

is on the line segment connecting $\bar{v}_1$ and $\bar{v}_2$ and, hence, is in the convex set T. Next

$$\bar{u}_3 = \frac{w_1 + w_2}{w_1 + w_2 + w_3}\bar{u}_2 + \frac{w_3}{w_1 + w_2 + w_3}\bar{v}_3$$

$$= \frac{w_1}{w_1 + w_2 + w_3}\bar{v}_1 + \frac{w_2}{w_1 + w_2 + w_3}\bar{v}_2 + \frac{w_3}{w_2 + w_2 + w_3}\bar{v}_3$$

is on the line segment connecting $\bar{u}_2$ and $\bar{v}_3$ and, hence, in T. Continuing in this fashion, we have

$$\bar{u}_n = \frac{w_1 + w_2 + \cdots + w_{n-1}}{w_1 + w_2 + \cdots + w_n}\bar{u}_{n-1} + \frac{w_n}{w_1 + w_2 + \cdots + w_n}\bar{v}_n$$

$$= \frac{w_1}{w_1 + w_2 + \cdots + w_n}\bar{v}_1 + \frac{w_2}{w_1 + w_2 + \cdots + w_n}\bar{v}_2 + \cdots$$

$$+ \frac{w_n}{w_1 + w_2 + \cdots + w_n}\bar{v}_n$$

$$= w_1\bar{v}_1 + w_2\bar{v}_2 + \cdots + w_n\bar{v}_n = \bar{u}$$

is on the line segment connecting $\bar{u}_{n-1}$ which is in T and $\bar{v}_n$ and, hence, is in T. Since any convex set containing A contains S, and S is convex, it follows that S is the smallest convex set containing A. That is, S is the convex set generated by A.

The reader may note that the above argument incidentally proves that *there* actually *is a smallest convex set containing* A. A simpler proof of this fact involves showing that the common part of all convex sets containing A is a convex set which contains A.

APPENDIX E_6

Relevance of Risks

Consider a problem where the expected losses are given by $L(\theta_1, s)$ and $L(\theta_2, s)$. Suppose you decide that you *prefer* some strategy s_0. Now let the problem be changed as follows. Mr. Breakwell tells you that he will toss a well-balanced coin. If it falls heads, he will see to it that you end up with k_1 or k_2 depending on whether θ_1 or θ_2 is the state of nature. If it falls tails, he will leave you to your own devices and you can proceed as before. Presumably if the coin falls tails, you should be satisfied to apply s_0.

However, the entry of Mr. Breakwell into the problem makes for some new expected losses. In this case, the expected losses corresponding to the use of s if the coin falls tails are

$$L^*(\theta_1, s) = (1/2)k_1 + (1/2)L(\theta_1, s)$$

and

$$L^*(\theta_2, s) = (1/2)k_2 + (1/2)L(\theta_2, s).$$

First let $k_1 = k_2 = 0$. (Mr. Breakwell is very kind.) Then we should still be satisfied to apply s_0. Thus, multiplying the expected losses by $\frac{1}{2}$ does not affect the nature of a good strategy. But if we interchange the L and L^*, it also follows that multiplying by 2 does not affect the nature of a good strategy. However, it is clear that by using a suitably biased coin we could prove that multiplying the $L(\theta, s)$ by any fixed positive number does not affect the nature of a good strategy.

Now let us return to the case where k_1 and k_2 are not necesarily zero. Multiply L^* by 2 and we obtain

$$L^{**}(\theta_1, s) = k_1 + L(\theta_1, s)$$
$$L^{**}(\theta_2, s) = k_2 + L(\theta_2, s).$$

Thus a problem with expected losses given by L^{**} should yield the same " good " strategies as the original problem. But this means that we can add an arbitrary constant to each row in the expected loss table without affecting what constitutes a good strategy. Geometrically this means we can shift S horizontally or vertically without affecting our choice. In particular, we can let k_1 and k_2 be the negative of the minimum losses under states θ_1 and θ_2 (i.e., shift S until it touches both axes). But this way L^{**} coincides with the risk R. Thus, an analysis of R should provide the "good" strategy.

For this reason, some statisticians often feel free to confine their attention to regrets and risks, and often do not bother computing with losses. Of course, this argument also applies if there are $k > 2$ states of nature.

APPENDIX E_7

Probabilities of Compound Sets

In the appendix we prove certain propositions stated in Section 2 of Chapter 6.

CONSEQUENCE 1.

$$P\{A \text{ and } B \text{ and } C\} = P\{A\}\, P\{B \mid A\}\, P\{C \mid A \text{ and } B\}.$$

Proof: $P\{A \text{ and } B\} = P\{A\}\, P\{B \mid A\}$

$$\begin{aligned} P\{A \text{ and } B \text{ and } C\} &= P\{(A \text{ and } B) \text{ and } C\} \\ &= P\{A \text{ and } B\}\, P\{C \mid A \text{ and } B\} \\ &= P\{A\}\, P\{B \mid A\}\, P\{C \mid A \text{ and } B\}. \end{aligned}$$

This argument extends easily to yield

CONSEQUENCE 1a. $P\{A_1 \text{ and } A_2 \text{ and } \cdots \text{ and } A_n\}$

$$= P\{A_1\}\, P\{A_2 \mid A_1\}\, P\{A_3 \mid A_1 \text{ and } A_2\} \cdots P\{A_n \mid A_1 \text{ and } A_2 \text{ and } \cdots \text{ and } A_{n-1}\}.$$

CONSEQUENCE 2. $P\{A \text{ or } B\} = P\{A\} + P\{B\} - P\{A \text{ and } B\}$.

Proof: $\{A \text{ or } B\} = \{E_1 \text{ or } E_2 \text{ or } E_3\}$

where E_1, E_2, E_3 are the nonoverlapping sets:

$$\begin{aligned} E_1 &= \{A \text{ and } \tilde{B}\} \\ E_2 &= \{B \text{ and } \tilde{A}\} \\ E_3 &= \{A \text{ and } B\}. \end{aligned}$$

Thus

$$\begin{aligned} P\{A \text{ or } B\} &= P\{E_1\} + P\{E_2\} + P\{E_3\} \\ &= (P\{E_1\} + P\{E_3\}) + (P\{E_2\} + P\{E_3\}) - P\{E_3\}. \end{aligned}$$

But

$$A = \{E_1 \text{ or } E_3\}$$

and thus

$$P\{A\} = P\{E_1\} + P\{E_3\}.$$

Similarly,

$$P\{B\} = P\{E_2\} + P\{E_3\}$$

and thus

$$P\{A \text{ or } B\} = P\{A\} + P\{B\} - P\{A \text{ and } B\}.$$

This result can be extended. We give the extension, without proof, for three sets.

CONSEQUENCE 2a. $P\{A \text{ or } B \text{ or } C\}$

$$= P\{A\} + P\{B\} + P\{C\} - P\{A \text{ and } B\} - P\{B \text{ and } C\} - P\{A \text{ and } C\} + P\{A \text{ and } B \text{ and } C\}.$$

APPENDIX E_8

Bayes Strategies Obtained by Using *A Posteriori* Probabilities[1]

For the discrete case we shall prove that the Bayes strategies are equivalent to Mr. Solomon's of computing *a posteriori* probabilities on the basis of the data $\mathbf{Z}$ and solving the corresponding no-data problem. We have

$$\mathbf{w}_i = \frac{w_i f(\mathbf{Z} \mid \theta_i)}{f(\mathbf{Z})}$$

$$f(\mathbf{Z}) = w_1 f(\mathbf{Z} \mid \theta_1) + w_2 f(\mathbf{Z} \mid \theta_2) + \cdots + w_k f(\mathbf{Z} \mid \theta_k)$$

$$B(\overline{\mathbf{w}}, a) = \sum_{i=1}^{k} \mathbf{w}_i \, r(\theta_i, a) = \frac{1}{f(\mathbf{Z})} \sum_{i=1}^{k} w_i f(\mathbf{Z} \mid \theta_i) \, r(\theta_i, a).$$

Mr. Solomon's strategy is to react to $\mathbf{Z}$ with the action which minimizes

$$B(\overline{\mathbf{w}}, a) \qquad \text{or} \qquad \sum_{i=1}^{k} w_i f(\mathbf{Z} \mid \theta_i) \, r(\theta_i, a).$$

The Bayes strategy as originally described is the strategy s which minimizes the average expected loss

$$\mathscr{R}(\overline{w}, s) = \sum_{i=1}^{k} w_i \, R(\theta_i, s).$$

If $\mathbf{Z}$ can take on values $z_1, z_2, \cdots$, we have

$$\begin{aligned} R(\theta, s) &= f(z_1 \mid \theta) \, r(\theta, s(z_1)) + f(z_2 \mid \theta) \, r(\theta, s(z_2)) + \cdots \\ &= \sum_j f(z_j \mid \theta) \, r(\theta, s(z_j)) \end{aligned}$$

$$\begin{aligned} \mathscr{R}(\overline{w}, s) = \quad & w_1 f(z_1 \mid \theta_1) \, r(\theta_1, s(z_1)) + w_1 f(z_2 \mid \theta_1) \, r(\theta_1, s(z_2)) + \cdots \\ & + w_2 f(z_1 \mid \theta_2) \, r(\theta_2, s(z_1)) + w_2 f(z_2 \mid \theta_2) \, r(\theta_2, s(z_2)) + \cdots \\ & + w_3 f(z_1 \mid \theta_3) \, r(\theta_3, s(z_1)) + w_3 f(z_2 \mid \theta_3) \, r(\theta_3, s(z_2)) + \cdots \\ & \quad \vdots \qquad\qquad\qquad\qquad \vdots \qquad\qquad\qquad\qquad \vdots \end{aligned}$$

[1] This appendix proves a remark made in Sections 1 and 4 of Chapter 6.

The expression $s(z_1)$ appears only in the first column. Thus to minimize $\mathscr{R}(\overline{w}, s)$, we should select $s(z_1)$ to minimize

$$w_1 f(z_1 | \theta_1)\, r(\theta_1, s(z_1)) + w_2 f(z_1 | \theta_2)\, r(\theta_2, s(z_1)) \\ + w_3 f(z_1 | \theta_3)\, r(\theta_3, s(z_1)) + \cdots .$$

This defines the Bayes strategy reaction to the possible observation z_1; similarly, if z_2 is observed, $s(z_2)$ should be that action which minimizes

$$w_1 f(z_2 | \theta_1)\, r(\theta_1, s(z_2)) + w_2 f(z_2 | \theta_2)\, r(\theta_2, s(z_2)) + \cdots .$$

In general, if an observation **Z** is recorded, the Bayes strategy should give for $s(\mathbf{Z})$ that action which minimizes

$$\sum_{i=1}^{k} w_i f(\mathbf{Z} | \theta_i)\, r(\theta_i, a).$$

But this is exactly the same as the rule advocated by Mr. Solomon.

APPENDIX E_9

Crossing Bridges One at a Time[1]

We shall prove that the *a posteriori probability* $\bar{\mathbf{w}}$ derived from the data $(\mathbf{X}, \mathbf{Y})$ together is the same as $\bar{\mathbf{w}}^{**}$ which is derived serially by considering the data $\mathbf{X}$ and $\mathbf{Y}$ one after the other. For simplicity, we assume that $\mathbf{X}$ and $\mathbf{Y}$ are *discrete* and *independent* for fixed θ and give a proof for that case only.

$$\begin{aligned}
f(x, y \mid \theta) &= P\{\mathbf{X} = x \text{ and } \mathbf{Y} = y \mid \theta\} \\
g(x \mid \theta) &= P\{\mathbf{X} = x \mid \theta\} \\
h(y \mid \theta) &= P\{\mathbf{Y} = y \mid \theta\} \\
f(x, y \mid \theta) &= g(x \mid \theta)\, h(y \mid \theta) \\
f(x, y) &= \sum w_i f(x, y \mid \theta_i) \\
g(x) &= \sum w_i\, g(x \mid \theta_i).
\end{aligned}$$

The *a posteriori probabilities* $\mathbf{w}_i$ are given by

$$\mathbf{w}_i = \frac{w_i f(\mathbf{X}, \mathbf{Y} \mid \theta_i)}{f(\mathbf{X}, \mathbf{Y})}.$$

On the other hand, the *a posteriori* probability derived from $\mathbf{X}$ alone is given by

$$\mathbf{w}_i^* = \frac{w_i\, g(\mathbf{X} \mid \theta_i)}{g(\mathbf{X})}.$$

Finally,

$$\mathbf{w}_i^{**} = \frac{\mathbf{w}_i^*\, h(\mathbf{Y} \mid \theta_i)}{\sum_{i=1}^{k} \mathbf{w}_i^*\, h(\mathbf{Y} \mid \theta_i)}$$

$$\begin{aligned}
\mathbf{w}_i^{**} &= \left[\frac{w_i\, g(\mathbf{X} \mid \theta_i)\, h(\mathbf{Y} \mid \theta_i)}{g(\mathbf{X})}\right] \Bigg/ \left[\frac{\sum_{i=1}^{k} w_i\, g(\mathbf{X} \mid \theta_i)\, h(\mathbf{Y} \mid \theta_i)}{g(\mathbf{X})}\right] \\
&= \frac{w_i f(\mathbf{X}, \mathbf{Y} \mid \theta_i)}{f(\mathbf{X}, \mathbf{Y})} = \mathbf{w}_i.
\end{aligned}$$

[1] This appendix refers to a statement in Section 4, Chapter 6.

APPENDIX E_{10}

Mean and Variance of $\overline{\mathbf{X}}$

We have already given some properties of expectation. Now we can apply them to the sample mean. First of all, suppose that $\mathbf{X}_1$, $\mathbf{X}_2, \cdots, \mathbf{X}_n$ are n random variables. Then,

$$E\left(\frac{\mathbf{X}_1 + \mathbf{X}_2 + \cdots + \mathbf{X}_n}{n}\right) = \frac{E(\mathbf{X}_1) + E(\mathbf{X}_2) + \cdots + E(\mathbf{X}_n)}{n}.$$

Thus, if $\mathbf{X}_1, \mathbf{X}_2, \cdots, \mathbf{X}_n$ all have the same probability distribution which has mean μ_{X}

$$(E_{10}\text{–}1) \qquad E(\overline{\mathbf{X}}) = \frac{\mu_{\mathrm{X}} + \mu_{\mathrm{X}} + \cdots + \mu_{\mathrm{X}}}{n} = \mu_{\mathrm{X}}.$$

To treat the variance of $\overline{\mathbf{X}}$, we use a property of expectation not previously considered. This is basic enough to warrant addition to the list of the four properties in Section 3, Chapter 4.

Expectation Property 5. *If* **X** *and* **Y** *are independent random variables, then*

$$(E_{10}\text{–}2) \qquad E(\mathbf{XY}) = E(\mathbf{X})\, E(\mathbf{Y}).$$

For discrete random variables, one may easily demonstrate this property. As a special case, suppose that **X** assumes the values x_1, x_2 with probabilities $f(x_1)$ and $f(x_2)$ while **Y** assumes the values y_1, y_2, and y_3 with probabilities $g(y_1)$, $g(y_2)$, and $g(y_3)$. Then **XY** can be x_1y_1, x_1y_2, x_1y_3, x_2y_1, x_2y_2, or x_2y_3. Since **X** and **Y** are independent, these outcomes have probabilities $f(x_1)\, g(y_1)$, $f(x_1)\, g(y_2)$, etc.

$$\begin{aligned} E(\mathbf{XY}) &= x_1y_1 f(x_1)\, g(y_1) + x_1y_2 f(x_1)\, g(y_2) + x_1y_3 f(x_1)\, g(y_3) \\ &\quad + x_2y_1 f(x_2)\, g(y_1) + x_2y_2 f(x_2)\, g(y_2) + x_2y_3 f(x_2)\, g(y_3) \\ &= [x_1 f(x_1)][y_1\, g(y_1) + y_2\, g(y_2) + y_3\, g(y_3)] \\ &\quad + [x_2 f(x_2)][y_1\, g(y_1) + y_2\, g(y_2) + y_3\, g(y_3)] \\ &= [x_1 f(x_1) + x_2 f(x_2)][y_1\, g(y_1) + y_2\, g(y_2) + y_3\, g(y_3)] \\ &= E(\mathbf{X})\, E(\mathbf{Y}) \,. \end{aligned}$$

Note that, if X and Y are independent and h_1 and h_2 are two functions, then $h_1(X)$ and $h_2(Y)$ are independent. Now let us compute σ^2_{X+Y} when X and Y are independent. First

$$\mu_{X+Y} = \mu_X + \mu_Y \tag{E$_{10}$–3}$$

whether or not X and Y are independent.

$$\begin{aligned}
(X + Y - \mu_{X+Y})^2 &= [(X - \mu_X) + (Y - \mu_Y)]^2 \\
&= (X - \mu_X)^2 + 2(X - \mu_X)(Y - \mu_Y) + (Y - \mu_Y)^2 \\
\sigma^2_{X+Y} &= E[(X + Y - \mu_{X+Y})^2] \\
&= E[(X - \mu_X)^2] + 2E[(X - \mu_X)(Y - \mu_Y)] \\
&\quad + E[(Y - \mu_Y)^2] \\
E[(X - \mu_X)^2] &= \sigma^2_X \\
E[(X - \mu_X)(Y - \mu_Y)] &= E(XY - X\mu_Y - \mu_X Y + \mu_X\mu_Y) \\
&= \mu_X\mu_Y - \mu_X\mu_Y - \mu_X\mu_Y + \mu_X\mu_Y = 0 \\
E[(Y - \mu_Y)^2] &= \sigma^2_Y.
\end{aligned}$$

Thus

$$\sigma^2_{X+Y} = \sigma^2_X + \sigma^2_Y \tag{E$_{10}$–4}$$

if X *and* Y *are independent*. Applying this result to n *independent* observations on X, we have

$$\sigma^2_{X_1+X_2+\cdots+X_n} = \sigma^2_{X_1} + \sigma^2_{X_2} + \cdots + \sigma^2_{X_n} = n\sigma^2_X \tag{E$_{10}$–5}$$

and

$$\sigma^2_{\overline{X}} = \sigma^2_{(X_1+X_2+\cdots+X_n)/n} = \frac{1}{n^2}\sigma^2_{X_1+X_2+\cdots+X_n} = \frac{n\sigma^2_X}{n^2}$$

$$\sigma^2_{\overline{X}} = \frac{\sigma^2_X}{n}. \tag{E$_{10}$–6}$$

The reader should note that the argument which leads to Equation (E$_{10}$–4) can be applied as well to $X - Y$ as to $X + Y$, which gives the result, for X and Y independent,

$$\sigma^2_{X-Y} = \sigma^2_X + \sigma^2_Y. \tag{E$_{10}$–7}$$

APPENDIX E_{11}

For Testing a Simple Hypothesis Versus a Simple Alternative, the Bayes Strategies Are Likelihood-Ratio Tests

Consider the problem of testing a simple hypothesis H_1: $\theta = \theta_1$ versus the simple alternative H_2: $\theta = \theta_2$. Suppose that the *a priori* probabilities $w_1 = 1 - w$ and $w_2 = w$ are given for θ_1 and θ_2. First we shall solve the no-data problem. Here there are only two strategies: take action a_1 or take action a_2. The risks associated with these are

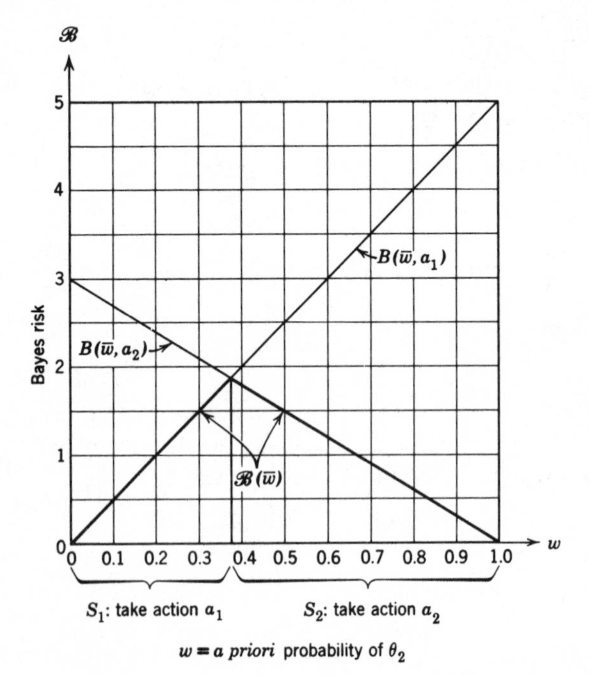

Figure E_{11}-1. Bayes risk for the no-data problem.

$$B(\overline{w}, a_1) = (1 - w)0 + w\,r(\theta_2) = w\,r(\theta_2)$$

and

$$B(\overline{w}, a_2) = (1 - w)\,r(\theta_1) + w \cdot 0 = (1 - w)\,r(\theta_1)$$

and the Bayes risk for the no-data problem is then

$$\mathscr{B}(\overline{w}) = \min\,[w\,r(\theta_2), (1 - w)\,r(\theta_1)]$$

and is represented in Figure E_{11}–1 for Example 9.1. The two risks are equal where the two lines $B(\overline{w}, a_1)$ and $B(\overline{w}, a_2)$ intersect, i.e., for $w = w_0$ where

$$w_0\,r(\theta_2) = (1 - w_0)\,r(\theta_1)$$

$$w_0 = \frac{r(\theta_1)}{r(\theta_1) + r(\theta_2)}.$$

Thus, the solution to the no-data problem is to take action a_1 if $w < w_0$, a_2 if $w > w_0$, and either action if $w = w_0$.

Now we consider the problem with data. Here the data transform the *a priori* probability w to $\mathbf{w}$ and the Bayes strategy leads to a_1 if $\mathbf{w} < w_0$, a_2 if $\mathbf{w} > w_0$, and either action if $\mathbf{w} = w_0$. But from Equation (6.7) we know that

$$\mathbf{w} = \mathbf{w}_2 = \frac{w f(\mathbf{Z}\,|\,\theta_2)}{(1 - w) f(\mathbf{Z}\,|\,\theta_1) + w f(\mathbf{Z}\,|\,\theta_2)}$$

$$= \frac{1}{\{[(1 - w) f(\mathbf{Z}\,|\,\theta_1)]/[w f(\mathbf{Z}\,|\,\theta_2)]\} + 1}$$

$$= \frac{1}{\{[(1 - w)/w][\lambda(\mathbf{Z})]\} + 1}.$$

As $\lambda(\mathbf{Z})$ increases, $\mathbf{w}$ decreases. Furthermore, $\mathbf{w} = w_0$ when

$$\frac{1}{\{[(1 - w)/w][\lambda(\mathbf{Z})]\} + 1} = \frac{r(\theta_1)}{r(\theta_1) + r(\theta_2)} = \frac{1}{1 + \{[r(\theta_2)]/[r(\theta_1)]\}}$$

or

$$\lambda(\mathbf{Z}) = \frac{w\,r(\theta_2)}{(1 - w)\,r(\theta_1)} = k.$$

Thus,

if $\lambda(\mathbf{Z}) > k$, $\mathbf{w} < w_0$, and our Bayes strategy calls for a_1;
if $\lambda(\mathbf{Z}) < k$, $\mathbf{w} > w_0$, and our Bayes strategy calls for a_2;
if $\lambda(\mathbf{Z}) = k$, $\mathbf{w} = w_0$, and our Bayes strategy calls for either a_1 or a_2.

This means that the Bayes strategy is the likelihood-ratio test for

$$k = \frac{w\, r(\theta_2)}{(1 - w)\, r(\theta_1)}.$$

APPENDIX E_{12}

Some Likelihood-Ratio Tests

We present the likelihood-ratio tests for three examples.

Example E_{12}–1. Binomial Distribution. Test the hypothesis H_1: $p = p_1$ versus H_2: $p = p_2$ where $p_1 > p_2$, p represents the probability of a head, and the experiment consists of n tosses of the coin.

The probability of observing $HHTHHTT\cdots$ is $pp(1 - p)\,pp(1 - p)(1 - p)\cdots$. Thus if the data $\mathbf{Z}$ consist of a sequence of n tosses of which $\mathbf{m}$ are heads,

$$f(\mathbf{Z}\,|\,p) = p^{\mathbf{m}}(1 - p)^{n-\mathbf{m}}$$

$$\lambda(\mathbf{Z}) = \frac{p_1^{\mathbf{m}}(1 - p_1)^{n-\mathbf{m}}}{p_2^{\mathbf{m}}(1 - p_2)^{n-\mathbf{m}}} = \left(\frac{1 - p_1}{1 - p_2}\right)^n\left(\frac{p_1/(1 - p_1)}{p_2/(1 - p_2)}\right)^{\mathbf{m}}.$$

The expression

$$\frac{p_1}{1 - p_1} = \frac{1}{1 - p_1} - 1$$

increases as p_1 increases. Since $p_1 > p_2$, it follows that

$$\frac{p_1/(1 - p_1)}{p_2/(1 - p_2)} > 1$$

and, thus, $\lambda(\mathbf{Z})$ increases as $\mathbf{m}$ increases. Then $\lambda(\mathbf{Z})$ will increase when $\hat{\mathbf{p}} = \mathbf{m}/n$ increases, and the likelihood-ratio test which consists of accepting H_1 if $\lambda(\mathbf{Z})$ is larger than some constant is equivalent to accepting H_1 if $\hat{\mathbf{p}}$ is larger than some related constant.

Example E_{12}–2. Normal Distribution, Known σ. Test the hypothesis H_1: $\mu = \mu_1$ versus H_2: $\mu = \mu_2$ where $\mu_1 > \mu_2$, and the data $\mathbf{Z}$ consist of n independent observations $\mathbf{X}_1, \mathbf{X}_2, \cdots, \mathbf{X}_n$ which are normally distributed with mean μ and known standard deviation σ.

In this problem

$$f(\mathbf{Z}\,|\,\mu) = \left[\frac{1}{\sqrt{2\pi\sigma^2}}\exp\left(-\frac{(\mathbf{X}_1 - \mu)^2}{2\sigma^2}\right)\right]\left[\frac{1}{\sqrt{2\pi\sigma^2}}\exp\left(-\frac{(\mathbf{X}_2 - \mu)^2}{2\sigma^2}\right)\right]$$

$$\cdots \left[\frac{1}{\sqrt{2\pi\sigma^2}} \exp\left(-\frac{(\mathbf{X}_n - \mu)^2}{2\sigma^2}\right)\right]$$

$$= (2\pi\sigma^2)^{-n/2} \exp\left[-\frac{1}{2\sigma^2}\left(\sum_{i=1}^{n} (\mathbf{X}_i - \mu)^2\right)\right].$$

Then

$$\lambda(\mathbf{Z}) = \exp\left\{-\frac{1}{2\sigma^2}[\sum (\mathbf{X}_i - \mu_1)^2 - \sum (\mathbf{X}_i - \mu_2)^2]\right\}.$$

Since $\exp(-x)$ decreases when x increases, the likelihood-ratio test consists of accepting H_1 if

$$\sum (\mathbf{X}_i - \mu_1)^2 - \sum (\mathbf{X}_i - \mu_2)^2 = -2n(\mu_1 - \mu_2)\bar{\mathbf{X}} + n(\mu_1^2 - \mu_2^2)$$

is small. But $2n(\mu_1 - \mu_2)$ is positive, and this expression is small when $\bar{\mathbf{X}}$ is large. Thus the likelihood-ratio test consists of accepting H_1 if $\bar{\mathbf{X}}$ exceeds some constant.

Example E_{12}*–3. A Peculiar Example Involving the Cauchy Distribution.* Suppose that the data consist of a single observation $\mathbf{X}$ with density given by

$$f(x \mid \theta) = \frac{1}{\pi[1 + (x - \theta)^2]}.$$

This density is symmetric and centered about θ and resembles the normal density. Test $H_1\colon \theta = \theta_1 = 1$ versus $H_2\colon \theta = \theta_2 = -1$. The likelihood-ratio is

$$\lambda(\mathbf{X}) = \frac{1 + (\mathbf{X} + 1)^2}{1 + (\mathbf{X} - 1)^2} = \frac{\mathbf{X}^2 + 2\mathbf{X} + 2}{\mathbf{X}^2 - 2\mathbf{X} + 2}.$$

To show the peculiarity of the likelihood-ratio tests for this example, let us illustrate with the test which accepts H_1 if $\lambda(\mathbf{X}) > 1/2$. Thus we accept H_1 if

$$2\mathbf{X}^2 + 4\mathbf{X} + 4 \geq \mathbf{X}^2 - 2\mathbf{X} + 2$$

or

$$\mathbf{X}^2 + 6\mathbf{X} + 2 \geq 0.$$

But

$$\begin{aligned} &\mathbf{X}^2 + 6\mathbf{X} + 2 = 0 && \text{when } \mathbf{X} = -3 \pm \sqrt{7} \\ &\mathbf{X}^2 + 6\mathbf{X} + 2 > 0 && \text{when } \mathbf{X} < -5.65 \quad \text{or} \quad \mathbf{X} > -0.35 \end{aligned}$$

and

$$\mathbf{X}^2 + 6\mathbf{X} + 2 < 0 \qquad \text{when } -5.65 < \mathbf{X} < -0.35.$$

Thus for $k = 1/2$, we accept H_1 when $\mathbf{X}$ is large enough (> -0.35), or *very small* (highly negative, i.e., < -5.65).

It should be remarked that, although an intuitively appealing strategy, such as, "accept H_1 if $\mathbf{X} > -0.35$," is not admissible, it is almost so, and its error point lies close to the error curve.

APPENDIX E_{13}

Admissibility of the Two-State Likelihood-Ratio Tests for Certain Problems Involving Composite Hypotheses

In this appendix we shall prove that, for the binomial problem of testing $H_1: p \geq p_0$ versus $H_2: p < p_0$, the admissible tests are the same as for the two-state problem. That is, accept H_1 if $\hat{\mathbf{p}}$ is large enough. More specifically let $\mathscr{T}$ be the class of tests which can be described as follows:

Take action a_1	if $\hat{\mathbf{p}} > c$
Take action a_2	if $\hat{\mathbf{p}} < c$
Take either action	if $\hat{\mathbf{p}} = c$.

We know that $\mathscr{T}$ is the class of admissible tests for testing H_1^*: $p = p_1$ versus $H_2^*: p = p_2$ $(p_1 > p_2)$. We shall prove that $\mathscr{T}$ is the class of admissible tests for H_1 versus H_2. The proof follows in two parts.

1. *Every Test in $\mathscr{T}$ is Admissible.* Let s be a dominated test of H_1 versus H_2. Then there is a test s_1 which dominates s and for which $R(p, s_1) \leq R(p, s)$ for all p and $R(p, s_1) < R(p, s)$ for some p, say p_3. To be specific, let us suppose $p_3 > p_0$. Select an arbitrary $p_4 \leq p_0$. Then s_1 dominates s for the two-state problem of testing $H_1^{**}: p = p_3$ versus $H_2^{**}: p = p_4$. But no strategy which is dominated for the two-state problem can be in $\mathscr{T}$. Hence s is not in $\mathscr{T}$ and every test in $\mathscr{T}$ is admissible.

2. *Every Admissible Test is in $\mathscr{T}$.* Suppose that s is not in $\mathscr{T}$. For this test, $\alpha_s(p_0, a_1)$ is some number. By suitably adjusting c, we find a test s_2 in $\mathscr{T}$ which has the same action probabilities for p_0, i.e.,

$$\alpha_{s_2}(p_0, a_1) = \alpha_s(p_0, a_1).$$

We shall now show that s_2 dominates s. Suppose $p_5 < p_0$. Since s_2 is admissible and s is not for testing $p = p_0$ versus $p = p_5$, and the action probabilities coincide at p_0, it follows that $\alpha_{s_2}(p_5, a_1) < \alpha_s(p_5, a_1)$. Thus $R(p_5, s_2) < R(p_5, s)$ unless $r(p_5) = 0$, in which case equality replaces inequality. Similarly, if $p_6 > p_0$, we consider testing $p = p_6$ versus $p = p_0$ and obtain $\alpha_{s_2}(p_6, a_2) < \alpha_s(p_6, a_2)$, giving $R(p_6, s_2) < R(p_6, s)$ unless $r(p_6) = 0$. Thus we have

$$R(p, s_2) \leq R(p, s) \qquad \text{for all } p$$

and

$$R(\text{p}, s_2) < R(p, s)$$

for some p provided that the problem is nontrivial and $r(p) > 0$ for some p other than p_0. Thus any test s not in $\mathscr{T}$ is dominated and, consequently, every admissible test is in $\mathscr{T}$.

Combining (1) and (2), it follows that $\mathscr{T}$ is the class of admissible tests.

This argument applies also to the normal distribution problem with known σ. The main condition which allows for the generalization of this result is that the class $\mathscr{T}$ of likelihood-ratio tests for $\theta = \theta_1$ versus $\theta = \theta_2$ ($\theta_2 < \theta_1$) does not involve the particular values of θ_1 and θ_2.

APPENDIX E_{14}

Sequential Likelihood-Ratio Tests

Here we shall show that, when the cost of taking N observations is $C(N) = cN$, the Bayes sequential tests of a simple hypothesis versus a simple alternative are the sequential likelihood-ratio tests.

First modify the problem so that the first observation is free and we can proceed optimally thereafter. The minimum average risk is given by some function $\mathscr{B}^*(\overline{w})$ which is no greater than

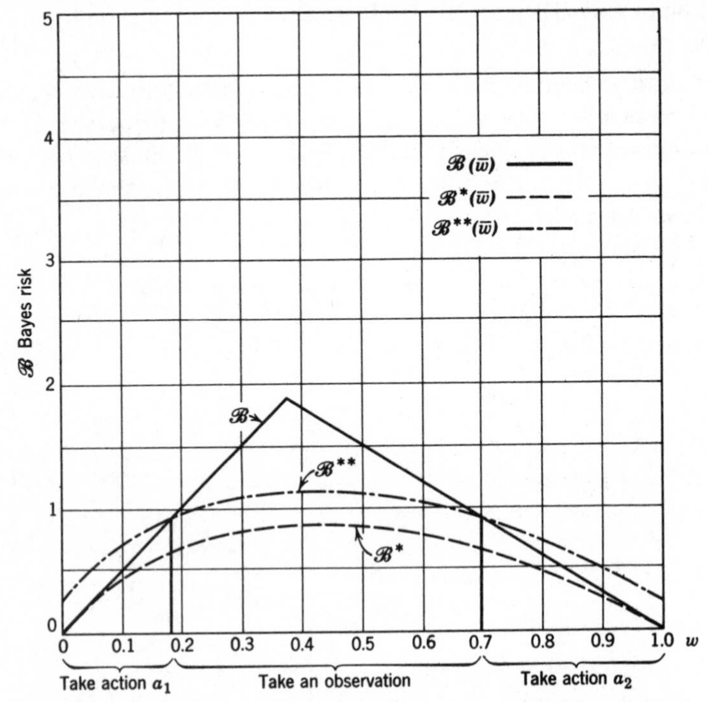

Figure E_{14}–1. $\mathscr{B}(\overline{w})$, the Bayes risk for the no-data problem; $\mathscr{B}^*(\overline{w})$, the minimum average risk for the sequential problem with the first observation free; and $\mathscr{B}^{**}(\overline{w})$, the minimum average risk for the sequential problem with the first observation costing c whether it is used or not.

the Bayes risk for the no-data problem, $\mathscr{B}(\overline{w})$. (See Figure E_{14}–1.) Let us assume for the time being that $\mathscr{B}^*(\overline{w})$ is concave (all chords lie below the curve). Now suppose that we were forced to take the first observation and to pay for it, and then were allowed to proceed optimally. The minimum average risk would be $\mathscr{B}^{**}(\overline{w}) = c + \mathscr{B}^*(\overline{w})$.

Now let us consider the original problem where we can choose whether to take the first observation (at a cost of c). If $\overline{w}$ is such that $\mathscr{B}(\overline{w}) < \mathscr{B}^{**}(\overline{w})$, it pays to take no observation but to take the appropriate action depending on the location of $\overline{w}$. If $\mathscr{B}(\overline{w}) > \mathscr{B}^{**}(\overline{w})$, it pays to take an observation.

It may occur that the cost of observation c is so large that $\mathscr{B}(\overline{w}) < \mathscr{B}^{**}(\overline{w})$ for all w. If this is the case, all Bayes strategies prohibit taking observations. Often this is not the case. Then we note that each flat section of $\mathscr{B}(\overline{w})$ will intersect $\mathscr{B}^{**}(\overline{w})$ at one point, altogether giving two points with abscissas v_1 and v_2. Then the Bayes strategies dictate: take action a_1 if $0 \leq w \leq v_1$, take an observation if $v_1 < w < v_2$, and take action a_2 if $v_2 \leq w \leq 1$.

Suppose that the problem is such that *each* observation taken costs c, and all successive observations are independently distributed with a common distribution depending only on θ. Suppose that $v_1 < w < v_2$ so that a first observation is taken. Digest it by computing $\mathbf{w}^*$, the *a posteriori* probability of θ_2. The problem now facing us is *exactly the same* as the one we had before except that w is replaced by $\mathbf{w}^*$. Hence, we should stop sampling unless $v_1 < \mathbf{w}^* < v_2$.

In other words, in this problem where each observation costs c, *and the observations are independent with the same distribution, each Bayes solution can be characterized by two numbers* v_1 *and* v_2. *If, after any observation, the a posteriori probability* $\mathbf{w}$ *is such that:*

(a) $0 \leq \mathbf{w} \leq v_1$, *take action* a_1

(b) $v_1 < \mathbf{w} < v_2$, *take another observation*

(c) $v_2 \leq \mathbf{w} \leq 1$, *take action* a_2.[1]

The problem of finding the two numbers v_1 and v_2 corresponding to $r(\theta_1)$, $r(\theta_2)$, and c is not a trivial one, but it has been treated successfully.[2]

[1] We have implicitly ignored some Bayes strategies. A Bayes strategy could call for another observation if $w = v_1$ or v_2.

[2] Incidentally, v_1 and v_2 are determined by $f(x \mid \theta)$ and the ratios of the three costs, $r(\theta_1)$, $r(\theta_2)$, and c.

What does this procedure have to do with the likelihood-ratio procedure ? To answer this question, let us compute $\mathbf{w}_n$ after the data $\mathbf{Z}_n = (\mathbf{X}_1, \mathbf{X}_2, \cdots, \mathbf{X}_n)$ have been taken. Referring to Appendix E_{11} we see that

$$\mathbf{w}_n = \frac{1}{\{[(1-w)/w][\lambda_n(\mathbf{Z}_n)]\} + 1}.$$

An increase in the likelihood-ratio increases the denominator and decreases the above fraction. In fact

$$\lambda_n(\mathbf{Z}_n) = \frac{1-\mathbf{w}_n}{\mathbf{w}_n}\frac{w}{1-w}.$$

Thus $\mathbf{w}_n \leq v_1$ if and only if $\lambda_n(\mathbf{Z}_n) \geq k_1 = [(1-v_1)/v_1][w/(1-w)]$. Hence we take action a_1 if $\lambda_n(\mathbf{Z}_n) \geq k_1 = [(1-v_1)/v_1][w/(1-w)]$. Also, we take action a_2 if $\lambda_n(\mathbf{Z}_n) \leq k_2 = [(1-v_2)/v_2][w/(1-w)]$, and we take another observation if $k_2 < \lambda_n(\mathbf{Z}_n) < k_1$. Note that, while v_1 and v_2 did not involve the original *a priori* probability w, the limits on the likelihood-ratio, k_1 and k_2, do involve w, although in a rather simple fashion. In any case, k_1 and k_2 depend on r_1, r_2, c and w.

It remains to show that $\mathscr{B}^*(\overline{w})$ is represented by a concave curve. The argument we shall give for $\mathscr{B}^*(\overline{w})$ is quite general and may be used to show the concavity of the Bayes risk for any decision problem.

For each strategy s, the risk $\mathscr{R}(s) = (1-w)R(\theta_1, s) + wR(\theta_2, s)$. In our present graphical representation, s is represented by a straight line segment from $(0, R(\theta_1, s))$ to $(1, R(\theta_2, s))$. The curve giving $\mathscr{B}^*(\overline{w})$ is the lower boundary of a set of line segments corresponding to the strategies being considered. Since each line segment is on or above the curve at $w = w_1$ and at $w = w_2$, each line must be on or above the chord connecting the points at $w=w_1$, and $w = w_2$. Hence the lower boundary of the line segments must lie on or above the chord. Thus the lower boundary is concave.

These results can be extended to an arbitrary finite number of states of nature and finite number of available actions. In fact, another point of view leads to a shorter proof. Suppose that the states of nature are $\theta_1, \theta_2, \cdots, \theta_k$ and the available actions are a_1, $a_2, \cdots, a_m$. The preceding argument for the concavity of $\mathscr{B}^*(\overline{w})$ applies equally well to show that $\mathscr{B}^{***}(\overline{w})$, the weighted average of risks for the optimal Bayes strategy, is concave. Let S_j be the jth stopping set, i.e., the set of $\bar{w}$ for which taking action a_j with no observation is a Bayes strategy. We shall prove that the *stopping sets are convex.*

Suppose $\bar{w}_{(1)} = (w_{11}, w_{12}, \cdots, w_{1k})$ and $\bar{w}_{(2)} = (w_{21}, w_{22}, \cdots, w_{2k})$ are in S_j and $\bar{w}_{(3)} = (1 - a)\bar{w}_{(1)} + a\bar{w}_{(2)}$, $0 \leq a \leq 1$. Then

$$\mathscr{B}^{***}(\bar{w}_{(1)}) = \sum_{i=1}^{k} w_{1i}\, r(\theta_i, a_j),$$

$$\mathscr{B}^{***}(\bar{w}_{(2)}) = \sum_{i=1}^{k} w_{2i}\, r(\theta_i, a_j),$$

and, by concavity, the best that can be done, if $\bar{w} = \bar{w}_{(3)}$, is

$$\mathscr{B}^{***}(\bar{w}_{(3)}) \geq (1 - a)\, \mathscr{B}^{***}(\bar{w}_{(1)}) + a\, \mathscr{B}^{***}(\bar{w}_{(2)}).$$

But if $\bar{w} = \bar{w}_{(3)}$ and action a_j is taken with no observation, the risk is

$$\begin{aligned}\sum_{i=1}^{k} w_{3i}\, r(\theta_i, a_j) &= \sum_{i=1}^{k} [(1 - a)\, w_{1i} + a\, w_{2i}]\, r(\theta_i, a_j) \\ &= (1 - a)\, \mathscr{B}^{***}(\bar{w}_{(1)}) + a\, \mathscr{B}^{***}(\bar{w}_{(2)}),\end{aligned}$$

which is the best that can be done. Thus $\bar{w}_{(3)}$ is in S_j, and S_j is convex.

In the special two-action, two-state case, it is clear that the stopping set for a_1 is a convex set containing $\bar{w} = (1, 0)$, i.e., $S_1 = \{\bar{w}\colon 0 \leq w \leq v_1\}$. Similarly $S_2 = \{\bar{w}\colon v_2 \leq w \leq 1\}$. For three-action, two-state problems, we may have three intervals which are stopping sets. For three-state problems, graphical representations are more complicated.

APPENDIX E_{15}

Some Sequential Likelihood-Ratio Tests

We consider the sequential likelihood-ratio tests for the sequential extensions of Examples E_{12}–1 and E_{12}–2.

Example E_{15}–1. Binomial Distribution. Here we have

$$\lambda_n(\mathbf{Z}_n) = \left(\frac{1 - p_1}{1 - p_2}\right)^n \left(\frac{p_1/(1 - p_1)}{p_2/(1 - p_2)}\right)^{n\hat{\mathbf{p}}_n}.$$

The relation $k_2 < \lambda_n(\mathbf{Z}_n) < k_1$ is equivalent to

$$k_2\left(\frac{1 - p_2}{1 - p_1}\right)^n < \left(\frac{p_1/(1 - p_1)}{p_2/(1 - p_2)}\right)^{n\hat{\mathbf{p}}_n} < k_1\left(\frac{1 - p_2}{1 - p_1}\right)^n$$

$$\frac{\log\,[(1 - p_2)/(1 - p_1)]}{\log\left(\dfrac{p_1/(1 - p_1)}{p_2/(1 - p_2)}\right)} + \frac{1}{n}\frac{\log k_2}{\log\left(\dfrac{p_1/(1 - p_1)}{p_2/(1 - p_2)}\right)}$$

$$< \hat{\mathbf{p}}_n < \frac{\log\,[(1 - p_2)/(1 - p_1)]}{\log\left(\dfrac{p_1/(1 - p_1)}{p_2/(1 - p_2)}\right)} + \frac{1}{n}\frac{\log k_1}{\log\left(\dfrac{p_1/(1 - p_1)}{p_2/(1 - p_2)}\right)}$$

$$a - \frac{b_2}{n} < \hat{\mathbf{p}}_n < a + \frac{b_1}{n}$$

where

$$a = \frac{\log\,[(1 - p_2)/(1 - p_1)]}{\log\left(\dfrac{p_1/(1 - p_1)}{p_2/(1 - p_2)}\right)}.$$

Example E_{15}–2. Normal Distribution, Known σ. Here we have

$$\lambda_n(\mathbf{Z}_n) = \exp\left[-\frac{1}{2\sigma^2}[\textstyle\sum(\mathbf{X}_i - \mu_1)^2 - \sum(\mathbf{X}_i - \mu_2)^2]\right].$$

The relation $k_2 < \lambda_n(\mathbf{Z}_n) < k_1$ is equivalent to

$$\frac{\log k_2}{\log e} < -\frac{1}{2\sigma^2}[\textstyle\sum(\mathbf{X}_i - \mu_1)^2 - \sum(\mathbf{X}_i - \mu_2)^2] < \frac{\log k_1}{\log e}.$$

Since

$$\sum (\mathbf{X}_i - \mu_1)^2 - \sum (\mathbf{X}_i - \mu_2)^2 = -2n(\mu_1 - \mu_2)\left(\overline{\mathbf{X}} - \frac{\mu_1 + \mu_2}{2}\right)$$

and

$$\mu_1 - \mu_2 > 0,$$

the preceding inequality is equivalent to

$$\frac{1}{n}\left(\frac{\sigma^2 \log k_2}{(\mu_1 - \mu_2) \log e}\right) < \left(\overline{\mathbf{X}} - \frac{\mu_1 + \mu_2}{2}\right) < \frac{1}{n}\left(\frac{\sigma^2 \log k_1}{(\mu_1 - \mu_2) \log e}\right)$$

or

$$a - \frac{b_2}{n} < \overline{\mathbf{X}} < a + \frac{b_1}{n}$$

where

$$a = \frac{\mu_1 + \mu_2}{2}.$$

APPENDIX F_1

Remarks About Game Theory

In the study of statistics as decision making under uncertainty, a strategy is evaluated in terms of its consequences, namely, the average loss. This average loss depends on the statistician's strategy and the unknown state of nature.

A closely related situation exists in game theory. We illustrate with two very simple games. These are called two-person zero-sum games because they involve two persons and the winning player collects from the loser. (One player's gain is the other's loss.) The gain of player A is called the *payoff*.

Example F_1–1. Matching Coins. Each player selects heads or tails. If they match at heads, player A wins $1 from player B. If they match at tails, he wins $3. If they do not match, he loses $2 or $3 as illustrated by Table F_1–1. Note that each player has two strategies available. These are "call heads" and "call tails."

TABLE F_1-1

PAYOFF (TO PLAYER A) FOR THE COIN-MATCHING GAME

		Strategy of Player B		
		H	T	min
Strategy of Player A	H	1	−3	−3
	T	−2	3	−2
	max	1	3	

Example F_1–2. There are two bags. Bag 1 has 3 white and 7 red balls. Bag 2 has 6 white and 4 red balls. Player A selects a bag and draws a ball at random from the bag. Player B observes the bag that player A selects and then calls white or red. If player B guesses the color, he wins $1 from A. Otherwise he loses $1.

In this example, player A has two strategies. These are "select bag 1" and "select bag 2." But player B has more strategies available. Each of his strategies must tell him how to call for each possible move of player A. A typical strategy would be (R, W), which means call red if player A picks bag 1 and white if player A picks bag 2. Suppose player A picks bag 2 and player B applies (R, W). Then player A will obtain a white ball and lose one dollar 60% of the time and obtain a red ball and win one dollar 40% of the time. On the average he will lose 20 cents, as shown in Table F_1–2.

TABLE F_1-2

PAYOFF TO PLAYER A FOR EXAMPLE F_1-2

		Strategy of Player B				
		(W, W)	(W, R)	(R, W)	(R, R)	min
Strategy of Player A	1	0.4	0.4	−0.4	−0.4	−0.4
	2	−0.2	0.2	−0.2	0.2	−0.2
	max	0.4	0.4	−0.2	0.2	

We analyze Example F_1–2 first. Note that if player A selects strategy 1, the worst that can happen to him is a loss of 0.4. If he selects strategy 2, the worst that can happen to him is a loss of 0.2. The strategy which minimizes the worst that can happen to him (called the *minimax strategy*) is strategy 2. Thus player A can assure himself he will lose no more than 0.2.

The worst that can happen to player B for his four strategies are 0.4, 0.4, −0.2, and 0.2 respectively. His minimax strategy is (R, W) with which he can make sure that player A will lose at least 0.2.

Player A can assure himself of a loss of no more than 0.2 and player B can prevent him from doing better. Then it seems reasonable for both players to select their minimax strategies and to call −0.2 the *value* of the game (to player A).

Now let us consider the coin-matching game. The minimax strategy for player A is to select T which assures him that, at the worst, he will lose \$2. The minimax strategy for player B is to select H. This assures him that at the worst he will lose \$1. If they both play their minimax strategies, player A will lose \$2 but player B will do considerably better than he was guaranteed. Should player A be satisfied with this situation? The answer is "no." For example, suppose he tossed his coin and called whatever fell. This *randomized* strategy will yield a payoff (on the average) of

$$1/2 \times (1) + 1/2 \times (-2) = -1/2$$

if player B calls H, and of

$$1/2 \times (-3) + 1/2 \times (3) = 0$$

if player B calls T. Then the worst that can happen to player A is that he will lose 50 cents on the average, which is a considerable improvement over the previous situation.

Is there a rational strategy for the players in the coin-matching game? There is, and this answer is based on a theorem which applies to two-person, zero-sum games with a finite number of non-randomized strategies. This theorem states that, if the players are allowed to use a random device to select strategies, then, by using the minimax randomized strategies, player A can assure himself an amount v and player B can prevent him from doing better. *Then these minimax randomized strategies seem to be reasonable for playing against an intelligent opponent, and v is called the value of the game.*

We illustrate with the coin-matching problem. Suppose player A selects H with probability 5/9. Then, if player B selects H, player A averages

$$5/9 \times (1) + 4/9 \times (-2) = -1/3.$$

If player B selects T, player A averages

$$5/9 \times (-3) + 4/9 \times (3) = -1/3.$$

No matter what player B does, player A averages a loss of 1/3. If A were to use a different probability, he could improve his performance for one strategy of player B at the expense of hurting it for the other, thereby incurring a larger maximum loss. Thus the above randomized strategy is the minimax strategy for player A. Similarly, the minimax strategy for player B consists of calling H with probability 2/3 and T with probability 1/3. This strategy for player B yields a payoff (to player A) of $-1/3$, no matter what he does. Thus player A can avoid losing more than 1/3, and player B can force him to lose 1/3. Then $-1/3$ is the value of the game, for which two intelligent opponents should be willing to settle.

The techniques studied in Sections 5 and 6 of Chapter 5, can be applied in computing the minimax strategies.

There is a striking similarity between the structure of the two-person, zero-sum game and the statistical problem. In fact, statistics is often referred to as a game against nature. However there is a major difference, in that it seems unreasonable to regard the unknown state of nature as the strategy selected by a malevolent opponent. For this reason, the applicability of the minimax criterion in the game problem does not necessarily imply that it is the "correct" criterion for statistical problems.

APPENDIX F_2

Outline of the Derivation of the Utility Function

This appendix represents a bare and informal outline of the main steps in a proof that the four assumptions of Section 2.1 of Chapter 4 imply the existence of a utility function with the utility function properties 1 and 2. As previously indicated, a complete and formal proof exists in *Theory of Games and Economic Behavior* by Von Neumann and Morgenstern. Incidentally, a slightly weaker version of our fourth assumption is used in that presentation. The complete proof is quite long and formal and requires some mathematical sophistication. This outline is presented merely as a handy reference for the use of readers with considerably more background in mathematics than is expected in the majority of the readers of this book.

In the trivial case where all prospects are liked equally well, there is no problem and the utility function is constant. Hereafter, assume that there are at least two prospects P_0 and P_1 which are not liked equally well. By assumption one, one of these is preferred to the other. Let us assume that P_1 is preferred to P_0. In general let $(P, Q; p, 1 - p)$ represent the mixed prospect of facing P, with probability p and Q otherwise. Let $P_p = (P_1, P_0; p, 1-p)$. Note that P_1 may be regarded as $(P_1, P_1; p, 1-p)$. Using Assumption 4, we obtain:

LEMMA 1. *P_1 is preferred to P_p which is preferred to P_0 for* $0 < p < 1$.

If $0 < q < p$, P_q may be regarded as the mixture $(P_p, P_0; q/p, 1 - q/p)$. Using Assumption 4 again, we have:

LEMMA 2. *P_p is preferred to P_q if* $0 \leq q < p \leq 1$.

Using Assumption 1, we derive a stronger form of Assumption 2:

LEMMA 3. *If P is preferred to Q, and Q is regarded at least as* *well as R, then P is preferred to R. If P is regarded at least as well as Q and Q is preferred to R then P is preferred to R.*

Now suppose that P_1 is preferred to P which is preferred to P_0. Applying Lemmas 2 and 3, and Assumption 3, there is a number p between 0 and 1 so that P_q is preferred to P for $p < q < 1$, and P is preferred to P_q for $0 < q < p$. Using Assumption 3 again, it follows that P and P_p are liked equally well. We have

LEMMA 4. *If P_1 is regarded at least as well as P, and P is regarded at least as well as P_0, then there is a single number p between 0 and 1 so that P and P_p are liked equally well.*

In the case described in Lemma 4, define $u(P) = p$. If P is preferred to P_1, Lemma 4 tells us that there is a number a, $0 < a < 1$, so that P_1 and $(P, P_0; a, 1 - a)$ are liked equally well. Then let $u(P) = 1/a$ (which exceeds one). If P_0 is preferred to P then there is a number b, $0 < b < 1$, so that P_0 and $(P_1, P; b, 1 - b)$ are liked equally well. Then let $u(P) = -b/(1 - b)$ (which is negative).

LEMMA 5. *The function u satisfies the utility function properties if we consider only prospects for which $0 \leq u \leq 1$.*

Proof: Property 1 follows from Lemmas 2 and 4. Property 2 follows because $(P_a, P_b; p, 1 - p)$ is essentially the same as $P_{ap+b(1-p)}$.

THEOREM 1. *The function u satisfies the utility function properties.*

Proof: We wish to show that these properties apply to two arbitrary prospects Q and R. Let P_1^* be the most preferred among P_1, Q, and R, and let P_0^* be the least preferred among P_0, Q, and R. Define u^*, as u was but based on P_1^* and P_0^*. By Lemma 5, u^* satisfies the desired properties for Q and R and for all prospects "between" P_1^* and P_0^*. Hence it suffices to show that, for such prospects, u and u^* are linearly related with positive slope.

If P is preferred to P_1, P_1 is regarded as well as $(P, P_0; u(P)^{-1}, 1 - u(P)^{-1})$. Applying Lemma 5 to u^*, $u^*(P_1) = u(P)^{-1} u^*(P) + [1 - u(P)^{-1}] u^*(P_0)$. Hence

$$u^*(P) = u^*(P_0) + [u^*(P_1) - u^*(P_0)] u(P)$$

which is a linear relation with positive slope. The same relation is obtained similarly in the cases where P_0 is preferred to P and where P is "between" P_1 and P_0.

In this book, we have been deliberately vague about the distinction between finite and countable additivity in expectation and utility. In Theorem 1, we have established $u(P, Q; p, 1-p) = pu(P) + (1-p)\,u(Q)$. This does not imply that $u(P) = E[u(\mathbf{P})]$ if P is a mixed prospect which yields the random prospect $\mathbf{P}$ if there are infinitely many possibilities for $\mathbf{P}$. To obtain this stronger result, Assumption 4 must be strengthened slightly. We shall apply

Assumption 4a. *If P_i is regarded at least as well as Q_i for $i = 1, 2, \cdots$, any mixture of the P_i is regarded at least as well as the same mixture of the Q_i.*

First we shall prove that utility is bounded. A new proof is required because the one in Section 5, Chapter 4, used countable additivity.

Theorem 2. *The utility function u is bounded.*

Proof: Suppose that u is unbounded from above. Then there are prospects $P_1, P_2, \cdots$, such that $u(P_1) \geq 2^{i-1}$. The St. Petersburg game gives P which may be regarded as a mixture of $P_1, P_2, \cdots, P_n$, and a "remainder" R_n which is itself a mixture of $P_{n+1}, P_{n+2}, \cdots$. Then

$$u(P) \geq (1/2)1 + (1/4)2 \cdots + (1/2^n)2^{n-1} + (1 - 1/2^n)\,u(R_n).$$

But Assumption 4a tells us that R_n is regarded at least as well as P_1 and hence $u(R_n) \geq 1$. Then $u(P) = \infty$ which is impossible. Hence u is bounded from above. Similarly, it can be shown that u is bounded from below.

Theorem 3. (*Countable Additivity.*) *If P is a mixed prospect which yields $\mathbf{P} = P_i$ with probability p_i, then*

$$u(P) = \sum p_i\,u(P_i) = E[u(\mathbf{P})].$$

Proof: As in the proof of Theorem 2

$$u(P) = \sum_{i=1}^{n} p_i\,u(P_i) + r_n\,u(R_n)$$

where $r_n = 1 - \sum_{i=1}^{n} p_i \to 0$ and $u(R_n)$ is bounded. Let $n \to \infty$ and the theorem follows.

Partial List of Answers to Exercises

CHAPTER 1

1. 1. (a_1, a_1, a_1, a_1)

1. 6.

	s_1	s_2	s_3	s_4
θ_1	3.2	6.4	4	3.2
θ_2	4.6	3.4	2	5.4
θ_3	1.0	0.8	1	2.1

; 256 strategies

CHAPTER 2

2. 5. 350; 1434; 1784; 7000; 16
2. 7. 3; 41; 218
2. 9. 11; −4; 7; 35; 24; 47; 405

CHAPTER 3

3. 6. 7/10; 3/10
3.10. 1/2
3.11. 1/2; 1/3
3.21. 0.5000; 0.3413; 0.8400; 1.000; $A = 69$; $B = 64.065$
3.23. 0.0548
3.24. 44.8; 50; 53.9
3.27. Straight lines, parabola; (a) overlaps (b) at one point and (c) at two points; (b) and (c) are nonoverlapping.
3.28. (a) is a circle of radius 2 with center at origin, (b) is a disk of radius 2 and center at origin, (c) is a disk minus its boundary with radius 1 and center at (2,1); all three of these sets overlap.
3.34. 1/2
3.35. 9/47
3.37. 4/47
3.39. $$F(a) = \begin{cases} \sqrt{a} & \text{for } 0 \leq a \leq 1 \\ 1 & \text{for } a \geq 1 \\ 0 & \text{for } a \leq 0 \end{cases}$$
3.40. 0.021
3.43. 1/2
3.45. $f(1) = 2/3$, $f(2) = 1/4$, $f(3) = 1/14$, $f(4) = 1/84$
3.47. $$f(y) = \begin{cases} \dfrac{1}{2\sqrt{y}} & \text{for } 0 \leq y \leq 1 \\ 0 & \text{otherwise} \end{cases}$$
3.49. $$F(a) = \begin{cases} 1 & \text{for } a \geq 1 \\ 1 - \dfrac{1}{\pi} \cos^{-1}(a) & \text{for } -1 \leq a \leq 1 \\ 0 & \text{for } a \leq 1 \end{cases}$$
$$f(x) = \begin{cases} (\pi\sqrt{1 - x^2})^{-1} & \text{for } -1 \leq x \leq 1 \\ 0 & \text{otherwise} \end{cases}$$

CHAPTER 4

4. 1. 9; 4; 43
4. 5. Yes
4. 7. 0.30 approximately
4. 8. 9.6
4.12. A_1 is a disk with radius 2 and center at origin, A_2 is a disk of radius 1 with center at (1,2); their union is the collection of all points in either or both disks.
4.14. 60%
4.15. 80%; 20%
4.18. 7 cents
4.20. 0.59
4.22. Win 4000 with probability 1/2 or lose 4000 with probability 1/2; lose 500 with probability 15/16 or win 7500 with probability 1/16.
4.26. Select rifle with least variance.
4.29. 0.27; 0.33; 0.27
4.32. 4, 12.4, 3.52; 28.4, 899.44
4.36. $p, p(1-p)$
4.38. 1/2; $\sqrt{1/12}$
4.40. 17.5
4.43. 2; $\sqrt{2}$

CHAPTER 5

5. 1.

	a_1	a_2	a_3
θ_1	0.4	0.6	0
θ_2	0.8	0.2	0

	Expected Loss
θ_1	0.6
θ_2	4.6

5. 6. 3/8; 1/4; 3/8
5.10. $(4.4, 2.4) = 0.4(6,1) + 0.2(2,2) + 0.4(4,4)$
5.11. Yes; no
5.13. 1; 2; 2; 3; 2; 2; 3
5.14. (*c*), (*d*)
5.18.

	s_1	s_2	s_3	s_4	s_5	s_6	s_7	s_8
z_1	a_1	a_1	a_1	a_1	a_2	a_2	a_2	a_2
z_2	a_1	a_1	a_2	a_2	a_1	a_1	a_2	a_2
z_3	a_1	a_2	a_1	a_2	a_1	a_2	a_1	a_2

$L(\theta, s)$

	s_1	s_2	s_3	s_4	s_5	s_6	s_7	s_8
θ_1	1	1.5	1.4	1.9	1.1	1.6	1.5	2
θ_2	4	3.2	1.6	0.8	3.2	2.4	0.8	0

s_1, s_5, s_7, s_8, and mixtures of s_1 and s_5, of s_5 and s_7, and of s_7 and s_8.

5.21. $(2/5)x + (3/5)y = 1$; $(2/7)x + (5/7)y = -3/7$; $(3/2)x - (1/2)y = 1/2$
5.28. s_8
5.32. Corresponding supporting line is $(1-w)L_1 + wL_2 = c$. At point (L_1^*, L_1^*) of intersection, we get $L_1^* = c$.

5.37. s_7 with probability 3/4 and s_5 with probability 1/4.

5.43. s_7 with probability 0.77 and s_8 with probability 0.23 approximately.

CHAPTER 6

6. 1. a_3; $\mathscr{B} = 1$

6. 5. a_2; $\mathscr{B} = 1/2$

6. 7. 1

6.11. 13/850

6.13. 16 cents

6.17. 45/1081

6.18. 1/5; 1/5

6.19. 2/5; 1/5

6.20. If z_1 is observed, $\mathbf{w}_1 = 1$, $\mathbf{w}_2 = 0$, and $\mathbf{w}_3 = 0$
If z_2 is observed, $\mathbf{w}_1 = 1/2$, $\mathbf{w}_2 = 1/2$, and $\mathbf{w}_3 = 0$
If z_3 is observed, $\mathbf{w}_1 = 0$, $\mathbf{w}_2 = 3/5$, and $\mathbf{w}_3 = 2/5$
If z_4 is observed, $\mathbf{w}_1 = 0$, $\mathbf{w}_2 = 0$, and $\mathbf{w}_3 = 1$

6.25. 1/9; 1/2; 2/9

6.26. 41/126

6.27. 7/41; 12/41; 22/41

6.28. 1/32; 3/16

6.29. (a_1, a_2, a_3, a_3)

6.35. 0.0864

6.36. 1/18

6.38. $f(n) = (1/2)^n$, $n = 1, 2, 3, \cdots$

6.39. $E(\mathrm{N}) = 2$, $\sigma^2_{\mathrm{N}} = 2$
In general, $E(\mathrm{N}) = x^{-1}$ and $\sigma^2_{\mathrm{N}} = x^{-1} + (1 - 2x)x^{-2}$, where $x = k2^{-(k-1)}$.

6.40. 0.008; 0.64; $(0.8)^5$; $1 - (0.8)^5$

6.42. $\binom{n}{r} p^r(1 - p)^{n-r}$

6.43. $\dfrac{e^{-\lambda}\lambda^r}{r!}$

6.48. Normal with mean $\mu_{\mathrm{X}} + \mu_{\mathrm{Y}}$ and variance $\sigma^2_{\mathrm{X}} + \sigma^2_{\mathrm{Y}}$
Normal with mean $\mu_{\mathrm{X}} - \mu_{\mathrm{Y}}$ and variance $\sigma^2_{\mathrm{X}} + \sigma^2_{\mathrm{Y}}$
Normal with mean $2\mu_{\mathrm{X}} - 3\mu_{\mathrm{Y}}$ and variance $4\sigma^2_{\mathrm{X}} + 9\sigma^2_{\mathrm{Y}}$

6.51. (a_1, a_2, a_3, a_3); 6/10;

	a_1	a_2	a_3
θ_1	1/2	1/2	0
θ_2	0	1/2	1/2
θ_3	0	0	1

;

$\mathrm{R}(\theta_1, s) = 1/2$, $R(\theta_2, s) = 1/2$, $R(\theta_3, s) = 0$

CHAPTER 7

7. 1.

θ	Regrets		s^*_{31} Action Probabilities		Risk	$s^*_{31.25}$ Action Probabilities		Risk
	$r(\theta, a_1)$	$r(\theta, a_2)$	$P\{\overline{X} \geq 31 \mid \theta\}$	$P\{\overline{X} \leq 31 \mid \theta\}$		$P\{\overline{X} \geq 31.25 \mid \theta\}$	$P\{\overline{X} \leq 31.25 \mid \theta\}$	
28.0	40.00	0.00	0.000	1.000	0.000	0.000	1.000	0.000
28.5	27.50	0.00	0.000	1.000	0.000	0.000	1.000	0.000
29.0	19.40	0.00	0.000	1.000	0.001	0.000	1.000	0.000
29.5	13.15	0.00	0.001	0.999	0.018	0.000	1.000	0.003
30.0	8.00	0.00	0.023	0.977	0.184	0.006	0.994	0.048
30.5	3.55	0.00	0.159	0.841	0.564	0.067	0.933	0.238
31.0	0.00	0.00	0.500	0.500	0.000	0.308	0.692	0.000
31.5	0.00	2.20	0.841	0.159	0.350	0.692	0.308	0.678
32.0	0.00	3.55	0.977	0.023	0.082	0.933	0.067	0.238
32.5	0.00	4.40	0.999	0.001	0.006	0.994	0.006	0.026
33.0	0.00	5.00	1.000	0.000	0.000	1.000	0.000	0.001
33.5	0.00	5.35	1.000	0.000	0.000	1.000	0.000	0.000
34.0	0.00	5.50	1.000	0.000	0.000	1.000	0.000	0.000

7. 3. Accept; accept

7. 4. $\mathcal{R}(s^*_{31}) = 0.119$; $\mathcal{R}(s^*_{31.25}) = 0.172$; $1.706 - 0.119 = 1.587$

7.12. $\mathbf{W}/100$; Normal with mean θ and standard deviation 10.

7.13. (*a*) $168/N$; (*c*) $N = 13$

7.14. $(\overline{\mathbf{X}} - 2.576, \overline{\mathbf{X}} + 2.576)$

7.15. $(\overline{\mathbf{X}} - 7.828, \overline{\mathbf{X}} + 7.828)$

7.16. (0.291, 0.309); (0.390, 0.410); (0.490, 0.510)

7.20. Reject if $\overline{\mathbf{X}} \leq 30.355$

7.22. Reject if $|\hat{\mathbf{p}} - 0.2| > 0.052$

CHAPTER 8

8. 1. Yes; no

8. 2. Yes; yes

8. 4. (1) $B = 0$; (2) $B = C = 0$; (3) $B = 0$

CHAPTER 9

9. 3. 268

9. 7. 174

9. 9. Accept H_1 if $\sum(\mathbf{X}_i - \mu)^2$ is large enough. (The action probabilities of a test of this form can be obtained by reference to Table D_2 since $\sum(\mathbf{X}_i - \mu)^2/\sigma^2$ is known to have the χ^2 distribution with n degrees of freedom if $\mathbf{X}_1, \mathbf{X}_2, \cdots, \mathbf{X}_n$ are independent and normally distributed with mean μ and variance σ^2.)

9.12. $n = 235$; reject if $\hat{\mathbf{p}} < 0.526$.

9.18. Reject if $|\overline{\mathbf{X}} - 500| \geq 19.60$

9.19. Reject H_1 if $|\overline{\mathbf{X}} - \mu_0| \geq 2.262 \mathbf{s}_{\mathbf{X}}/\sqrt{10}$

CHAPTER 10

10. 1. $\overline{\mathbf{X}} + (3/n) \sum_{i=1}^{n} (\mathbf{X}_i - \overline{\mathbf{X}})^2$; $4\overline{\mathbf{X}}$

10. 3. $(p^*)^{\mathbf{m}_1 + \mathbf{m}_2}(1 - p^*)^{\mathbf{m}_2 + 2\mathbf{m}_3}$, $(\mathbf{m}_1 + \mathbf{m}_2)/(\mathbf{m}_1 + 2\mathbf{m}_2 + 2\mathbf{m}_3)$

10. 4. $\hat{\mu} = 6.5$

10. 5. 1, −1

10. 6. 1, 0

10. 7. 0, 0; 0, 0; 1, −1; 0, 0; 1, −1

10.10. 25

10.12. $\dfrac{n\hat{\mathbf{p}}^2}{n-1} - \dfrac{\hat{\mathbf{p}}}{n-1}$

10.14.

	95%	98%
(a)	(0.352, 0.448)	(0.343, 0.457)
(b)	(0.071, 0.129)	(0.065, 0.135)
(c)	(0.402, 0.598)	(0.383, 0.617)

10.15. (19.18, 20.82); (19.59, 20.41); (19.73, 20.27)

Index

25 KITES THAT FLY, Leslie Hunt. Full, easy-to-follow instructions for kites made from inexpensive materials. Many novelties. 70 illustrations. 110pp. 5⅜ × 8½. 22550-X Pa. $2.50

PIANO TUNING, J. Cree Fischer. Clearest, best book for beginner, amateur. Simple repairs, raising dropped notes, tuning by easy method of flattened fifths. No previous skills needed. 4 illustrations. 201pp. 5⅜ × 8½. 23267-0 Pa. $3.50

EARLY AMERICAN IRON-ON TRANSFER PATTERNS, edited by Rita Weiss. 75 designs, borders, alphabets, from traditional American sources. 48pp. 8¼ × 11. 23162-3 Pa. $1.95

CROCHETING EDGINGS, edited by Rita Weiss. Over 100 of the best designs for these lovely trims for a host of household items. Complete instructions, illustrations. 48pp. 8¼ × 11. 24031-2 Pa. $2.25

FINGER PLAYS FOR NURSERY AND KINDERGARTEN, Emilie Poulsson. 18 finger plays with music (voice and piano); entertaining, instructive. Counting, nature lore, etc. Victorian classic. 53 illustrations. 80pp. 6½ × 9¼. 22588-7 Pa. $1.95

BOSTON THEN AND NOW, Peter Vanderwarker. Here in 59 side-by-side views are photographic documentations of the city's past and present. 119 photographs. Full captions. 122pp. 8¼ × 11. 24312-5 Pa. $7.95

CROCHETING BEDSPREADS, edited by Rita Weiss. 22 patterns, originally published in three instruction books 1939-41. 39 photos, 8 charts. Instructions. 48pp. 8¼ × 11. 23610-2 Pa. $2.00

HAWTHORNE ON PAINTING, Charles W. Hawthorne. Collected from notes taken by students at famous Cape Cod School; hundreds of direct, personal *apercus*, ideas, suggestions. 91pp. 5⅜ × 8½. 20653-X Pa. $2.95

THERMODYNAMICS, Enrico Fermi. A classic of modern science. Clear, organized treatment of systems, first and second laws, entropy, thermodynamic potentials, etc. Calculus required. 160pp. 5⅜ × 8½. 60361-X Pa. $4.50

TEN BOOKS ON ARCHITECTURE, Vitruvius. The most important book ever written on architecture. Early Roman aesthetics, technology, classical orders, site selection, all other aspects. Morgan translation. 331pp. 5⅜ × 8½. 20645-9 Pa. $5.95

THE CORNELL BREAD BOOK, Clive M. McCay and Jeanette B. McCay. Famed high-protein recipe incorporated into breads, rolls, buns, coffee cakes, pizza, pie crusts, more. Nearly 50 illustrations. 48pp. 8¼ × 11. 23995-0 Pa. $2.00

THE CRAFTSMAN'S HANDBOOK, Cennino Cennini. 15th-century handbook, school of Giotto, explains applying gold, silver leaf; gesso; fresco painting, grinding pigments, etc. 142pp. 6⅛ × 9¼. 20054-X Pa. $3.50

FRANK LLOYD WRIGHT'S FALLINGWATER, Donald Hoffmann. Full story of Wright's masterwork at Bear Run, Pa. 100 photographs of site, construction, and details of completed structure. 112pp. 9¼ × 10. 23671-4 Pa. $7.95

OVAL STAINED GLASS PATTERN BOOK, C. Eaton. 60 new designs framed in shape of an oval. Greater complexity, challenge with sinuous cats, birds, mandalas framed in antique shape. 64pp. 8¼ × 11. 24519-5 Pa. $3.75

KEYBOARD WORKS FOR SOLO INSTRUMENTS, G.F. Handel. 35 neglected works from Handel's vast oeuvre, originally jotted down as improvisations. Includes Eight Great Suites, others. New sequence. 174pp. 9⅜ × 12¼.
24338-9 Pa. $7.50

AMERICAN LEAGUE BASEBALL CARD CLASSICS, Bert Randolph Sugar. 82 stars from 1900s to 60s on facsimile cards. Ruth, Cobb, Mantle, Williams, plus advertising, info, no duplications. Perforated, detachable. 16pp. 8¼ × 11.
24286-2 Pa. $2.95

A TREASURY OF CHARTED DESIGNS FOR NEEDLEWORKERS, Georgia Gorham and Jeanne Warth. 141 charted designs: owl, cat with yarn, tulips, piano, spinning wheel, covered bridge, Victorian house and many others. 48pp. 8¼ × 11.
23558-0 Pa. $1.95

DANISH FLORAL CHARTED DESIGNS, Gerda Bengtsson. Exquisite collection of over 40 different florals: anemone, Iceland poppy, wild fruit, pansies, many others. 45 illustrations. 48pp. 8¼ × 11. 23957-8 Pa. $1.95

OLD PHILADELPHIA IN EARLY PHOTOGRAPHS 1839-1914, Robert F. Looney. 215 photographs: panoramas, street scenes, landmarks, President-elect Lincoln's visit, 1876 Centennial Exposition, much more. 230pp. 8⅜ × 11¾.
23345-6 Pa. $9.95

PRELUDE TO MATHEMATICS, W.W. Sawyer. Noted mathematician's lively, stimulating account of non-Euclidean geometry, matrices, determinants, group theory, other topics. Emphasis on novel, striking aspects. 224pp. 5⅜ × 8½.
24401-6 Pa. $4.50

ADVENTURES WITH A MICROSCOPE, Richard Headstrom. 59 adventures with clothing fibers, protozoa, ferns and lichens, roots and leaves, much more. 142 illustrations. 232pp. 5⅜ × 8½. 23471-1 Pa. $3.95

IDENTIFYING ANIMAL TRACKS: MAMMALS, BIRDS, AND OTHER ANIMALS OF THE EASTERN UNITED STATES, Richard Headstrom. For hunters, naturalists, scouts, nature-lovers. Diagrams of tracks, tips on identification. 128pp. 5⅜ × 8. 24442-3 Pa. $3.50

VICTORIAN FASHIONS AND COSTUMES FROM HARPER'S BAZAR, 1867-1898, edited by Stella Blum. Day costumes, evening wear, sports clothes, shoes, hats, other accessories in over 1,000 detailed engravings. 320pp. 9⅜ × 12¼.
22990-4 Pa. $10.95

EVERYDAY FASHIONS OF THE TWENTIES AS PICTURED IN SEARS AND OTHER CATALOGS, edited by Stella Blum. Actual dress of the Roaring Twenties, with text by Stella Blum. Over 750 illustrations, captions. 156pp. 9 × 12.
24134-3 Pa. $8.50

HALL OF FAME BASEBALL CARDS, edited by Bert Randolph Sugar. Cy Young, Ted Williams, Lou Gehrig, and many other Hall of Fame greats on 92 full-color, detachable reprints of early baseball cards. No duplication of cards with *Classic Baseball Cards*. 16pp. 8¼ × 11. 23624-2 Pa. $3.50

THE ART OF HAND LETTERING, Helm Wotzkow. Course in hand lettering, Roman, Gothic, Italic, Block, Script. Tools, proportions, optical aspects, individual variation. Very quality conscious. Hundreds of specimens. 320pp. 5⅜ × 8½.
21797-3 Pa. $4.95

THE RIME OF THE ANCIENT MARINER, Gustave Doré, S.T. Coleridge. Doré's finest work, 34 plates capture moods, subtleties of poem. Full text. 77pp. 9¼ × 12. 22305-1 Pa. $4.95

SONGS OF INNOCENCE, William Blake. The first and most popular of Blake's famous "Illuminated Books," in a facsimile edition reproducing all 31 brightly colored plates. Additional printed text of each poem. 64pp. 5¼ × 7. 22764-2 Pa. $3.50

AN INTRODUCTION TO INFORMATION THEORY, J.R. Pierce. Second (1980) edition of most impressive non-technical account available. Encoding, entropy, noisy channel, related areas, etc. 320pp. 5⅜ × 8½. 24061-4 Pa. $4.95

THE DIVINE PROPORTION: A STUDY IN MATHEMATICAL BEAUTY, H.E. Huntley. "Divine proportion" or "golden ratio" in poetry, Pascal's triangle, philosophy, psychology, music, mathematical figures, etc. Excellent bridge between science and art. 58 figures. 185pp. 5⅜ × 8½. 22254-3 Pa. $3.95

THE DOVER NEW YORK WALKING GUIDE: From the Battery to Wall Street, Mary J. Shapiro. Superb inexpensive guide to historic buildings and locales in lower Manhattan: Trinity Church, Bowling Green, more. Complete Text; maps. 36 illustrations. 48pp. 3⅞ × 9¼. 24225-0 Pa. $2.50

NEW YORK THEN AND NOW, Edward B. Watson, Edmund V. Gillon, Jr. 83 important Manhattan sites: on facing pages early photographs (1875-1925) and 1976 photos by Gillon. 172 illustrations. 171pp. 9¼ × 10. 23361-8 Pa. $9.95

HISTORIC COSTUME IN PICTURES, Braun & Schneider. Over 1450 costumed figures from dawn of civilization to end of 19th century. English captions. 125 plates. 256pp. 8⅜ × 11¼. 23150-X Pa. $7.50

VICTORIAN AND EDWARDIAN FASHION: A Photographic Survey, Alison Gernsheim. First fashion history completely illustrated by contemporary photographs. Full text plus 235 photos, 1840-1914, in which many celebrities appear. 240pp. 6½ × 9¼. 24205-6 Pa. $6.00

CHARTED CHRISTMAS DESIGNS FOR COUNTED CROSS-STITCH AND OTHER NEEDLECRAFTS, Lindberg Press. Charted designs for 45 beautiful needlecraft projects with many yuletide and wintertime motifs. 48pp. 8¼ × 11. (EDNS) 24356-7 Pa. $2.50

101 FOLK DESIGNS FOR COUNTED CROSS-STITCH AND OTHER NEEDLECRAFTS, Carter Houck. 101 authentic charted folk designs in a wide array of lovely representations with many suggestions for effective use. 48pp. 8¼ × 11. 24369-9 Pa. $2.25

FIVE ACRES AND INDEPENDENCE, Maurice G. Kains. Great back-to-the-land classic explains basics of self-sufficient farming. The one book to get. 95 illustrations. 397pp. 5⅜ × 8½. 20974-1 Pa. $5.95

A MODERN HERBAL, Margaret Grieve. Much the fullest, most exact, most useful compilation of herbal material. Gigantic alphabetical encyclopedia, from aconite to zedoary, gives botanical information, medical properties, folklore, economic uses, and much else. Indispensable to serious reader. 161 illustrations. 888pp. 6½ × 9¼. (Available in U.S. only) 22798-7, 22799-5 Pa., Two-vol. set $16.45

519.5 C423 91-172
Chernoff, Herman.
Elementary decision theory

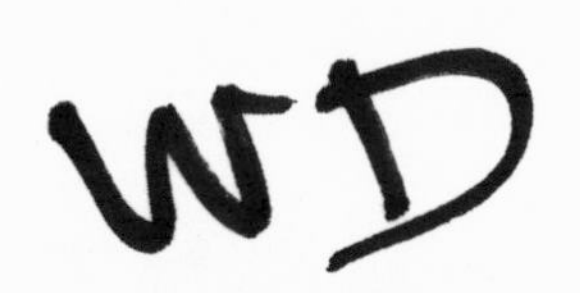